An Exaltation of Parks

AN EXALTATION OF PARKS

JOHN D. ROCKEFELLER JR.'S CRUSADE TO SAVE AMERICA'S WONDERLANDS

STEVE KEMP

THE UNIVERSITY OF UTAH PRESS
Salt Lake City

The Defiance House Man colophon is a registered trademark of the University of Utah Press. It is based on a four-foot-tall Ancient Puebloan pictograph (late PIII) near Glen Canyon, Utah.

Library of Congress Cataloging-in-Publication Data
Names: Kemp, Steve, author
Title: An exaltation of parks : John D. Rockefeller Jr.'s crusade to save America's wonderlands / Steve Kemp.
Other titles: John D. Rockefeller Jr.'s crusade to save America's wonderlands
Description: Salt Lake City : The University of Utah Press, 2025. |
Includes bibliographical references and index.
Identifiers: LCCN 2024056918 | ISBN 9781647692230 hardback | ISBN 9781647692247 paperback | ISBN 9781647692254 ebook
Subjects: LCSH: Rockefeller, John D., Jr. (John Davison), 1874-1960 | National parks and reserves--United States--History | United States. National Park Service--History--20th century | Rockefeller family | Philanthropists--United States--Biography
Classification: LCC SB482.A4 R635 2025 | DDC 333.78/0973--dc23/eng/20250214
LC record available at https://lccn.loc.gov/2024056918

Portions of this work related to the Great Smoky Mountains and park museum associations previously appeared in *Smokies Life Journal*, a biannual magazine published by Smokies Life, a nonprofit dedicated to supporting the scientific, historical, and interpretive activities of Great Smoky Mountains National Park by providing educational products and services. Learn more at SmokiesLife.org.

Cover photos by Ansel Adams (Grand Tetons) and Schuyler U. Bunnel (redwoods).

Errata and further information on this and other titles available at UofUpress.com

For Janet

Our public lands—whether a national park, monument, wildlife refuge, forest, or prairie—make each one of us land-rich. It is our inheritance as citizens of a country called America.

—Terry Tempest Williams, *The Hour of Land*

It sort of flattens the notion that it's a terrible thing for anybody to have money.

—Columnist Dave Boone in *New York Sun,* Dec 13, 1946

CONTENTS

PREFACE

I believe America's national parks are in deep trouble. While I recall decades ago when the media presaged every Memorial Day weekend with the obligatory "Are We Loving Our National Parks to Death?" story, and while the answer even back then was "yes," recently the problem of too much love for our fragile national treasures has become untenable.

Today's visitors to crown jewel national parks are often greeted, even before they reach the rustic wooden entrance sign, by miles-long lines of exhaust-spewing vehicles.[1] Trails that are little more than four feet wide and in poor condition host thousands of hikers per day. From nine a.m. on, there is no parking available at the more popular trailheads, scenic overlooks, visitor centers, restaurants, or points of departure for shuttlebuses.[2] Even in the backcountry, hikers stand in line for their turn to photograph a waterfall.[3] Reservations are required for campsites, some hiking trails, backcountry areas, river access, firefly viewing, lodging, and even to enter Glacier, Mount Rainier, Yosemite, Rocky Mountain, and other national parks.[4] A former park superintendent at Mount Rainier reported that visitors were getting into fights over parking spaces at her park prior to the reservation system. Some employees quit due to repeated verbal abuse from frustrated tourists.[5] The eleven-mile scenic drive on the Cades Cove Loop Road in Great Smoky Mountains National Park has become a three-hour-long, blood-pressure-spiking, bumper-to-bumper motorized slog. On a busy summer afternoon, many of our best national parks look more like Day Three at Woodstock than the pristine wildlands visitors seek and park managers are charged with protecting.

Since the early 2000s, as social media evolved to focus on sharing images, visitation to many already overcrowded national parks arched skyward. Visits to tiny Acadia on the coast of Maine increased by 1.7 million (over 75 percent) between 2004 and 2024. Nearly four million people now tour the small island park during a tourist season that stretches barely five months. Grand Teton National Park visitation jumped by more than one million (44 percent) over the same period. Yellowstone surged over 50 percent, and visits to Great Smoky Mountains National Park swelled from 9.1 million in 2004 to more than 12 million in 2024 (35 percent).[6] Bear in mind these hefty increases have befallen already maxed-out parks with stagnant budgets, staff reductions, failing infrastructure, and finite space.

While other tourist destinations such as theme parks and beach towns generally have the flexibility to expand parking and other infrastructure in response to rising attendance (and revenue), national park budgets are not tied to visitation and superintendents are mandated to protect their landscapes, not pave them. Even a 10 percent increase in visitation to a popular national park can be suffocating.

Potential park visitors are now inspired by both social media and amazingly comprehensive travel websites when in the planning stages of their trips.[7] This new type of marketing, whether intentional or not, is at least part of the reason why visitation to national parks (not all NPS sites) nationwide jumped from 69 million in 2013 to 94 million in 2024, an increase of better than 35 percent.[8] The number of annual visitors to all NPS sites (including national monuments, historic sites, seashores, etc.) is now roughly equivalent to the population of the United States.[9] It's a conundrum: parks need all the admirers they can muster to continue to be recognized for funding from Congress, but there's no longer enough room to shoehorn them into the more popular "crown jewels" of the national park system.

In our capitalist system, too much demand and not enough supply generally causes either prices to spike or supply to increase. Raising the cost of visiting a national park, however, undermines the egalitarian goals of our park system.[10] The parks are, after all, America's superior system of managing scenic places; they are sanctuaries open to all as opposed to the Old World model of exclusive private estates owned by nobles and other wealthy persons and available only to the elite.[11] Unfortunately, in iconic parks like Yosemite, Grand Teton, and Yellowstone, basic in-park concessionaire lodging often goes for $325 to more than $600 per night.[12]

Planning a trip to a busy national park today also involves a teeth-grinding amount of time spent (usually during the wee hours) battling apathetic online reservation systems. In a recent study by researchers at the University of Montana, William L. Rice and colleagues noted that one area received nineteen thousand online reservation requests for fifty-seven NPS campsites for the same date! Most reservations to drive the Going-to-the-Sun Road in Glacier National Park fill up months in advance.[13]

That leaves supply.

*

Because I grew up as an outdoorsy, free-range kid on the outskirts of a medium-sized city in Iowa, I may have a special appreciation for John D. Rockefeller Jr.'s public lands munificence. Although my school-age friends and I had a strong desire to hike, camp, hunt, and fish, for the most part we lacked the space to do so. Suburban homeowners had little reason to welcome a pack of subadult males onto their properties and farmers had even less. Consequently, our expeditions into the former wild prairies and woodlots of southeastern Iowa were often thwarted by No Trespassing signs or landowners understandably suggesting our immediate departure. Or the discovery that another open space had been transformed into a housing subdivision, like the one we lived in.

Within seventy-two hours of high school graduation, some of the same friends and I were off for a two-month tour of the western parks and wilderness areas. It was an almost unbelievable feeling of freedom to have hundreds of thousands of acres of mountains, forests, and valleys to explore, without the fences or discouraging signs. As an adult, I spent nearly my entire career working in Grand Teton, Yellowstone, and Great Smoky Mountains National Parks. My wife and I met and were married in the Smokies and we honeymooned in Acadia, where we hiked and bicycled the Rockefeller carriage roads. Yet, well into my career, what I knew of John D. Rockefeller Jr.'s contributions to national parks would have fit on one side of a three-by-five-inch card.

This book began as a magazine article about John D. Rockefeller Jr.'s contribution to the creation of Great Smoky Mountains National Park. I soon realized I would never discern Junior's motives for coming to the rescue of the Smokies without investigating his contributions to other parks, including Acadia, the Blue Ridge Parkway, George Washington's Birthplace, Grand Canyon, Grand Tetons, Mesa Verde, Redwoods, Shenandoah, Yellowstone, and Yosemite. Early on, I came to believe that if more people knew this story of a humble, insecure, pious, and very rich man who became a conservation superhero through his partnering with gung-ho NPS leaders and a whole lot of eccentric hikers and other grassroots conservationists, they might be inspired to help our parks in their immediate twenty-first-century crisis.

We need more national parks, and many of the parks we now have need to be expanded. This declaration echoes goals set by President Joe Biden's America the Beautiful Initiative to conserve 30 percent of our nation's lands and waters by 2030 and biologist Edward O. Wilson's proposal to preserve 50 percent of the planet.[14] These clarion calls for more wild places may not be as pie-in-the-sky as you imagine. Not long ago in the Smokies, the NPS had to turn down an opportunity to purchase some three thousand acres of beautiful mountain lands adjacent to the park because of lack of funds. The people making the offer loved their land, they loved the park, and would have loved to have seen the two combined.[15] The sum of money it would have taken to add miles of hiking trails, outstanding trout streams, dramatic vistas, and scenic roadways would have been less than one of the scores of America's digital-age billionaires makes in a morning. To date, the good people with the three thousand acres still own most of that property.

In 2015 the state of Wyoming offered to sell its 640-acre Kelly Parcel to Grand Teton National Park for $46 million, but the park couldn't scrape together the coin.[16] The square-mile tract is the largest unprotected land within the park boundary and is critical range for antelope, elk, bison, and moose. During the closing hours of 2024, Wyoming accepted an offer to sell its Kelly Parcel to the national park for $100 million. Roughly $62.4 million of the cost came from the federal government's Land and Water Conservation Fund (supported by royalties from offshore oil and gas production), and the $37.6 million balance was raised by the philanthropic Grand Teton National Park Foundation. Had the astonishingly successful private fundraising campaign failed, there is little doubt the land would have become an ecologically disruptive inholding packed with extravagant homes. While the Kelly Parcel was a very expensive piece of land, it will feel like a bargain in twenty years.

During the 1990s, a wealthy family from California opted to give their priceless mountaintop land to Great Smoky Mountains National Park. As a result, America's most visited national park grew by more than 570 gorgeous and biologically diverse mountaintop acres. The donors even contributed to developing a science center for area high school students and other researchers on the property and worked with the local Friends group to establish a large endowment to help cover operational expenses. Our national parks have many neighbors who love their landscapes and their national parks and don't want to see their property desecrated after they're gone.[17]

As in other endeavors, the first step in solving our parks' overpopularity problem is recognizing we have a problem. The second is accepting that the solution is more parkland. The next is making an expanded national park system a priority of the federal government, citizen groups, and philanthropists.

Look at a map of our scenery-rich country and you'll see many places with national park potential. Why are there so few national parks in New England? (And, yes, the Katahdin Woods and Waters National Monument in Maine—a highly controversial yet successful public-private conservation partnership—not to mention the proposed North Woods megapark, are exactly what I'm talking about.)[18] Why are there not massive national parks in the increasingly nonarable Great Plains, where land is relatively inexpensive and natural prairie could be restored and repopulated with its native megafauna—elk, bison, antelope, golden eagles, wolves, and grizzly bears? There could be more and larger national parks in the Cascades (NPS director Stephen T. Mather proposed the expansion of Crater Lake National Park in 1923), the Sierras, the Rockies, and the Appalachians. America's frequently scoured flood plains and hurricane-hammered coastal areas that regularly require billion-dollar bailouts need to be surrendered to Mother Nature to become parks.

Many new national parks can be created by the relatively noninvasive process of redesignating existing public lands. Others might require at least some acquisition of private lands, a more contentious and painful process but one we have learned to do more fairly and amicably than in the past. As the chapters in this book reveal, creating national parks is rarely easy, but well worth the effort. I doubt if many people who visit Acadia, the Grand Tetons, Great Smoky Mountains, Yosemite, or the Redwoods today wish the parkland would revert to commercial enterprises and private ownership.

It's difficult to overstate what a great idea "America's Best Idea" has turned out to be. National parks are excellent carbon sinks, absorbing and storing carbon dioxide and exhaling oxygen, day in and day out, free of charge (while the estimated value of this function is over half a billion dollars per year).[19] They are arks of biological diversity, providing permanent homes for tens of thousands of native plants and animals.[20] They help improve the physical and mental health of millions of Americans annually.[21] As conservationist Harvey Broome put it, national parks are places where "happiness is not only pursued, it is oft times found."[22] Support for national parks and other public lands is also one of the few government programs that enjoys bipartisan political support. Passage of the beneficial Land and Water Conservation Fund is evidence of this concurrence.[23]

Parks are drivers of economic growth. The taxpayers' annual bill for operating Great Smoky Mountains National Park is between $15 and $20 million. According to the NPS, revenue generated by the park's 12 million visitors is around $2.1 billion annually (with a cumulative effect of $3.3 billion). Nationwide, parks generate more than $50 billion and support 380,000 jobs; every $1 invested in national parks produces $10 for the nation's economy.[24] Compare that proven return on investment to that of your nearest multibillion-dollar professional sports stadium.

Next time lawmakers in Washington fail to pass a federal budget and the government goes into a "partial shutdown," watch how national parks play into the gambit. Generally, if the administration in power expects to gain from the continuation of the shutdown, they will find a way to keep the parks open, despite the lack of a budget. Those who wish the gridlock to end quickly will push to close the parks. If closure becomes imminent, city, county, tribal, and state governments will often become so desperate they will step in and offer to write checks to the federal government to keep the parks open. Of all the federal government's myriad services, the national parks are one of the things citizens, local communities, and voters will not tolerate losing. Ironically, the NPS consumes a measly one-fifteenth of one percent of the federal budget.[25]

This book is not intended to be a how-to guide for expanding or creating new national parks, though the story includes almost every conceivable scenario for doing so. Rather, I hope this story illuminates how positive things can be accomplished when well-intentioned philanthropists, government agencies, and passionate citizens join forces for a worthy cause. I hope it also grants long-overdue recognition to one of America's greatest conservationists, John D. Rockefeller Jr. While I realize boosting recognition for the son of an infamous Gilded Age tycoon is not always considered a virtuous effort, it is important for twentieth-century conservation history to set the record straight, not only on Junior, but also for his cohorts—Horace M. Albright, Arno B. Cammerer, George B. Dorr, Stephen T. Mather, Aileen and Jesse Nusbaum, and many other national park Hall of Famers.

Finally, during what some are calling "America's second Gilded Age,"[26] it may be beneficial for Americans to be reminded of one of the senior Rockefeller's sayings: "The very rich are just like all the rest of us; and if they get pleasure from the possession of money, it comes from their ability to do things which give satisfaction to someone besides themselves."[27]

ACKNOWLEDGMENTS

During the dark days of the Covid-19 pandemic, when archives across the country were temporarily closed, I was fortunate to receive the generous help of several subject matter experts across the country. They shared their knowledge over the phone as well as copies of documents they had gleaned during their own important research projects. Thanks to Ronald H. Epp for sharing his nearly infinite knowledge of all things Acadia National Park and for gently guiding me along on my writing journey. Thanks to Patti Bell and Kathy Fiero for their vast knowledge and enthusiasm for Mesa Verde superintendent Jesse Nusbaum and his overachieving wife, Aileen. Betsy Engle's knowledge of the Grand Tetons, Harold Fabian, and Jackson Lake Lodge proved invaluable and is very much appreciated.

I am extremely grateful for the help of knowledgeable and patient archivists, including Mike Aday (Great Smoky Mountains National Park archives), Michele H. Beckerman and Monica Blank (Rockefeller Archive Center), Samuel Denman and Tara Travis (Mesa Verde National Park Archives), Nora Haskell (Jackson Hole Historical Society), Paul F. Rogers (Yosemite National Park archives), and all the diligent and hardworking professionals at the National Archives Center in Maryland. You are the keepers of all the world's wisdom.

I would also like to give a special shout-out to the unsung public historians who compile unheralded works such as National Register of Historic Places Nomination Forms, compliance documents, roads and bridges engineering records, administrative histories, and cultural landscape inventories. Your painstaking and well-written works will survive the test of time and endure as some of our most valuable public records. This book would have been impossible without your contributions.

Same goes for all the developers, contributors, and history hoarders behind npshistory.com. Wow. So many of us are in awe of what you have accomplished.

Many thanks to reviewers, advisors, encouragers, and supporters Claire Ballentine, Dave Ballentine, Robert Beatty, Theodore Catton, Kent Cave, Katie Corrigan, Dale Ditmanson, Tim Davis, Frances Figart, Jonathon Foster, Phil Francis, Stewart Harris, Virginia Hoffman, Pamela Hogan, Rose Houk, Clay Jordan, Valerie Kemp, Jeffrey Kimball, Richard Miller, Richard Powers, Laurel Rematore, Janet Rock, Laurie and John Rush, Mark Smith, Dana Sohen, Jill Stevens, Greg and Kenlyn Stewart, Susan Wegener, Christopher Wuthmann, and everyone who suffered through my lectures on park philanthropy at The Swag. And because of the consistent enthusiasm and optimism of editor Jedediah Rogers, publication of this book has been delightful. What a deft touch you have in guiding hyperventilating writers along the steep and rocky path to the finish line!

INTRODUCTION

All the great philanthropic deeds that John D. Rockefeller Jr. accomplished were made possible by the wealth his father created. And of all the Horatio Alger rags-to-riches tales that enliven nineteenth-century American history, none is more unlikely and dramatic than the one lived by John D. Rockefeller Sr.

Senior's father, William Avery Rockefeller, was nicknamed "Devil Bill" for good reason. While Devil Bill could be measured as a decent husband to Eliza Davison Rockefeller for around half the year, a severe case of wanderlust carried him far from hearth and home for the remainder. According to family recollections, Devil Bill would vanish without prelude into the night and quietly return months later, sometimes with new clothes, fast horses, and a fat wad of cash to show for his efforts, whatever those efforts may have been.[1]

Eliza Davison was a devout, beautiful, even saintly woman before she married Devil Bill at age twenty-four (he was twenty-seven). She was from good New England farming stock with thrifty, hard-working parents who were considered wealthy by the rural western New York standards of the day. Although Eliza and Devil Bill did not part until death, Eliza no doubt spent a considerable amount of her time questioning her judgment on the decision to marry the dashing, athletic, smooth-talking con man.

Trials started early on in their marriage, particularly when Devil Bill invited his former fiancée, Nancy Brown, into the couple's home as a "housekeeper." Between Eliza and Nancy, Devil Bill fathered four children in two years—one a boy named John Davison Rockefeller—born on July 8, 1839.[2]

Eliza and the, by now, five children grew up poor, wearing tattered clothes and feeling anxious about when, or whether, their breadwinner would return. Posterity reveals Devil Bill wandered mostly west, to the fringes of the mid-nineteenth-century frontier, where his vocations included trader, sharpshooter, trinket peddler, and, most prominently, snake-oil salesman. His "herbal medicines," which he proclaimed would cure cancer and most other serious maladies, fetched a premium price of $25 ($780 in 2025 dollars), a steep enough levy to drain the bank accounts and mattresses of many a soon-to-be bereft family.[3]

In 1853, Devil Bill moved the family west to Cleveland, Ohio, to be closer to the frontier (and farther from the law), and assumed the new traveling identity of Dr. Levingston. While peddling his wares in Ontario, Canada, he met and married a seventeen-year-old Canadian girl named Margaret Allen. Margaret took Devil Bill's alias of Levingston and neither family was aware of their mirror clans living in distant households.[4]

Devil Bill's eventual abandonment of his first family affected the teenage John D. Rockefeller in several ways. No doubt the economic insecurities he experienced as a child, with the family living off store credit and the generosity of neighbors for months at a time, influenced his adult hunger for money and stability. He may have also inherited his father's enthusiasm for a thick wad of cash. By describing his mother as "widowed" to acquaintances, John's shame and frustration were evident.[5] In 1855 John was informed by his father that he lacked the funds to allow John to attend college. Feeling the need to become the substitute breadwinner of the family, John dropped out of school at age sixteen and sought gainful employment. "There were younger brothers and sisters to educate and it seemed wise to go into business," John recalled.[6]

John was never a standout in his academic pursuits—with the exception, perhaps, of debate, math, and public speaking—yet, as a job hunter he beat the streets with confidence and lofty ambitions. He called upon only the most successful Cleveland enterprises—banks, railroad offices, and wholesale merchants. Unfortunately, after knocking on doors from morning until night that fateful summer, John was rewarded with nary a flicker of interest.

Patience, attention to detail, and persistence were to become John's defining business attributes. He called upon all respectable businesses a second time, seeking only a bookkeeper's position, but still to no avail. "No one wanted a boy," Rockefeller recalled, at least not a tall, string bean, high school dropout with mediocre grades and a somewhat dopey expression on his childish face.[7]

The future plutocrat's big break came in 1855 when a wholesale produce and commodities supplier took pity on the hapless youth. Within weeks, the meticulous sixteen-year-old John was discovering costly inefficiencies within Hewitt and Tuttle procedures and fingering swindlers and bamboozlers within the corporate supply chain. The firm's "casual way of conducting business did not appeal to me," Senior later recalled. After only three months, the baby-faced bookkeeper was rewarded with a meaningful promotion and significant pay raise.[8]

The teen with the twenty-five-dollar-a-month salary quickly threw his energy and financial resources into the activities of the lower crust Erie Street Baptist Mission Church. His punctual tithing to the church was in addition to his support for mother and siblings. Such benevolence came at a time when clergy advised their flock to "gain all they can [and] save all they can [to] give all they can [and thereby] grow in grace."[9] Rockefeller's lack of self-indulgence even allowed for small contributions to an African American church and Catholic orphanage in Cleveland as well as the inauguration of a lifelong practice of handing out coins to less fortunate parishioners on Sunday mornings. In 1859, when his carefully kept personal ledger recorded charitable giving in excess of 10 percent of his gross earnings, he helped a man buy his wife out of slavery. Historian Allan Nevins noted, "He had not waited to become rich before he became generous."[10]

As Rockefeller continued to climb the few rungs of the merchant's corporate ladder, his eyes began to scan the horizon for bolder enterprises. Not far away, a boom similar to a gold rush was sweeping western Pennsylvania and overflowing into Cleveland: rock oil. The pesky effluent had been seeping from the ground around Titusville, Pennsylvania, since time immemorial, but no one paid it much mind other than a few flimflam men who pawned it off to the unwell as a natural panacea. Then some American and British chemists stumbled upon a method of refining oil by mixing it with sulfuric acid to produce kerosene.

Kerosene was revolutionary. In the gloomy days before kerosene's mass production, the citizens of the world lit their homes mainly with whale oil lamps, but also with lard-based candles, coal oil, or turpentine oil. As the number of whales still swimming the oceans inevitably declined, the price of the commodity rose until only the well-to-do could afford the bright, clean-burning blubber derivative. The less affluent households had to settle for the smoky, dim, lung-tarnishing light from inferior combustibles.[11]

Therefore, while most consumers today consider the oil industry the antithesis of good environmental stewardship, in the 1860s it was viewed differently, especially if you were a whale. Cheap, nearly smokeless kerosene allowed the family, farm, and business to be more productive after sunset, a long leap for humanity toward the affluent, intellectually stimulating lives we've now accepted as birthright.

Rockefeller soon came to believe the fledgling oil refining business was the big opportunity he had always been seeking. Even as a lowly Cleveland bookkeeper, he was not only determined to be rich, he also trusted that God's plan for him was to become rich and accomplish many great things in His name. As business partner Maurice Clark stated, "John had abiding faith in two things—the Baptist creed and oil."[12]

Rockefeller's parsimonious nature gave him an advantage over his numerous Cleveland competitors. Not only did he save some 50 percent on the cost of a Pennsylvania forty-two-gallon oil barrel (a standard of measure set in the wooden-barrel period that still stands today) by coopering them inhouse, Rockefeller also prescribed that his oak staves be kiln dried, making the barrels lighter and cheaper to transport. Later, he was an early adopter and builder of railroad tank cars, the next big step forward in efficient shipping. Rather than dumping the by-products of his refining into the Cuyahoga River (yes, the one that burned), Rockefeller used leftover gasoline to fuel his refineries and found markets for paraffin, benzene, and petroleum jelly. He also grasped early on the long-term advantage of hiring associates who possessed both sound moral character and exceptional business skills.

The twenty-five-year-old Rockefeller's sudden rise to power in Cleveland coincided perfectly with the end of the Civil War and the advent of one of the greatest economic boons America would ever see. In hindsight, writers like Mark Twain (who would eventually become a family friend) pinpointed the end of the war in 1865 as the dawn of the Gilded Age (a term coined by Twain).

Still, the oil business was no place for the meek or anxious. Many feared the honey hole in western Pennsylvania represented the sum of all crude oil on our planet. Other aggravations included volatility in the supply of crude—every gusher caused the price to plummet while a series of dry holes created a spike. Prices fluctuated between ten cents and ten dollars a barrel in a given year, resulting in kerosene too cheap to profit from or too expensive for consumers to purchase.

For the young Rockefeller, the opportunistic solution to these erratic supply swings was to consolidate and control. In 1871, when a glut of oil caused prices and profits to dive, Rockefeller and his associates went on a buying spree in the sludgy industrial districts of Cleveland. With the threat of Rockefeller and the railroads colluding to create a theoretically formidable cartel, twenty-two of Cleveland's twenty-six refiners sold out to "the Standard" oil company.

Rockefeller was polite and perhaps generous when he acquired a competitor's facility. The smaller companies were paid a price determined by an assessor, and, considering the circumstances (the Standard's competitors in Cleveland had a slim prognosis for survival), Rockefeller paid fair value or better to acquire (and often dismantle) their marginal operations and regularly integrated the staff of the overtaken companies into the Standard.[13] Better yet, the titan frequently offered to compensate the smaller refiners with Standard Oil stock, an option which, if accepted, would eventually make them wealthy.[14]

Not that Rockefeller was benevolent in his cornering of the oil market. His participation in a scheme to create a massive cartel called the South Improvement Company harried all operators not included in the fold and forcefully pushed many toward abandoning their businesses. When details of the cartel's onerous pricing pressures leaked, there was such an outcry that the Pennsylvania legislature yanked the company's charter. The intensity of Rockefeller's zeal for absolute control of everything oil, as well as the intensity of his Christian faith, have intrigued historians for over a century and a half. Yet Rockefeller's preferred style of Christianity actually looked quite favorably upon wealth. In Senior's own words, "I believe it every man's duty to get all he honestly can, and to give all he can."[15]

When the dust had settled in Cleveland, Rockefeller had not only eliminated most of the competition (as well as middlemen) in town, he had attained an economy of scale that allowed him enormous leverage when negotiating shipping rates, bulk oil purchases, pipeline access, and other links in the kerosene supply chain. The commodity was no longer simply a national product, either; households in Europe and Asia were beginning to use kerosene. Standard Oil, so named for its goal of providing a product of consistent quality, had become a colossus.

*

No one was better at informing the world of Standard Oil's monopolistic practices than Ida Tarbell, America's first and foremost muckraking journalist. Tarbell grew up in the petroleum-slickened streets of Titusville, Pennsylvania, home of Oil Creek and the epicenter of the mid-1800s rock oil rush. Her father, who progressed from a cooper of oil barrels to an independent oil producer and refiner, was a typical Titusvillian caught up in the boom. Ida Tarbell recounted as her halcyon days those bustling times when gushers flooded the hollows of Titusville with black torrents and her homeland was known around the world for its rough-and-tumble gold rush mentality. From her perspective, the best part of the bonanza was the large number of small, independent drillers and affiliated entrepreneurs merrily vying with one another across a level, if slippery, playing field. It was an egalitarian (almost agrarian) view that was certain to be challenged as the age of great industrialists dawned.

Tarbell was no impartial observer when she accepted the assignment to write an exposé on Standard Oil in the prestigious *McClure's Magazine*. Not only had "the Standard" tanked her father's livelihood, her brother was founder of Pure Oil, one of Standard's few surviving competitors. Yet, as a work of journalism, "The History of Standard Oil" is admirably researched and extremely well written and organized. It helped catapult *McClure's* from a well-regarded monthly to a highly acclaimed—and wildly popular (circulation of 350,000)—leader in the field of progressive journalism.

Tarbell's criticism of Standard Oil was not for their success and massive size but the evidence that they cheated to get there. Much of her disparagement zeroed in on Standard's sneaky deals with the railroads, which were also run by monopolist robber barons like Cornelius Vanderbilt, Edward Harriman, Jay Gould, and Leland Stanford. For the most part, however, Standard's preferential deals were perfectly legal and even initiated by the railroads, though they did pave the way to monopolization.[16]

Even more serious were Tarbell's accusations of bribery of elected officials. Rockefeller and his key associates became increasingly interested in politics when legislation to regulate interstate commerce and oil pipelines began to circulate in Congress. The great danger fomented by political corruption during the zenith of the Gilded Age lay in the fact that it was almost ubiquitous. Lobbyists strolled Washington, DC with bags of cash destined for crooked politicians' campaign funds. Big business had grown to become greater than big government, and compared with the cost of building a railroad or acquiring a competing refinery, politicians were favorably priced.[17]

The fact that America had become populated with some two dozen plutocrats, each so rich from their monopolies that they could stack state houses, the U.S. Congress, even the presidency in their favor, threatened our nation's institutions and opened a pathway to a future that looked more like an oligarchy of tycoons in steel, railroads, oil, banking, sugar, real estate, shipping, tobacco, copper, and the stock market than a functioning democracy. As Ron Chernow expressed in his definitive Rockefeller biography, *Titan*, "Modern industry not only menaced small-scale commerce but appeared to constitute a sinister despotism that endangered democracy itself as giant corporations overshadowed government as the most dynamic force in American society."[18]

Considering these dangers, Tarbell and the press (there were some ten thousand newspapers in America in the late nineteenth century) did citizens a great service by beginning to reveal and unravel the monopolies that Rockefeller and the other tycoons had so masterfully woven. *McClure's* and Tarbell, however, were not the originators of social reform. By the 1880s, Progressive Era "trustbusting" was becoming an almost universal cry among voters and within the ranks of both Democratic and Republican political parties. America by then was crowded with robber baron mansions—the Vanderbilt family had over thirty, one of which was the 250-room Biltmore House in North Carolina, the largest privately owned home in the country. Working-class people, many of whom toiled up to seven days a week and might live in dank, windowless tenements, were stirring.

Not every Rockefeller biographer (and there were many) was as critical of Senior's business practices as Tarbell. Most objective historians have consistently acknowledged Senior for his important innovations in business, including creating efficiency and economy in the chaotic oil business, setting a new high standard for quality of personnel, and making corporate decisions by consensus rather than decree. Business writer Charles R. Morris called him "not only the first great corporate executive but one of the greatest ever."[19] They also concur his vertically integrated monopolies were somewhat inevitable because of the nation's rapid industrialization and lack of regulation within a capitalistic system. There is little disagreement, however, that Senior's monopolies needed to be unraveled by government action to preserve fair business practices and democratic institutions.[20]

*

Ironically, even though Senior's monopolistic practices and concentration of wealth were catalysts for Progressive Era reforms, Rockefeller philanthropy was a model of progressivism. The Rockefellers' gargantuan philanthropic endeavors included the General Education Board, Laura Spelman Rockefeller Memorial, and Rockefeller Institute, all devoted to the betterment of humankind, mostly through improvements in health care and education.[21] Author Karl Zinsmeister recently wrote that the foundations Senior begat led to thousands of breakthroughs and accelerated research in medicine and biochemistry for sixty Nobel laureates.[22]

The simplest explanation for this apparent contradiction is that while Senior's fixation on relentlessly expanding the Standard Oil empire could certainly be labeled as greed, he did not create the company out of malice. Rather, as historians Nevins and Ralph W. and Muriel E. Hidy contend, it was Senior's "innovations in administrative techniques" as well as "methods of production and manufacturing" that made him so successful.[23] As we shall see throughout this book, Rockefeller philanthropy was largely motivated by their Christian morals and sense of duty. Senior viewed himself as a "steward of wealth which God had placed in his trust."[24] John D. Rockefeller Jr. (referred to hereafter—in the family tradition—as "Junior") was carrying out a family custom of charitable giving established by his parents. Philanthropic duty was instilled in Junior and his siblings at a young age by Senior and Laura. Junior recalled his mother frequently raising the same question, "Is it right, is it duty?"[25] It was reinforced by corporate officers like Frederick T. Gates, a former Baptist clergyman who served as Junior's mentor in business and giving. Senior described Gates as possessing a "passion to accomplish some great and far-reaching benefits to mankind, the influence of which will last."[26]

Both John D. Rockefellers aligned themselves with the progressive view that most of society's major ills could be solved through business-like and scientific methodologies. Senior became a pioneer in both business and philanthropy by hiring a large network of the best and brightest professionals and effectively delegating most of the day-to-day work to their desks. Junior was very much aligned with the progressives in support of women's suffrage and a prohibition on alcohol. He joined with the progressives and a growing number of conservationists and conservation organizations in feeling that God could be found in nature, and that nature offered relief from the stresses of city life.[27]

Junior's work as a full-time philanthropist began in the middle of the Progressive Era (1890–1920). The progressive crusade was an optimistic movement characterized by campaigns to uplift the poor, educate the illiterate, break up the monopolies, fight political corruption, cure the sick, and create public parks, libraries, and museums. Most concisely, it was an effort to place the public interest over special interests. Junior embraced the spirit of the movement with unprecedented energy and devotion, even serving on a Progressive Era hallmark—a New York grand jury formed to crack down on prostitution and white slavery in the city. As wife Abby Aldrich Rockefeller once confessed to Junior's advisor and biographer, Raymond B. Fosdick, "I never know where John is anymore, but he is out saving the world somewhere!"[28]

Unfortunately, siding with the progressives occasionally landed Junior in the same bandwagon as some of the era's most prominent eugenicists, including Madison Grant, Henry F. Osborn, and Richard Corwin. It's likely the concept of eugenics interested Junior and other philanthropists because it promised to treat problems like crime, mental illness, disease, and poverty at the source, rather than simply addressing the symptoms. Prior to the rise of the Nazis in Germany during World War II, other conservationist-progressives, including President Theodore Roosevelt and U.S. Forest Service head Gifford Pinchot, erroneously saw potential in the flawed science.

Efficiency was a hallmark of progressivism, and that quality also steered leading progressives, including Theodore Roosevelt and Pinchot, toward conserving America's forests, waterways, wildlife, and other resources. Wildlands could be used more efficiently if they were protected from fire and disease and in other ways scientifically managed.[29] The peak of Theodore Roosevelt's national forest and wildlife refuge crusade was 1908–1909, only three or four years before Junior's transformation from an owner of a large private estate to an advocate for sharing the great outdoors with the public.

Conservation was certainly in the air when some prominent nature lovers on Mount Desert Island in Maine, Charles W. Eliot and George B. Dorr, recruited Junior to help them protect their wilderness paradise from tacky tourist developments and the loggers' saws. By then, Junior was experienced in philanthropy, had a strong emotional bond with the beauty of Mount Desert Island, and possessed a lifelong passion for landscape architecture. All three of these ingredients prepped him for the transition from lord of the manor to patron of the parks. It was also during the opening decades of the twentieth century that the western frontier was officially closed, and many Americans were noticing that scenery-rich wildlands were being replaced by farms and towns. Ann Rockefeller Roberts believed Junior's (her grandfather's) conservation work was motivated by his "deep personal feelings for the earth and its beauty, his conviction that the Divine Presence is revealed in nature, and his belief that nature plays an important role in our lives."[30]

While Junior's conservation legacy is rightly overshadowed by the unprecedented accomplishments of his contemporary, President Theodore Roosevelt, one should consider that most of the president's scores of national forests, monuments, and refuges were created by the stroke of a pen, while the president was on the clock with the federal government. Junior, to accomplish his arduous conservation good deeds, had to dig into his own pockets and toil as a volunteer. He could have been island hopping in Greece or betting on the ponies at Saratoga, but preferred to spend his hours hunched over maps, reports, proposals, ledgers, blueprints, and unending correspondence, stoically accomplishing all the things he considered duty and worthwhile.

One might presume, after reading of Senior's vilification by Tarbell and others in the press, that much of the Rockefellers' conservation philanthropy was an attempt to redeem the family name. And while Junior was deeply wounded by Senior's humiliation and was ever conscious of the family's legacy, his primary motives for his national park projects did not include what we would call today "greenwashing." To the contrary, Junior is little known and certainly underappreciated as a public lands conservationist. This is largely due to Junior's humble nature, but it also supports the contention that he wasn't mainly out to cleanse the family name. The fact that he conducted much of his philanthropy anonymously and always eschewed dedications and other such ceremonies and celebrations also supports this assertion. When the rare plaque recognizing Junior's philanthropy was erected in a park, it was usually done posthumously, or without Junior's knowledge or consent.[31]

Further, many if not most of Junior's conservation efforts elicited firestorms of controversy and criticism, making them poor choices as public relations ploys. Some of the campaigns even mushroomed into highly publicized congressional investigations. Parks like Great Smoky Mountains and Shenandoah required the displacement of thousands of people, creating embers of resentment that continue to smolder today. More importantly, Junior's widely accepted motives for his conservation philanthropy are well documented and ring true to his upbringing and psychological complexion.

Junior's national park philanthropy was extraordinary in many ways. It was an odd and pleasantly congenial collaboration between the world's richest person (who happened to be a nature lover) and several woodsy bureaucrats from the federal government. Among the latter, Stephen T. Mather, Horace M. Albright, Arno B. Cammerer, Aileen and Jesse Nusbaum, and George B. Dorr are deservedly ranked as National Park Service legends. They were brilliant, workaholic, courageous, tenacious leaders who made painful sacrifices for the national parks cause. The ease in which a Rockefeller melded with these famous park service icons and granted their conservation wishes—almost like a genie coaxed from a magic lamp—is one of the most remarkable parts of the story.

It is unusual and noteworthy that, because of their public-private nature, none of these philanthropic endeavors became monetary black holes. None requires massive influxes of private cash on an annual basis to keep the good work going. Quite the contrary. Because of the well-deserved popularity of America's crown jewel national parks, the taxable tourism revenue they generate each year is more than enough to cover the government's annual operating costs.[32]

There are, happily, a lot of philanthropists in the world and the sums they have given to great causes deserve our gratitude. Yet, if you have ever watched a sunrise over the Atlantic from Cadillac Mountain in Acadia, gazed upward to the crown of a three-hundred-foot-tall redwood tree in California, visited Linville Falls along the Blue Ridge Parkway in autumn, watched the sun set from Morton Overlook in the Great Smoky Mountains, or traipsed through a sugar pine forest in Yosemite, you are likely to concur that no conservation philanthropist has ever accomplished more than Junior and his allies.

Every year, an astonishing thirty-five million people or more visit the national parks which Junior played a major role in establishing, expanding, or improving.[33] No other philanthropic projects have ever been more enduring or economically viable. Our nation is fortunate indeed that John D. Rockefeller Jr. resisted the temptation of other Gilded Age tycoons to litter the globe with yachts and mansions, choosing instead to toil and bear the slings and arrows of conflict to bequeath for us a legacy of incomparable natural beauty, open to all, forever.

ISLAND IDYLL

Acadia, 1908–1913

It was the unborn Nelson A. Rockefeller who led Abigail (Abby) Aldrich Rockefeller and husband John Davison Rockefeller Jr. to Mount Desert Island, Maine. Abby was pregnant with her third child in 1908 and was having a hard time of it. Her biographer, Bernice Kert, reported Abby was "in and out of bed all winter with a cough and rheumatism." When Abby's physician, Dr. Thomas, announced that he would be summering in Maine on Abby's due date, Abby and Junior resolved they would be doing likewise.[1]

The expecting parents, along with youngsters Abby (Babs), age five, and John D. Rockefeller III, age two, supported by several staff members and a stable of fine horses, assembled at Grand Central Station at eight p.m. on June 19 and boarded the comfortable all-Pullman-car night train from New York City to Hancock, Maine. The following day they caught a ferry across Frenchman's Bay and arrived in Eden (now Bar Harbor).[2]

A healthy Nelson made his perfectly timed debut shortly after noon on July 8, the same day his famous grandfather, John Davison Rockefeller Sr., was celebrating his sixty-ninth birthday. Nelson, however, showed definite signs of favoring the extroverted, self-confident Aldrich side of the family (as opposed to the much more reserved Rockefellers) almost immediately, and therefore his Christian name, after Abby's father, Senator Nelson Aldrich, proved providential. Perhaps because of the coincidental birthday, Nelson also became one of Senior's favorite grandchildren.

Records of doctor's bills and fees for nurses indicate that Abby continued to struggle physically during and after the birth of the couple's third child. Despite these hardships, the growing family developed a favorable opinion of Mount Desert Island by summer's end. They had rented a handsome "cottage" on the water constructed for the real estate–savvy Sears family, one of the richest clans in New England. The well-situated abode—"The Briars"—had the pedigree of being designed by one of Ralph Waldo Emerson's cousins, William R. Emerson, purported originator of shingle-style architecture. Tragically, a mere five weeks after Nelson's birth, the Sears's only son, Joshua "Monty" Sears was killed while driving his new roadster "at high speed late at night" in Rhode Island.[3] Had the family elected to summer in Eden, as was their custom, Monty might well have been spared the untimely death; automobiles were still prohibited on Mount Desert Island.

By the 1908 summer season, Eden had more in common with Newport, Rhode Island, and the Hamptons than the secluded destination for bird watchers and other nature lovers it had once been. There were dances and entertainments almost nightly in the ballrooms of the grand hotels, and young singles were drawn to the shores to mingle, flirt, and cavort.[4] Rumors circulated that some of the well-to-do men had even opted to bring their mistresses, in lieu of wives, to the prestigious watering hole. Consequently, when the Rockefellers returned for the summer of 1909, sans pregnancy, they chose the much more staid and studious Seal Harbor, some nine miles distant.

"Bar Harbor was too flashy and ostentatious," surmised David, Junior and Abby's last child. As a youth, he recalled observing Gatsby-esque parties given by ultrawealthy families like the Dorrances (of Campbell Soup fame) "with bands playing on yachts...and dancing all through the night. Speedboats carried guests back and forth, and champagne flowed for all ages."[5] By contrast, Seal Harbor was downright scholarly—known for its summering professors from Yale, Harvard, and Johns Hopkins, including Harvard president and Rockefeller associate Charles W. Eliot. Other summer people included authors James Ford Rhodes and Winston Churchill, and a plethora of accomplished physicians.

Toward the end of their second summer, Abby and John were invited to dinner at a stately Tudor summer "cottage" built in 1900–1901 by a professor from Massachusetts. The home overlooked Seal Harbor and was being rented by the Miltons, family friends from Pocantico Hills, New York, whose son, David Milton, would one day fall in love with "Babs," the most rebellious of John and Abby's six children. The cottage's front terraces commanded views of the wild Atlantic, including Sutton Island, the Cranberry Isles, and the Eastern Way. From the opposite side, one's eyes were drawn upslope to a forested mountain wilderness, including The Bubbles and Penobscot Mountain. The cottage contained sixty-five rooms and sat on sixteen acres. Rockefeller grandchild Eileen described it as "perched like an eagle's nest on a granite ledge overlooking the approach to Seal Harbor." Like all Maine cottages, it had a name: the Eyrie.[6]

Junior was immediately taken with the place. "He totally adored it," granddaughter Ann Rockefeller Roberts recalled.[7] He inquired of the Miltons if they planned to return the following summer? They did not. Perchance did they have a desire to buy? None whatsoever. Junior quickly arranged its purchase for $26,000 ($900,000 in 2025 dollars) through Stebbins Realty. It was the first of scores of land purchases that Junior would arrange through George L. Stebbins over the succeeding decades.[8]

With the purchase of the Eyrie (likely accomplished with funds from Senior), Junior continued the pattern of country estate acquisition and development established by his family at their beloved Forest Hill property on the bluffs above Cleveland, Ohio, and Pocantico Hills, New York, overlooking the Hudson River Valley. The "Rockefeller Way" was to purchase an existing home with acreage far enough from the hustle and bustle of the city to afford open space to ride horses and wander through the woods. The homes themselves were not particularly interesting, and Laura Spelman Rockefeller, Junior's pious mother, showed her thrift and disdain for material things by opting not to refurnish or redecorate most of their second-hand domiciles.[9]

The true purpose of the estates was to encourage healthy outdoor recreations, build carriage roads to scratch the Rockefeller family amateur landscape architect itch, and gain a measure of privacy. For the entire Rockefeller clan, Forest Hill evoked sentimental feelings of family closeness and youthful vigor while providing opportunities for outdoor play and safe harbor from the adversities of business and city life. Senior always had an eye out for available land and continually acquired adjacent properties until the estates sprawled for thousands of acres and offered adequate breadth for the careful design and construction of miles and miles of horse and walking trails and carriage roads. Graciously, Senior allowed the public to use most of these privately owned facilities.[10]

Time spent in the out-of-doors, including doing rote physical labors like planting trees, chopping firewood, clearing brush, sawing rails, and tapping trees to make maple syrup had long been a prescription for Junior's many psychosomatic ailments. He had been tormented by nervousness, depression, and insecurities since childhood. He suffered frequent head-popping migraines, sinus infections, and even temporary deafness; he slept poorly, experienced dyspepsia and a variety of other digestive disorders, and had had the rug yanked from beneath his feet by lengthy nervous breakdowns. During middle age, while enjoying a stretch of improved health, Junior reported he was sleeping better and for the first time in many years had gone four consecutive days without a headache.[11] Junior's future interest in national parks was no doubt entwined with his perception of nature as a tonic for stress and anxieties.

Abby shared Junior's affection for Mount Desert Island. "She loved the beauties of nature: flowers, the song of the wood thrush in the forest, and the crashing of waves on the beach," her youngest son, David, recalled.[12] She was less absorbed by religion than the Rockefellers but much more infatuated with art and architecture. By 1911 she and architect Duncan Candler had initiated a five-year expansion and remodeling of the Eyrie which produced a respectable Tudor-style cottage of 107 rooms staffed by 22 uniformed cooks, butlers, maids, and gardeners. Its walls and tables soon became tastefully adorned with art, with several rooms dominated by Abby's lovingly collected buddhas. The Asian influence eventually spread outdoors into the famed Oriental (now Rockefeller) Garden. Thanks to Abby, Candler, and landscape gardener Beatrix Farrand, the Eyrie became a postcard-pretty estate that was in fact featured on several postcards available for purchase at tourist shops in Eden.[13]

True to his Rockefeller roots, Junior, from 1910 to 1914, worked discreetly with realtor Stebbins, acquiring small parcels of land in the vicinity of the Eyrie as they became available. At the time Stebbins was treasurer of the Hancock County Trustees of Public Reservations (HCTPR), a group described by Park Service historian Richard H. Quin as "summer residents who wanted to protect the island's mountain summits and other scenic spots from development...and the island's diverse forest from logging."[14] The Rockefeller estate gradually grew with acquisitions near the Eyrie and west to Long Pond. The pond surely reminded Junior of boyhood days at Forest Hill where he swam and boated and enjoyed ice skating with his father and sisters on the estate's small manmade ponds. Such sentimental memories no doubt provided emotional thrust for the half-century of conservation work Junior was soon to embark upon.

However, Junior's memories of growing up very wealthy and living on large estates were bittersweet. He recalled having "no childhood friends, no school friends."[15] Friday was prayer meeting night and Sunday was church and Bible study and rest—no work or play; even cooking a meal was forbidden. A former tutor described the Rockefeller home as lacking "the laughter of youth, the light heartedness, the romping about." The tutor recalled a spooky old house filled with "silence and gloom."[16] There was no card playing, no dancing, no theater going, and little if any socializing with those from outside the church. In 1918, Senior confessed to one of his biographers that he "had never had time to become really acquainted" with his only son.[17] Junior got along well with his three sisters, Elizabeth, Alta, and Edith, but they were older, and they were girls.

Although the Rockefellers' wealth eventually eclipsed that of the other great industrialists and financiers of the period, the family escaped the mold of typical Gilded Age plutocrats. The puritanical Rockefellers feared great wealth and its accompanying temptations like it was the dark lord himself. They considered the flamboyance and extravagance of the Vanderbilts and Morgans as not only foolish but sacrilege. Junior, for example, wore his sisters' secondhand clothing until age eight. "Their dresses were always being made over for me," he recalled.[18] Later, on the occasion of his twenty-first birthday, the son received from his father—the richest man in America—a gift of twenty-one dollars. Included with the check was a letter filled with gratitude and encouragement from dad and stern warnings from mom and grandma Spelman. They dourly reminded the heir to the Rockefeller throne of his Christian duty and the temptations one must be wary of while away at college, chief among them dancing. The son's prompt thank you note gushed genuine kindness and appreciation. He called the sum of the bank draft as "perfect."[19]

*

Junior's very first carriage roads on Mount Desert Island connected the Eyrie to his stables and circled Long Pond at the bottom of Barr Hill. Like all Rockefeller riding trails, they were painstakingly designed to lay lightly on the landscape and offer the user a series of experiences such as scenic views of the ocean and glimpses of meadows, streams, and noteworthy trees. He continued his land acquisition with significant purchases stretching all the way from Long Pond to Jordan Pond House, an idyllic lakeside teahouse established in 1895.

By journeying from New York City and their sprawling Pocantico Hills estate to Mount Desert Island, the Rockefellers were following in the physical and intellectual footsteps of the Hudson River School of American landscape painters, including Thomas Cole and Frederic Church. During the mid-nineteenth century, the artists' romantic paintings of mountains and seashore, illuminated by divinely golden glows of light, turned the eyes of the world toward places like the White Mountains, Catskills, Adirondacks, and the stony summits of Maine. "Prophets of old retired into the solitudes of nature to wait the inspiration of heaven," Cole wrote in his 1835 "Essay on American Scenery."[20]

Yet Cole and Church, along with their contemporaries Durand and Bierstadt, were capturing more than scenery in their oils; they were expressing subliminal feelings about the tension between wilderness and civilization. From the artists' perspectives, America's true riches lay in its wild places, a radical departure from the attitudes of the Old World where art and culture reigned supreme, and nature was relegated to hedgerows and highly manipulated gardens. The American painters, who were something like pop stars in the era before landscape photography, depicted the wilderness as dangerous—yet Godly and gorgeous—and sadly vulnerable to the utilitarian obsessions of the European psyche. Artists like Cole also hinted that Nature would have her way in the end. [21]

Likewise, the writers Ralph Waldo Emerson, Henry David Thoreau, and others with deep New England roots had trumpeted the importance of self-reliance and nurturing a close connection with nature seventy-five years before the Rockefellers began falling in love with the wildness of Maine. Writers and philosophers of that day preached that by immersion in nature, one could forge their own personal path to God and a better understanding of the universe. As Emerson put it, "In the woods, we return to reason and faith."[22] As a deeply religious man and a person who sought healing from nature, Junior likewise perceived God in nature. "To see a tree coming out in spring was enough to impress me with the fact that God existed," Junior told his biographer, Raymond B. Fosdick.[23] This association of God and nature was undoubtedly a catalyst for his eventual blooming as a conservationist.

However, the Mount Desert Island of the early twentieth century was radically different from the period when Cole and Church had set up their easels, due largely to the influence of Cole and Church. The painters' awesome landscapes had attracted the first wave of sleep-on-the-ground travelers, mostly other artists and nature enthusiasts. Whether camping or boarding with locals, these pioneer tourists reveled in the island's abundance and diversity of natural wonders and were invigorated by the clean air and rigorous lifestyle. College professors, students, clergy, physicians, and lawyers were quick to follow and stake their seasonal claim to Eden.

Once the travel writers discovered the coast of Maine and inevitably bellowed its praises, the floodgates opened and bourgeoisie and some of the better-off proletariats sailed off to Mount Desert Island. The journalists emphasized the island's rare views which encompassed both mountains and sea. George W. Nichols, writing for *Harper's Magazine*, declared, "You will gaze upon the most beautiful views you have ever seen."[24] Regular steamship service, and later the railroad, made the once-secluded haven increasingly accessible from Boston, New York, and Philadelphia. Locals, who viewed the exotic tourists as profuse sources of income as well as amusement, expanded their boarding houses into hotels and set up tintype shops and eateries along the downtown streets.

The rising tide of "rusticators" (well-to-do families leaving the city for country estates) arrived long before Nelson Rockefeller's birth in Eden. By the mid-1880s, many of the families who were the engineers and benefactors of the Gilded Age—including such eminent names as Vanderbilt, Pulitzer, Archbold, Stotesbury, Procter, Bowler, Howard, Sears, and Dorr—had built mansion-sized summer places along the harbor in Eden. These old-money rusticators likely considered themselves somewhat aloof from the Rockefellers and other newcomers to the island. The well-established Brahmins were, perhaps, a bit more refined than the party crowd and more likely to be critical of any change to the island that might upset their staid rustication.

The presence of both the old and new guard attracted others seeking influence and notice, including princes, senators, and presidents. Yet, since the mansions were clad in redwood shingles and were for summer use only, the owners, brimming with false modesty, referred to them as "cottages." Dozens and dozens of these wood-clad castles sprang up between 1878 and the first World War, most boasting over fifty rooms and with quarters for as many as thirty staff. They were designed by the best architectural firms the Northeast had to offer—Andrews, Jaques & Rantoul, W. H. Day, Emerson—and were landscaped for the brief summer season by the world's best professional landscape designers, including Beatrix Farrand and the Olmsted brothers.

*

The Rockefeller habit of enlarging estates and constructing carriage roads might have continued indefinitely (at its peak, Pocantico Hills sprawled to over 3,400 acres with 52 miles of carriage roads) if not for a letter written by Harvard president Charles W. Eliot to Junior on February 25, 1915. As a fellow Seal Harbor summer resident, advisor to Abby, and Rockefeller General Education Board (GEB) trustee, Eliot and Junior were well acquainted. The letter, like most correspondence addressed to a Rockefeller, sought money; specifically, $15,000 ($467,000 in 2025 dollars) "to see a large part of its [Mount Desert Island's] hills, valleys, and ponds preserved for the enjoyment of the present and future generations." From his work with the GEB, Eliot knew the Rockefellers preferred to work as partners in their philanthropy. Consequently, he provided a long list of others who had already given money and lands to protect the area. Eliot even mentioned that George B. Dorr "had already put more money into this enterprise on behalf of the public than he should have done; and his estate was seriously embarrassed in consequence."[25]

The concept of purchasing private lands and converting them to public use, as Eliot's letter suggested, would not have been entirely foreign to Junior. As stated in the introduction, the Rockefellers were always generous donors to the church, the needy, civic organizations, and myriad other causes. Senior had contributed an enormous sum, $500,000 (over $16.5 million in 2025 dollars) for the development of Palisades Interstate Park, along New York's Hudson River, around 1910. According to Junior's biographer, the donation was "largely stimulated by the younger Rockefeller [Junior], who had known and loved the Palisades since his boyhood days." The park was also supported by J. P. Morgan and was cosigned into existence by New York Governor Theodore Roosevelt and New Jersey Governor Foster Voorhees.[26]

The senior John D. Rockefeller had been a generous charitable giver even back in his younger days as a lowly mercantile clerk, but the supercharging of Rockefeller philanthropy is often attributed to Frederick T. Gates. An office mate and mentor from Junior's earliest days at Standard Oil headquarters, Gates was a former Baptist clergyman with a highly energetic, passionate form of charisma who also possessed extraordinary business acumen. This unusual combination of business smarts and moral authority had proven invaluable to Senior during his Standard Oil years, and the elder Rockefeller was wise to position Gates as a surrogate father figure to Junior. On one very important topic—which pertained to both Junior and Senior—Gates was evangelical: the destructive powers of great wealth. Gates espoused the urgent need for organized philanthropy on a grand scale. "Your fortune is rolling up, rolling up like an avalanche!" Gates admonished Senior in a 1906 letter. "You must keep up with it! You must distribute it faster than it grows! If you do not, it will crush you and your children and your children's children!"[27]

The former preacher's great epiphany was to use his employer's bank account to spawn a new breed of philanthropy, one larger in scale and more thoughtful in execution than almost anything previously imagined. Rather than the Rockefellers and their staff randomly combing through the thousands of requests for money that arrived by mail and in person every single day (up to fifty thousand per month), Gates argued for organizing the Rockefellers' charitable giving by creating powerful foundations that would outlive Senior, Junior, and their heirs. The philanthropies would be governed by the Rockefellers and trustees who were chosen from the top experts in the fields. The institutions would exist for the purpose of promoting "human well-being" and their activities would be transparent and open to scrutiny from the best-respected thinkers and analysts of the day. The more specific causes undertaken by the crusade would include education, medical research, the arts, scientific agriculture, and citizenship and civic virtue.[28] Senior voted his acceptance of the proposal with $42 million (over $1.4 billion in 2025 dollars) in gifts to the General Education Board.[29] Later scholars described the pioneering transformation as "large-scale giving on a businesslike basis."[30] Junior, who would lead many of the foundations, expressed his consent in a letter to his father. "I believe this is a great thing, far greater than any of us can conceive today."[31]

Despite his handicaps of introversion and social awkwardness, Junior very much enjoyed meeting social crusaders, many of whom had a religious bent, and learning about their projects. Junior was not naturally chatty or quick with the one-liners, but he was curious about everything, from beet farming in Kansas to Holy Land archaeology. Thus, he moved cautiously but with efficacy from one diverse charitable project to the next. If nothing else, Junior was duty-bound in a way that rarely manifests itself in the genetic soup of humanity. "The training of my parents was to the effect that a duty could not be evaded," Junior recalled.[32] As Gates described him in his private autobiographical papers in 1929: "I have known no man who entered life more absolutely dominated by his sense of duty, more diligent in the quest for the right path, more eager to follow it at any sacrifice." Junior's friends and colleagues also recalled that he always finished what he started.[33]

Still, the conversion of Junior's beloved summering grounds into a public park would have required a significant change in attitude on Junior's behalf. Before the letter, Junior's long-term plan was likely to continue acquiring properties as they became available and building more carriage roads. Lots more carriage roads. By owning the land, he would be an unencumbered autocrat, free to pursue perfection in his hobby of landscape architecture, following in the footsteps of Frederick Law Olmsted, his sons, and others in the exciting new field. Instead, the letter, and subsequent meetings with the genteel Dorr and Eliot, nudged Junior toward one of the more fruitful collaborations in American conservation history. The success of the alliance is especially remarkable when you consider Eliot was nearly eighty years old and there have rarely been two men as dissimilar in temperament as Mr. Rockefeller and Mr. Dorr.

Sporting bushy eyebrows and a moustache as voluminous as some men's beards, Dorr resembled a police inspector from a nineteenth-century English murder mystery. He was an eccentric lifelong bachelor who was full of passion. Eliot described Dorr as an "impulsive, enthusiastic, eager person, who works at high tension, neglects his meals, sits up too late at night, and rushes about from one pressing thing to another."[34] Ann Rockefeller Roberts, Nelson Rockefeller's daughter, described him as "highly intuitive, enormously impulsive, very fluid." She added that Dorr "had a kind of fire."[35] Though born into a wealthy Boston mercantile family and educated at Harvard, Dorr struggled to discover a discernable career path other than dabbling in horticulture and a bit of writing, until he took up the cause to preserve Mount Desert Island. Conservation of Maine's mountains and coastline, as well as its cultural history, thus became his true life's calling.

Junior, on the other hand, was by both nature and nurture dispassionate, fastidious, and parsimonious—a man whom office mates claimed would toil for half an hour over the language in a telegram so as not to waste a nickel on an extraneous word. His accounting ledgers, mandated by John D. Rockefeller Sr., tracked his every expenditure, to the penny, dating back to childhood. The famed ledgers also tallied the kids' pennies given to charities, which was obligatory. Junior's college chums at Brown once chided him for "trimming frayed edges on his cuffs or standing in rapt concentration over a teakettle" attempting to separate two conjoined postage stamps.[36]

It's worth noting, however, that Rockefeller thrift was never practiced with the aim of purchasing larger yachts or more brilliant jewels, since neither Junior nor his parents considered such luxuries alluring. Money and its conservation were important to the Rockefellers because they believed their money had a purpose. Junior's many obsessions (with the possible exception of his pricey Chinese porcelain collection) orbited in the galaxy of lofty ideals and ambitious projects. During his myriad and protracted philanthropic undertakings, Junior frequently used words like "perfect," "ideal," and "taking pains."[37]

Eliot completed what would become known as the Acadia National Park "triumvirate."[38] His personality—formal, patient, thorough—was more akin to Rockefeller than Dorr. "Junior and Eliot were meticulous, orderly, planning types; very linear, very cognitive," Roberts summarized.[39] Yet Eliot and Dorr were both upper-crust Bostonians who had been visiting Mount Desert Island with their families for generations and had been collaborating on its preservation for at least a dozen years. One occasionally senses in correspondence that Dorr and Eliot looked upon Junior as somewhat of an outsider who lacked the thorough understanding and abiding love of the place that the old guard enjoyed.

Just as an unborn child steered the Rockefellers to Eden, a deceased son directed the triumvirate to protect the island. In conservation, Eliot reversed the typical pattern of generational succession by following in the footsteps of his son, also Charles Eliot, who died in 1897 at age thirty-seven of spinal meningitis. On Mount Desert Island, the younger Eliot was troubled by a real estate boom that seemed destined to plop summer homes, camps, and tourist facilities on the island's most scenic (and fragile) sites—particularly mountaintops, lakeshores, and the coastline. In essays for *Garden and Forest* magazine, a journal pioneered by Charles Sargent and Frederick Law Olmsted, Eliot frowned over "the small amount of thought and attention given to considerations of appropriateness" by the builders of summer houses. He likewise noted that "nowhere is lack of taste quite so conspicuous as on the sea-shore."[40]

Atop Green (Cadillac) Mountain, the summit was already obtrusively crowned with a couple of clumsily sited, rickety hotels. Access was provided by a screeching, smoke-belching, coal-powered cog railway system. The future seemed to promise shoulder-to-shoulder hotels, boarding houses, roads, utility corridors, billboards, hotdog stands, souvenir hawkers, and "absurdly pretentious" summer homes.[41]

Eliot's was one of the region's earliest (and youngest) voices for preserving scenery as public parks and came at a time when most Americans were still content with subduing the wilderness and plundering its resources. And unlike the establishment of Yellowstone and most other early national parks, the motivation was not primarily to create a tourist attraction to stimulate the local economy (since tourism was already well established). Eliot helped forge one of America's very first public land trusts, the Massachusetts Trustees for Public Reservations. "Just as the Public Library holds books and the Art Museum pictures," Eliot wrote, the trust would hold parcels of land. Trustees were presumed to donate their own lands to the cause and to acquire additional lands for expanding the reservations.[42]

Although there is a heaping measure of hypocrisy in summer people declaring that additional vacationers and summer home development will ruin the island, their concern is not without precedent. Places like Niagara Falls, celebrated by Church and other landscape painters, had long before been fouled by indiscriminate development and tacky tourist amenities. Critics pointed out that painters now had to intentionally exclude the souvenir stands and carnival barkers from their masterpieces, thereby skewing the reality of the place. Once a wonder of the world, Niagara had devolved into a county fair atmosphere with "rope-walking, diving, brass bands, fireworks," Frederick Law Olmsted lamented.[43] Ironically, in the nineteenth century as well as today, commercial tourism always seems determined to annihilate the features that attracted sightseers in the first place.

The real estate boom on Mount Desert Island also created a precarious situation in which the future teetered on a fulcrum between Old World and

New World sensibilities. The Old World model of countryside management, which made most freedom-loving, egalitarian-minded Americans cringe, was a fiefdom of estates owned by noblemen and patrolled by wardens. Upon such properties, according to national parks historian Alfred Runte, "trespassers were severely punished, and poachers could be put to death."[44] The New World model was being tentatively drafted in places like Yellowstone, Mesa Verde, Glacier, and Yosemite, where scenically and culturally rich lands were cordoned off from the creep of development and commercialization for the benefit of the public. Mount Desert Island could have tipped either way since the Vanderbilt, Kennedy, Rockefeller, and Sears families and others possessed ample wealth to sprawl their estates from one end of the fifteen-mile-long island to the other and erect fences and gate houses to exclude tourists, locals, and other riffraff.

Also vexing many Mount Desert Island inhabitants was a technological innovation referred to as the "portable sawmill." While a fair amount of logging had been happening on Mount Desert Island for decades, and some families' livelihoods depended on it, the new machinery allowed timber companies to seek out the big pines and hardwoods within the island's wild interior. In addition, Maine science writer Catherine Schmitt notes, the early 1900s saw loggers "targeting smaller spruce and fir suitable for making pulp and paper."[45] In other words, technology was advancing to allow the Maine forests to be sheared like sheep.

Over the winter of 1900–1901, while still grieving the death of his son, the elder Eliot took the time to sort through the younger Charles's publications and papers. A line from his "Waverly Oaks" piece in *Garden and Forest* must have been especially inspirational: "The establishment of a small public park at this place . . . would protect the trees from the dangers which now threaten them, and would make a valuable and interesting public resort within walking or driving distance . . . of a large number of people."[46] In late summer 1901, Eliot Sr. recruited Dorr, George L. Stebbins, John S. Kennedy, George W. Vanderbilt, and others to meet and take up the conservation banner; the result was the formation of the Hancock County Trustees of Public Reservations (HCTPR). Their mission, codified by the state legislature, certainly would have sounded familiar to the younger Eliot: to "acquire, hold and maintain and improve for free public use lands in Hancock County" for scenic beauty, historical interest, and sanitary proposes.[47]

*

Despite the prominence and resources of the trustees, their land acquisition ball was slow to get rolling. It was not until 1908 that Bostonian Eliza Homans donated 140 beautiful acres of mountains and lake—the Bowl and the Beehive—to the HCTPR. A close friend of President Eliot, Homan declared that she feared if she didn't protect the land, her grandchildren might find a "Merry-Go-Round established there."[48] Also in 1908, Dorr and Kennedy managed to buy the summit of Green Mountain, sparing the highest point on the Atlantic seaboard from a possible subdivision of summer homes.

Beginning in 1910 and continuing for several years thereafter, the trustees rushed to protect the shores of Eagle Lake from extensive development. Their string of land purchases preserved not only the scenery, but the public water supply as well, thereby benefitting permanent residents and summer people alike. By 1912 the HCTPR had added some three thousand acres of public land to their reservation. Most of the purchases were funded by Dorr, Kennedy, Henry L. Eno, the local water utility, and several wealthy cottagers from Seal Harbor.[49]

Pushback came in 1913 when realtors, and perhaps logging interests, convinced their state representatives to go to Augusta and introduce a bill to revoke the HCTPR's charter. Dorr wasted no time catching a fast train from Boston to the Maine capital and used his considerable charm and aristocratic notoriety to persuade the speaker of the house to drop the bill. Dorr's quick victory, however, was brittle. The mere existence of the bill in the state legislature eroded the permanence of the trust upon which Dorr and so many others had poured their faith, money, and an entire decade of their lives. As the sleeper train sped back to Boston, the sixty-year-old Dorr tossed and turned, tortured by the revelation that Mount Desert Island's ecological future twisted on erratic economic and political winds. By morning, Dorr recalled he had "decided that the only course to follow to make safe what we had secured would be to get the Federal Government to accept our lands as a National Park."[50]

Upon his return, Dorr immediately consulted Eliot and laid out his proposal to seek federal, not state, protection for their three thousand acres. This was not only a turning point for Dorr and Eliot, it was a prerequisite for Junior's evolution from lord of private estates and philanthropic generalist to munificent benefactor of America's national parks.

At first, Eliot strongly objected to Dorr's plan to seek federal assistance. However, after persistent persuasion by the charismatic Dorr, Eliot acquiesced.

"When will you go on to Washington?" Eliot inquired.[51]

AN ALTRUIST GOES TO WASHINGTON

Acadia, 1914

In the late spring of 1914 George Dorr took the train from Boston to Washington, DC, the second of many courageous solo trips to win federal protection for a "wild park" in Maine. At first, many of the federal bureaucrats whom Dorr called upon were unsure of what to make of the mustachioed Boston Brahmin and his unusual desire to donate land on an island in Maine to the government. Likewise, Dorr was unsure of what to make of Washington as he searched in vain for a bureau of national parks where he could plop down his deeds. Despite the fact that parks like Yellowstone, Yosemite, Sequoia, Crater Lake, Mount Rainier, Mesa Verde, and others had existed for many years, tracking down the people charged with running them proved elusive. In fact, there was no National Park Service in 1914, though there was a "Confidential Clerk to the Secretary" of the Interior, assigned to explore the possibility that someday such an agency might be born.[1]

That petty bureaucrat was a recent law school student from the University of California at Berkeley named Horace M. Albright. Dorr tracked him down in a labyrinth of offices and corridors in the Department of the Interior building, adjacent to the office of Secretary of Interior Franklin K. Lane. Albright had mixed feelings about a career in government. Upon graduation, his humble life plan had been to marry Grace Noble, his college sweetheart, and work for a mineral rights law practice near his alma mater in Berkeley, California. He had recently almost accepted a job with an old established law firm in San Francisco but turned it down after Lane gave him a bump in pay and a perfunctory promotion. Grudgingly, he remained a lovelorn bachelor making $133 per month (approximately $49,000 per year today) and living at the YMCA off G Street.

Albright had been recruited for his nebulous job with Secretary Lane by one of his UC law professors, Adolf C. Miller. Lane was also a University of California alum who often turned to his home state when seeking talent to fill jobs at the Department of the Interior. The professor who recruited Albright also loaned the shy student from a modest background train fare to Washington and an old suit which, Albright supposed, was "probably... the one he [Miller] had worn for his graduation from the university in the class of 1887."[2] Colleagues described Albright as a "very delightful, likable man" who "loved the outdoors" and made friends easily.[3] Although Albright had never personally visited a national park, shortly after arriving in Washington in 1913 he was informed that his purpose was to help create a bureau that would muster all thirty-one of the national parks and monuments under a single management umbrella. Albright agreed to give it a year.

Of his first meeting with Dorr, Albright recalled that Dorr's "earnestness and devotion to his cause impressed me very much." Dorr shared with Albright his portfolio of photographs from Mount Desert Island in Maine.[4] Albright also remembered that on that humid day in May, Dorr "looked like the Washington heat had worn him out. I brought back a pitcher of water. He gratefully drank several glasses and then related his reasons for wishing to see Lane. It was a fascinating story."[5]

Albright, who was more open to creating national parks and monuments in the East than others at the Department of Interior, gave Dorr an excellent (though daunting) piece of advice: convince President Woodrow Wilson to declare the Hancock County Trustees for Public Reservations lands a national monument, thereby avoiding the Congressional scrutiny required for designation as a national park. Albright also made Dorr an appointment to talk to Secretary Lane. Fortunately for Dorr, Lane was a Californian who was aware of a very small but precedent-setting national monument near San Francisco called Muir Woods. The six hundred acres of virgin redwood forest had been donated to the interior secretary in 1907 by William Kent, a politician, businessman, conservationist, and reformed slumlord. President Theodore Roosevelt created Muir Woods National Monument in 1908, the first such preserve fashioned from a donation of private lands. By the time the compelling Dorr left Lane's office, he had artfully gained the secretary of interior's pledge of support.

However, it was the president's support that Dorr needed, and that assistance was impeded by Wilson's friends in the U.S. Forest Service (part of the Department of Agriculture). There was an inherent competitiveness between overseers of parks and overseers of forests as new parklands were sometimes carved out of national forests. Creating parks therefore might mean the Forest Service needed to surrender some of its most scenic and prestigious lands to a government entity with which it vied for congressional funding each year. Even when there was no Forest Service land at stake, as with Mount Desert Island, the Forest Service was sometimes stingy with its support for the Interior's projects. Way back in 1911, Gifford Pinchot, the nation's premier forester and national forests advocate, protested the possible creation of a National Park Bureau by stating it was "no more needed than two tails on a cat."[6]

While Dorr looked for ways to warm Wilson's heart in favor of a national monument on the coast of Maine, he was informed of several steps he could take to make monument status more palatable. As it turned out, the federal government did not abide by the "never look a gift horse in the mouth" adage. In fact, they were extremely skeptical of anything resembling a free horse, an inclination that would vex Junior for decades to come. They likewise had an aversion to paying good taxpayer dollars for lands that would become national parks or monuments, even those for "the benefit and enjoyment of the people."[7] (However, Congress would open its pocketbook to purchase lands for national forests, a policy that made President Theodore Roosevelt's conservation legacy possible.) United States Speaker of the House Joseph Cannon vowed to spend "not one cent for scenery" [to acquire or improve lands for national park purposes], and more than a few government officials balked at the responsibilities and liabilities of receiving gifts of land.[8] The federal government even had the pluck to require potential land donors to absorb as many irksome preacquisition costs as possible.

In 1914 Dorr returned to Maine with a laundry list of prerequisites for the Feds to accept lands from private donors. The stipulations were numerous, but not altogether unreasonable considering a map of HCTPR lands looked like a jagged jigsaw puzzle with several key pieces missing. The government was especially cautious when it came to verification that the HCTPR actually possessed legal ownership of the lands they planned to surrender. Such certification required some deeds to be traced back to the original grant of Donaquet to Cadillac in the seventeenth century. Various knots that had to be untied in this process included "oral wills, unrecorded deeds," and "warranties given by men long dead,"[9] not to mention the ordinary problems of ancient surveys in rugged topography that delineated property with points like "blaze on big maple tree" or "dry streambed." More importantly, several major properties needed to be acquired to smooth the boundaries and fill the gaps.

Dorr had already sold his homeplace at 18 Commonwealth Avenue in Boston and exhausted much of his family fortune purchasing lands for the preserve. Kennedy had passed away in 1909. Fulfilling the government's wish list of requirements for achieving national monument status was going to be tedious and expensive. Hence the letter to John D. Rockefeller Jr.

In 1914 Junior was more focused on buying lands adjacent to the Eyrie and expanding his network of carriage roads for friends and family than helping create a national monument. Yet, Junior was from a long line of philanthropists, and by 1914 he had largely forsaken business to focus on charitable giving. For at least three generations the Spelman family (Junior's mother's side) had been crusaders for abolition, women's suffrage, prohibition, and religion.[10] By 1914, Senior had already established and endowed the Rockefeller Institute for Medical Research, the General Education Board, and the gargantuan Rockefeller Foundation. Consequently, the letter from Eliot was undoubtedly of interest to Junior as both a philanthropist and summer resident of the island.

The letter, however, arrived at a true low point in Junior's adult life. It was also a turning point—most would agree *the* turning point—in the forty-year-old son of a tycoon's existence. The harsh realities of being a Rockefeller in the early twentieth century were about to manifest themselves in the bleak coal fields of southern Colorado.

COAL, CATASTROPHE, AND A ROAD TOWARD REDEMPTION

Ludlow, Colorado, 1913–1915

The trouble in southern Colorado began in the autumn of 1913 when nine thousand miners in the region's shallow coal fields went on strike to demand the freedom to join a union and achieve better hours, wages, and living conditions.[1] Certainly for the miners and their families—most of whom were poor immigrants from Mexico, Italy, Greece, Spain, Poland, Austria, and twenty-six other countries, with limited experience with the English language—there was plenty not to like about what labor historians call the mine owner's "system of peonage."[2] At an elevation of 6,200 feet, winters were long and nights were brutally cold. The miners toiled in the dark, dangerous mines while the Colorado Fuel and Iron Company (CFI), which employed them, owned their housing, stores, schools, and even churches. Women especially found living in company houses within company towns stifling and oppressive.[3]

Management of CFI and bosses from adjacent mines responded to the strikes by booting the miners and their families out of company housing, just as the season's first snowflakes began drifting from the leaden skies. The union organizers built a tent city for the strikers, and the miners dug large basement bunkers beneath their canvas roofs to shield their families from sporadic gunfire. The southern Colorado mining companies steadily ramped up their state-supported militias and eventually armed them with machine guns and an armored vehicle. Union organizers and sympathizers helped provide weapons for the striking miners. In late October of 1913, "a general battle between strikers on the one side, and mine guards and Deputy Sheriffs on the other was waged for more than twelve hours."[4] Contradictory reports stated "several men" were killed.[5]

Although John D. Rockefeller Sr. had semiretired from Standard Oil and the day-to-day affairs of corporate management around 1900, and Junior had decided to delegate most of his business management activities to an accomplished staff and focus on philanthropy in 1910, the Rockefellers maintained an office full of lieutenants who devoted their days to adding to the Rockefeller fortune by investing in a wide range of companies and endeavors from oil refining and pipelines to real estate, banks, steel manufacturing, agricultural commodities, bonds, railroads, and mining companies like CFI. Senior controlled about forty percent of CFI stock and Junior was on the company's board of directors. Although Junior served on the boards of numerous nonprofit charitable organizations, with missions ranging from helping prostitutes escape the white slave trade to eliminating hookworm in the South, the sole corporate board upon which he continued to serve in 1913 was, unfortunately, CFI.

Junior maintained his seat on the CFI board only because his father's acquisition of the company's stock had proven to be a mistake. Senior had purchased the stock at the insistence of his most trusted advisor, Frederick T. Gates (who had a relative in CFI management), in 1902. Soon afterward the company revealed itself to be plagued by incompetent managers, phantom profits, and declining production. When dealing with troubled companies, Senior and Gates were determined "never to allow a company in which we had an interest to be thrown into bankruptcy court." Their preference was to use superior Rockefeller management practices, patience, and ample capital to rescue the enterprise.[6] Junior likewise felt compelled to nurse CFI back to health, if only to absolve his father's error by setting the company straight before selling it off. Junior's loyalty to his father was complex, but Senior was a good father who rarely had a harsh word for his children. No doubt watching his father's villainization by the press took a toll on the sensitive Junior. And while Senior and his children had a good, steady relationship, it was always more formal than warm.

Junior, regrettably, was also inclined to follow his father's lead toward labor issues and most other business matters. Junior was, after all, in awe of his father's professional talents—which had allowed the old man to rise from the humblest of rags to unfathomable riches—and was painfully aware those talents were largely absent from his own DNA.

When Junior was twenty-five years old and fresh out of college, attempting to follow in his father's gigantic footsteps (without the advantages of business experience, instincts, or training), he lost $1 million ($38 million in 2025 dollars) to the original "Wolf of Wall Street" David Lamar. The Wolf worked connivingly through one of Senior's staff, persuading Junior to invest heavily in the U.S. Leather Company. The more stock Junior bought, the higher the price rose, allowing Lamar to sell his considerable stake in the company at a handsome profit. When the bottom fell out, Junior was forced to ask his father to cover the loss.[7]

"I would rather have my right hand cut off than to have caused you this anxiety," Junior told his father. "My one thought and purpose since I came into the office has been to relieve you in every way possible of the burdens which you have carried for so long."[8] Upon learning of the duping, Senior, as always, was patient and calm. He asked his son many pointed questions to get to the bottom of the con, then concluded the conversation with, "All right, John, don't worry. I will see you through."[9]

Naturally the press latched on to the news that Junior, "The Babe of Finance,"[10] had been "Skinned in Leather."[11] They duly noted that many Wall Streeters were laughing over his rookie blunder.[12] Yet, in the years before Ida Tarbell's muckraking crusade against Standard Oil, many in the media somehow resisted an all-out roast of the unpretentious Junior. Some writers compared him to legendary investors George Gould and William K. Vanderbilt in their early years and applauded his character. "He works harder than any laborer" and "allows himself but half an hour for lunch," reported the *Deseret News*. "His wealth seems to bring him little satisfaction. He has been heard to say he looked upon his immense fortune as a responsibility."[13]

Senior did not create or manage Colorado Fuel and Iron, but he did purchase, sight unseen, a controlling share of its stock. And while conditions for CFI workers were unquestionably miserable, records show that Senior paid his own Standard Oil employees "moderately high salaries, to which he added bonuses for specially efficient officers" and even provided hospitalization and retirement benefits.[14] Senior was also known as an inspiring and engaged corporate leader. According to one Standard Oil refinery worker, Senior "always had a nod and a kind word for everybody. He never forgot anyone." In *Titan*, Chernow concluded Senior was "reasonably generous in wages, salaries, and pensions, he paid somewhat above the industry average." His employees "tended to revere" him and "vied to please him."[15] Senior also invited employees to send their complaints and suggestions directly to him.

However, Senior, like most business owners, was adamantly opposed to labor unions, conveniently maintaining that unions were nothing but "frauds perpetuated by feckless workers."[16] In Senior's view, the ultimate goal of union workers was to "do as little as possible for the greatest possible pay."[17]

Junior echoed the prejudices of Senior, Gates, and onsite CFI managers as the violence at the Colorado mines intensified: "The failure of our men [the miners] to work is due simply to their fear of assault and assassination [from union organizers]," he related to U.S. Secretary of Labor William Wilson.[18] This narrow and erroneous opinion by Junior could partially be blamed on Lamont M. Bowers, the chairman of the CFI board, who was Junior's chief informant on the strike. Bowers described the strike leaders as "agitators, socialists, and anarchists" and declared that 90 percent of the workers were satisfied with their jobs and opposed to the strike.[19] Bowers's credibility was enhanced by his being the uncle of Gates, Junior's mentor.

As the situation in Colorado worsened, Junior erred by continuing to accept the sanitized and fallacious reports from CFI staff. He never traveled to the site or sought out other perspectives and he continued to blindly support his on-site managers. Tragically, on October 6, 1913, Junior reassured Bowers, "We will stand by you to the end."[20]

When the U.S. House of Representatives began an investigation of the October killings in Colorado, Junior was invited to testify. His presence in the politically charged review was requested not so much because he was a CFI board member, but because he was a Rockefeller. In 1914, public opinion of the family and their fellow monopolists was at a low ebb. Muckraking journalists continued to make hay by exposing Senior's sketchy business ethics, and the growing number of bloody labor disputes—including even a strike by newsboys—highlighted the contrast between the aloofness of the ultrawealthy and the plight of the working poor.

The *New York Times* reported that during Junior's four hours of testimony before Congress, he remained "polite and thoroughly suave" but failed to take responsibility for the deaths and violence, arguing that the company's operations were controlled by on-site managers, not by absentee stockholders like himself. He said he and his father believed in "employing the best men to look after the details" and would support their managers as long as they were "capable and worthy." Junior stated that "he was not opposed to unions as such" but did not believe employees should be forced to join them.[21] Although the press seemed to react favorably to Junior's poise, his critics rightly accused him of defending "absentee capitalism" and for shirking his responsibilities as director.[22]

On April 20, 1914, a mere fourteen days after Junior's public testimony, unholy hell broke loose at one of the strikers' tent camps. State militiamen, some of whom were former guards and hooligans employed by the mining companies, strafed the tents with machine gun fire. The militia then descended on the camp with torches and burned some two hundred of the temporary homes. Eighteen strikers and family members died for every militaman.[23] In the morning, the bodies of two women and eleven children were discovered in a large bunker beneath one of the tents. They had apparently either suffocated while huddling together to save themselves during the machine gun fire or been choked by the militia-ignited fire.

As newspapers across the country headlined the catastrophe, the whole world gasped. The *New York Times* declared the fourteen-hour battle at Ludlow "a story of horror unparalleled in the history of industrial warfare."[24] The revulsion and anger of the strikers exploded into open warfare against mining companies throughout Colorado. For the next ten days, mineworkers engaged in "the fiercest, deadliest uprising since the Civil War," killing more than thirty strikebreakers, mine guards, and militiamen. They also destroyed railroad bridges, a dam, six mines, and several company towns. The anarchy ended only after President Woodrow Wilson ordered the U.S. Cavalry to the state and a thousand women marched on Denver calling for an end to the war.[25]

As the public struggled to make sense of the tragedy, the name Rockefeller was foremost in their minds. In fact, the *New York Sun* reported that a Mrs. Mary Bedford had erroneously announced to the Women's Peace Association that Junior's "statements before the Congress committee" had emboldened the soldiers and caused the miners to arm for war.[26]

In no time, crowds were picketing outside Standard Oil headquarters at 26 Broadway in Manhattan and at the Rockefeller home in Pocantico

Hills. For the first time, Senior, then seventy-five, felt he must erect fences and station guards to protect his family. This fear was reinforced when a homemade time bomb exploded prematurely in a tenement building on Lexington Avenue in New York City, killing four members of the Industrial Workers of the World (Wobblies) and injuring several other occupants. Police investigators revealed the bomb was likely intended for the Rockefeller home on West 54th Street.

*

The Rockefeller family's many advisors concluded it was time for damage control. They sought counsel from public relations pioneer Ivy Lee, a young, former newspaperman who had done some PR work for the railroads. Lee advised Junior to tell his side of the story through a series of bulletins which presented the "facts in an entirely good natured, attractive, and impressive manner" and then distribute said messages to influential individuals and organizations. Lee's entrance into the Rockefeller realm marked a timely and abrupt departure from Senior's long-time strategy of responding to most critics with stony silence.

Even more impactful was Charles W. Eliot's suggestion of enlisting the help of William Mackenzie "Mac" King, Canada's former minister of labor.[27] King arrived bearing a new perspective on labor-management disputes and, after some convincing, accepted the position as head of the Rockefeller Foundation Department of Industrial Relations. Junior considered Mac's arrival as "heaven sent" and their friendship grew over the years to the extent that "King would soon regard John D. Jr. as one of his greatest friends." Likewise, King became "the closest friend John D. Jr. ever had."[28]

Seven months later, Junior was hauled before the U.S. Commission on Industrial Relations by Senator Frank Walsh, a prolabor attorney from Missouri. It was clear early on that Walsh's goal was to implicate Junior in the massacre at Ludlow and gain acclaim for putting a Rockefeller head on a pike. Yet the Junior for whom Walsh had prepared his inquiry was not the Junior who took the stand. Despite the heat and what reporters called the "stifling air," and large crowds stretching out into the hallways, the new Junior, at age forty, was finally ready to step from his father's shadow.[29] He was not there to shirk. Having gained fresh perspectives from the counsel of King and Lee, Junior now professed his new belief that he and other absentee stockholders should "be constantly progressing to something higher, better," and accept more responsibility for the actions of their companies and welfare of their employees. Junior also acknowledged labor's right to organize and have a voice in company management: It is "just as proper and advantageous for labor to associate itself into organized groups for the advancement of its legitimate interests, as for capital to combine for the same object."[30] Junior's comment on combinations and capital no doubt pleased his father as it endorsed the monopoly that Senior and Standard Oil built and muckraking journalists and progressive-era trustbusters tore down. It also placed labor and management on the same plain with equal goals and equivalent adversaries!

Junior's summary at the end of the three-day grilling even elicited some applause in the packed hearing room. King later called the testimony the turning point in Junior's life. Others, including Senior, Abby, and the children, agreed.

Between hearings, Junior was introduced to a most remarkable woman: eighty-three-year-old Mary Harris Jones—"Mother Jones"—the patron saint of exploited workers and persecuted labor organizers. Jones had been jailed in Trinidad, Colorado, for several months without just cause simply because she had traveled there to support the striking workers around Ludlow. Although most journalists and others present expected Jones to give Junior a thorough tongue lashing, the reality was quite different. After Junior invited her to his office for a meeting, Mother Jones related that she had never believed Junior understood the full extent of the abuses "those hirelings out there were doing."[31]

The meeting with Mother Jones occurred in the office building Jones had recently denounced as the "well of evil from which most of the woes of humankind emanated." This was another turning point in Junior's life. After Junior and Jones met privately and discussed the specific issues plaguing the miners in southern Colorado, the doors were opened to the press for an informal news conference, the first in Standard Oil history. No doubt most in the room were shocked when the usually tempestuous and vulgar Jones launched into an elaborate and seemingly sincere exculpation. "John D. Rockefeller, Jr. has been misunderstood," she said. "I see now the young man has been misrepresented. He's frank and he's open and he wants to do right." Jones went on to craftily extract the following commitments from Junior: "The Colorado miners should be allowed to organize ... and not be organized by the company." Junior also agreed to a key demand of the strikers, to have their own "check weigher" to weigh the coal upon which compensation is determined.[32]

Junior, whom the *New York Sun* described as a "pale young man with keen eyes and softly modulating voice," reciprocally agreed with Jones that "it is my duty as a director to know more about actual conditions in the mines. Of course there should be free speech, free assembly and independent, not company owned schools, stores, and churches." Junior then went on to pledge to visit the mines with Mother Jones at his side. The next day's front-page headline in the *Sun* read "John D. Jr. and Mother Jones Join Forces."[33]

In May, Senator Walsh once again attempted to haul Junior to the woodshed, this time equipped with correspondence between Junior and Bowers subpoenaed from Rockefeller files. Walsh badgered Junior for three long days on the stand, not unjustly hammering away at Junior's inaction and indifference during the growing labor unrest, yet he never managed to crack Junior's earnest poise. The labor commission was unable to reach any coherent conclusions at the end of the inquiry and many in the press painted Walsh as an opportunistic politician looking to boost his popularity by scapegoating a legitimate businessman and philanthropist.[34]

Besides winning public approval for his conversion to champion of improving relations between labor and management, Junior's other windfall from the hearings was, unexpectedly, impressing the old man. "They tried so hard to badger my son, to harrow him into saying something that they could use against him, against us." Senior later told one of his biographers, William O. Inglis: "He surprised us all. I believe his sainted mother must have inspired him."[35] Somehow, even while defying his father's staunch antiunion bias and breaking Senior's cardinal rule of never admitting wrongdoing, Junior had finally proven to his father that he had grown a backbone and was capable of handling the responsibilities of Rockefeller-size wealth. Over the next several years, starting with forty thousand shares of CFI stock, Senior transferred over $450 million ($20 billion in 2025 dollars) in stock, bonds, and cash to his son.[36]

One can imagine then, in September of 1914, when the letter from Eliot seeking financial help for the establishment of a public nature preserve in Maine was received by Junior, that the Junior who opened the envelope was a changed person. The Junior reading Eliot's graceful script had perhaps matured a decade since the previous early autumn on Mount Desert Island.

After spending his entire life sequestered with capitalism's winners, he'd been confronted with the plight of the multitudes who made the gears of industry turn, yet who were relegated to the opposite side of the economic system's coin.

Junior had also learned that ignorance is not bliss. His failure with CFI and the Ludlow tragedy was not malicious, but he was guilty of negligence. He knew (or should have known) the militia's effort to break up the strike would result in significant violence. Being a Rockefeller, he had possessed the power to possibly avert the tragedy had he been less trusting of the status quo and more attentive to the root causes of the conflict. If he didn't fully comprehend it before, he knew now that being John D. Rockefeller Jr. was going to be a treacherous task requiring every bit of his energy and attention. Eliot's letter, then—describing how a pristine New England environment could be preserved as a tonic for the increasingly harried, industrialized, subjugated, and urbanized peoples of the world—must have seemed to Junior as refreshing as a balsam-scented breeze across a crystal-clear mountain lake.

FRIENDS, FOES, AND FAVORS IN OUR NATION'S CAPITAL

Acadia, 1915–1916

Before sending the letter to John D. Rockefeller Jr. requesting $15,000 for the national monument project, Charles W. Eliot had passed the hat to colleagues at Harvard and within the Hancock County Trustees for Public Reservations (HCTPR).[1] Unfortunately, after completing its rounds, the hat remained all but empty. Even Dorr was tapped out.

Although a final tally of Dorr's contributions to the public reserve are impaired by the eccentric bachelor's unusual record-keeping system, it is believed he purchased at least fifty-three properties on Mount Desert Island, some tracts larger than one thousand acres. Once each purchase was completed, Dorr dutifully donated the land to the HCTPR. After his considerable estate was exhausted, he mortgaged his family home, Oldfarm, and dove into debt to purchase additional lands for the reserve.

Unlike Dorr, Junior entered into new philanthropic endeavors prudently, usually after consulting his battery of well-paid and trusted associates, and often with a small initial pledge to test the waters. In this case, the ante from Junior was $100 to the HCTPR. His first significant contribution was $34,500 in 1916 (over $1 million in 2025 dollars) to cover the costs of two land purchases and the extensive survey and title work required for national monument status by the federal government. The Rockefellers also preferred that their philanthropy be a team effort;

therefore Junior added some stipulations to his pledge: matching funds must be found to obtain other properties vital to park establishment, and said lands, once procured, needed to be deeded to the HCTPR, just in case the whole national monument/park thing went bust.[2]

One of Senior's credos was, "It is easy to do harm in giving money," a warning not against philanthropy per se (since Senior in his lifetime gave more money to charitable causes than perhaps any other human being who came before), but against impulsive and poorly managed philanthropy.[3] Back in 1889, when Senior discovered he'd been overcharged a relatively small sum by a railroad, he demanded a refund and defended his parsimony by explaining he needed the money to fund mission churches in the West.[4]

And while the $34,500 contribution might seem like small potatoes coming from a Rockefeller, it was a huge windfall for the HCTPR, without which the entire national monument effort in Maine might well have collapsed. It was also a significant sum to Junior, coming during a period of transition from Junior living off an allowance or salary provided by his parents and Senior transferring the bulk of his wealth (excluding monies committed to charitable foundations) to Junior.

The research on deeds going back to the 1600s was completed by local attorneys Luere B. Deasy and Albert H. Lynam and was compiled into six hundred pages of documentation. Junior continued to buy land on the island, with the intention of donating it to the HCTPR. He also continued to lace his holdings and those of the HCTPR with artfully designed carriage roads.

It's hard to calculate which activity Junior enjoyed more, building carriage roads or using them. Junior and Abby often indulged in late-afternoon rides, pausing at scenic spots to appreciate a view of the Atlantic, or glacier-sculpted mountains, or wildflowers nodding in a meadow. They usually stopped at the Jordon Pond House for a bland snack of tea and toast and to absorb the long view across the lake. On other occasions, rather than instructing their staff to hitch up the horses, the Rockefellers would travel the bridle paths on foot, Junior's mind no doubt reeling with thoughts of better ways to engineer drainage along the routes and where to employ native stone in guardrails and retaining walls.

Carriage road construction proceeded intermittently for decades and became a major employer on Mount Desert Island, even drawing tradesmen from off-island communities. Junior employed local engineers, surveyors, contractors, stone masons, and laborers sometimes in crews of up to thirty. He even established an office in Seal Harbor from which to direct design and construction.

*

However, before Junior could devote himself exclusively to carriage roads and the national monument effort, he had unfinished business in Colorado. The journey he promised Mother Jones was postponed first by the death of his mother, Laura Spelman Rockefeller, on March 12, 1915, at age seventy-six. His virtuous mother, who devoted herself so fervently to the Christian religion that she even alienated some members of her own puritanical family, had been in poor health for several years. One month later, Abby's father, U.S. Senator Nelson Aldrich, died of a stroke. Then on June 12, after another difficult pregnancy, Abby gave birth to her and Junior's sixth and final child, David.

The promise of the Colorado trip was finally fulfilled in September of 1915. According to Junior's friend and biographer, Raymond Fosdick, Junior and his people took "every precaution against publicity." When reporters showed up anyway, he made them swear not to intrude on his personal talks with individual miners and not to leak his plans for any of the forthcoming days. Although there were undoubtedly multiple motives for the trip, one was an earnest effort to avert similar Ludlow tragedies in the future. The adventure was gutsy, too. Against Senior's better judgment, Junior and his labor relations sidekick Mac King refused to employ any form of security as they mingled with those who had witnessed and survived the Ludlow Massacre.[5]

Not surprisingly, the *Denver Labor Bulletin* was skeptical of the event. They maintained that Junior and King were "personally conducted by [CFI] President Welborn and Mine Manager Weitzel," thereby discouraging the workers from voicing any significant complaints. Junior and King visited various sites where the massacre had unfolded and paused to reflect at the places where workers like Louie Tekas had been beaten and shot and women and children had suffocated. The newspaper reported that Junior shook the "grimy hands" of coal diggers and patted them on the back.[6]

Junior and King then donned miners' coveralls and hard hats and headed eleven hundred feet underground where they swung pickaxes "for eleven minutes" and harvested chunks of coal.[7] The less hostile *Raymer*

(Colorado) *Enterprise* reported that Junior "asks pointed questions concerning the character of the work, and seeks suggestions from miner, fire boss, the clerk behind the store counters, the housewife and the camp doctors."[8] The enquirers sampled the drinking water and toured every school, church, store, garden, and clubhouse they could find.

Because of his bashful nature, Junior was very good at asking questions (thereby deflecting attention from himself) and listening to answers. His insecurity also made him naturally humble, something the workers seemed to recognize and appreciate. Gone was the absentee New York CFI board member and stockholder who had never laid eyes on an actual Colorado mining camp or steel mill and who deferred all responsibility to on-site managers. King's years of experience negotiating disputes between labor and management in Canada lent the pair credibility.

Although no doubt aware of the presence of top CFI brass, it is remarkable how the CFI rank and file seemed to welcome Junior into their hardscrabble world. After waging war to persuade CFI's incompetent and corrupt managers that labor deserved a seat at the conference table, they finally had the owner of the table standing beside them in their squalid hovels with open ears and eyes. They seemed to perceive in Junior an earnest man with some level of genuine interest in their welfare, even if he was overly serious and formal.

Some of the starch came out of Junior's shirt following an evening meeting with workers and managers at the Cameron Camp schoolhouse. Seeing some musicians preparing to play, Junior suggested they push the desks aside and kick up their heels. It wasn't long until the press corps that covered the entire three-week visit was fighting for the few available telephones. The front page of the next day's *New York Tribune* declared "John D. Jr. Trips to 'Tipperary' with Mine Maids." The subhead reported, "Spryly Does the Hesitation with Girls in Calico—All Got a Chance." With the four-piece band (accordion, snare drum, clarinet, and trombone) laying down the beat, the *Tribune* reported Junior spent two hours on the dance floor and "danced with practically every woman and girl in the room. Officials wives' and daughters in graceful summer silks and coal diggers' wives in calico were included in his invitations." After the dance, Junior pledged to buy the workers a bandstand.[9]

Abby Rockefeller followed her husband's progress in Colorado from his daily correspondence and the frequent newspaper reports. Her sympathies for the miners and their families are apparent in a letter to her husband. "If you can help bring about a solution of the labor problem...I shall die satisfied." And though flirtations and sexual jealousy were not facets of their enduring marriage, she did add "from the papers I gather that your dancing has been one of your greatest assets. I will never demur again."[10]

While there was minimal media attention given to the actual business conducted during the tour, there was progress made between labor and management in CFI's mining and steel empire in southern Colorado. Junior and King floated a plan that granted most of the workers' wishes and which was agreed to by CFI management (after Junior gave CFI vice president Bowers the boot) and approved by a majority of the employees who consented to vote. To the surprise of many, Junior went on the record several times in favor of "open shops" where workers were free to join or not join unions. He also claimed to have an open mind to the value of unions and collective bargaining.[11]

Junior and King were in fact cooking up a fresh (though not precisely pro-union) approach for facilitating cooperation between labor and capital that would develop into Industrial Relations Counselors, Inc., a strategy which, though imperfect, was a small step toward better communications between labor and management worldwide. Junior developed a nonpaternalistic creed which emphasized that labor and capital are partners, not enemies, and that every worker "is entitled to an opportunity to earn a living, to fair wages, to reasonable hours of work and proper working conditions, to a decent home, to the opportunity to play, to learn, worship and to love as well as toil." He further stated his belief in "stockholder responsibility" for the management of a company.[12]

More importantly, Junior declared "there shall be no discrimination by the company or any of its employees on account of membership or nonmembership in any society, fraternity, or union." The plan also formalized procedures for making and addressing grievances, prohibited most discharges without notice, and called for regular meetings between representatives of labor and management.[13]

When Junior, King, and even Abby returned to the CFI operations in subsequent years, they always reported improvements in employee living and working conditions and higher morale. Even though CFI managers issued "warnings of likely violence," Abby and Junior returned to Ludlow in 1917 for the memorial service marking the three-year anniversary of the tragedy.[14] Abby connected so well with the workers' families that she decided to prolong her stay and visit every camp. "In each camp she has met large groups of the women and has chatted with them about the Red Cross and other matters while she knitted. She also made it a point to gather together as many of the children as possible and treat them to candy, sodas and ice cream," Junior reported to his father. In subsequent years Abby would make donations from her own purse to trade unions and striking workers.[15]

Colorado historian Jonathan H. Rees wrote an entire book on the "Rockefeller Plan" and its impact on CFI. Rees characterized the plan as a "sincere" middle ground between paternalistic company control of workers and full unionization. The historian included Junior among the industrialists who believed "treating workers better might pay benefits in the long run." Rees maintains that workers gained a surprising number of benefits through the plan, but it ultimately lacked the teeth (and independence from the companies) of trade unions, things essential to the workers. The Rockefeller Plan did provide a useful stepping stone to unionization, however, and Junior pressured other companies within the Rockefeller orbit to liberalize their employee policies and spoke out for fair employee representation at numerous conferences.[16]

"The Colorado strike," Junior told Fosdick, "was one of the most important things that ever happened to the Rockefeller family."[17] Not only did it open Junior's eyes to the people and institutions responsible for creating the fortune he would soon inherit, the conflict influenced the course for how that fortune would be expended. Many of Junior's upcoming conservation efforts, especially those in the East, were tailored to benefit working class families, rather than being exclusive to nature-loving leisure class elites. And because the creation of new national parks and monuments can pit the aforementioned elites against working families tied to the lands being obtained by the government, Junior's experience working with the rank and file should have increased his empathy for their plights.[18]

So great was Junior's transformation in the aftermath of Ludlow, some of the family's industrialist peers began to view Junior as a "dangerous liberal."[19] Rees writes that labor advocates and critics determined that Junior had become a "very modest liberal" or a "Christian liberal" and conceded his employee representation effort should be considered part of the "Progressive movement."[20] In 1916 Junior met with a union organizer and assured him he was free to circulate among the workers of CFI "for the purpose of organizing them." When a CFI manager balked, Junior suggested he post notices reminding both workers and management that CFI was an "open camp."[21]

As for Mother Jones, in 1930, in recognition of "Mother's" one hundredth birthday, Junior sent a telegram of congratulations and praise. "Your loyalty to your ideals, your fearless adherence to your duty as you have seen it, is an inspiration to all who have known you." Upon receipt of the message, Mother remarked, "He's a damned good sport. I've licked him many times, but now we've made peace."[22]

A few weeks later, on Senior's ninety-first birthday, Mother Jones wrote Senior a note which likely reflected her concern about the Great Depression and its effect on working families. "Congratulations on the arrival of your 91st birthday. Thank God we have some men in this world as good as you. We never needed them as much as we do today. Most sincere wishes that you may be blessed with many more."[23]

*

Dorr, a Harvard-educated scholar well versed in Classical mythology, must have felt like he had successfully completed the twelve labors of Hercules as he approached the White House for his appointment with President Wilson in the spring of 1916. In accordance with requests from the Public Land Commission, Dorr, backed by Junior, with the painstaking efforts of Luere Deasy, Albert Lynam, and several mosquito- and briar-bitten surveyors and other native Mainers, had managed to verify the deeds all the way back to the French occupation of northern North America. Money had been raised and missing links of lands had been purchased, connecting isolated mountaintops and clarifying boundaries. Plans for further acquisitions that would extend the proposed national monument's borders to the seacoast were underway. Dorr had secured the publication of major articles in *National Geographic* magazine highlighting the natural wonders of Mount Desert Island. He had also amassed an arsenal of letters of support for the national monument from the likes of Secretary of Interior Lane and future NPS director Stephen T. Mather.

Although Dorr felt the fateful meeting with the president had gone even better than expected, it was followed by an extended period of anxious silence. Events were taking place far from the coast of Maine that were distracting attention from the national monument campaign. While the United States would not enter World War I until 1917, England, Germany, and much of the rest of Europe were already embroiled when Dorr chatted with Wilson. Horace Albright, who had been most helpful when Dorr made his first sojourn to Washington in 1914, was equally distracted by his and Mather's urgent fight to create the NPS, an excruciating political process that coincided with Dorr's epic quest. By far the largest obstacle, however, was the objection of the U.S. Forest Service, specifically the resistance of Secretary of Agriculture Franklin D. Houston, a cabinet member and powerful ally of the president.

In all likelihood, Houston was motivated more by the bureaucratic tug of war between the Departments of Agriculture and Interior than by any concern for the public good, especially as Albright and Mather drew closer to their goal of creating a brand new agency to unify the management of national parks and monuments. Yet Dorr, owing to his pedigree as a member of Boston's gentry, could pull a remarkable number of strings in Washington. In fact, his records reveal he could hardly take a dinner at the Metropolitan Club or a brandy at the Cosmos Club without running into an acquaintance who could open any number of politicians' or executives' doors.[24]

Dorr attempted to unsettle the president's inertia by rallying support from a squad of D.C. heavy hitters, including U.S. Forest Service Chief Henry Graves, Secretary of Interior Lane, Senator Charles F. Johnson from Maine, and Governor of the Federal Reserve Charles Hamlin. All contacted Wilson on Dorr's behalf and expressed their support for a national monument in Maine. In the end, however, it was fellow triumvirate member Charles W. Eliot who punched the right button to end Houston's adamant opposition. Dorr suddenly recalled "Secretary Houston to be indebted [to Eliot] for kindness shown him earlier in his academic career at Harvard as professor" (perhaps steering a fellowship his way).[25] At Dorr's urging, Eliot hand-wrote Houston a simple letter explaining why he and Dorr were so passionately committed to gaining protection for Mount Desert Island's sublime landscape.

Three days after receiving Eliot's letter, Houston wrote the president: "I have changed my view in regard to the proposed reservation on the coast of Maine and now think it highly desirable that you accept."[26] On July 8, 1916, three days after receiving the letter from Houston, President Wilson signed the proclamation creating Sieur de Monts National Monument. The odd (and short-lived) moniker was suggested by Dorr as a tribute to the French nobleman who ruled a swath of North America in the early 1600s. The presumption that the French and Americans would become allies in the Great War may have also been a factor in the general acceptance of the French name for an American park. Dorr also offered to work as park superintendent for a salary of one dollar per month to allay Wilson's and Houston's objections to spending a lot of taxpayer money on the new reservation. His offer was accepted.[27]

A celebration of this monumental achievement was held on the afternoon of August 22, appropriately at the Building of the Arts in Bar Harbor. The imposing structure with solid stone exterior and towering Greek columns was another of Dorr's charitable projects which he completed with considerable help from Catherine A. B. Abbe, George W. Vanderbilt, and other Mount Desert Island summer people. The highbrow facility hosted performances of music, theater, and dance. According to the report in the *Bar Harbor Times*, despite the facility's four hundred seats, the "capacity of the building was taxed." Eliot served as the master of ceremonies and started it off with a moving speech in which he promised, "The old lovers of the island expect to welcome many new lovers." Congratulatory letters were read from President Wilson and Maine governor Oakley C. Curtis. Maine secretary of state John E. Bunker, representing the governor, emphasized "the cordial good feeling" between summer people and residents.[28]

One of the more prophetic speakers was Alfred G. Mayer, director of the Department of Marine Biology of the Carnegie Institution. Tipping his hat to the efforts of other great philanthropists, including James Smithson of the Smithsonian, Mayer tantalized the crowd with the prospect for a natural science and outdoor education center on Mount Desert Island, especially one focused on marine biology. The science center was obviously agreeable to Dorr, who presented the afternoon's finale, a look-to-the-future, this-is-only-the-first-step speech calling for further land acquisitions for the monument and the establishment of a world-class science research center. The crowd responded with an ovation.[29]

Although Junior and Abby were likely "on island" on the day of the ceremony, there is no record they attended. It's probable they did not. Junior disliked crowds and abhorred being the focus of praise. At the rare tributes he did attend, he usually diverted attention by stating that his father was the one who amassed the great fortune and that "I am only the son."[30] Nor is there record that the Rockefeller name was mentioned at the ceremony, which is also not shocking since Junior's contributions to the preserve were still in the early stages and Junior almost always preferred to operate from behind the curtain rather than in the spotlight.

Achieving national monument status and the slew of federal protections that went with it was a huge victory for the triumvirate and the residents of Mount Desert Island who favored conservation over development. It was also a lunge forward for national parks in the East. Secretary of Interior Dr. Ray Lyman Wilbur stated, years later, at the commemoration of another major park, that scenic lands in the East had been passed over during westward expansion and before the concept of national parks had caught on. The only way to correct this error was to acquire parklands "by gift" from private individuals. Wilbur cited Sieur de Monts as one such example. "This demonstrated that by extraordinary efforts other lands could, if the right kind of and amount of energy and effort were put behind the movement, be added to the national park system."[31]

Wilbur's insightful words offer promise for similar conservation efforts today. After all, the creation of the new national monument in Maine was largely accomplished by just three mortals with their own flaws, idiosyncrasies, and very busy lives. Junior, with his astonishing wealth, was certainly no Everyman, yet the enormity of his other philanthropic projects, as well as commitments to business, church, and family, took an oftentimes debilitating toll on his health. As president of Harvard, Charles W. Eliot gave the campaign credence and wise guidance, but he never made significant contributions of funds. And though Dorr was almost comically disorganized and impulsive, he clearly possessed the heart and the charisma to push Sieur de Monts over the finish line. The triumvirate was a rare team, but not a unique one. Many, many more lovers of nature who are also persons of great wealth, wisdom, and passion share the planet with us today. Doug and Kris Tompkins, the American outdoor clothing entrepreneurs who helped create one-million-acre Pumalin Douglas R. Tompkins National Park in Chile, are two fine examples. Greg Carr, who parted with $100 million to help restore war-ravaged Gorongosa National Park in Mozambique, is another.

Sieur de Monts is also a reminder of one of the president's true superpowers, the ability to use the 1906 Antiquities Act to create a national monument (even without congressional support). As we shall see in Maine, national monument status is often a stepping stone to full national park designation and a very real and efficient way to address some of the over-popularity problem afflicting today's parks. When President Joe Biden announced his executive order in 2021 to "conserve 30 percent of America's lands and waters by 2030," supporters of the initiative strongly suggested the president use the powers of the Antiquities Act to make good on his bold commitment. Soon after taking office, Biden used the tool to restore lost portions of Bears Ears and Grand Staircase–Escalante National Monuments in Utah.[32] In 1978, President Jimmy Carter wielded the act to create seventeen new national monuments encompassing fifty-six million acres in Alaska.

A NEW PARK AND A NEW NATIONAL PARK SERVICE

Acadia, 1916–1924

On August 15, 1916, five weeks after President Woodrow Wilson signed the papers creating Sieur de Monts National Monument, Congress and the president spawned a spanking new federal agency called the National Park Service. Up to that moment, the nation's fourteen national parks and numerous monuments were managed by a hodgepodge of agencies ranging from the U.S. Forest Service to the military, none of which were organized specifically to preserve scenic and culturally significant lands. Credit for creating the NPS, an agency that was destined to become one of the most popular U.S. federal agencies of all time, rightfully went to Secretary of Interior Franklin Lane and the two fellow Californians he hired for the task: Stephen T. Mather and Horace M. Albright.[1]

Mather's experience as a journalist and public relations and advertising man would serve him well in the campaign to mint a new federal agency. Especially one whose creation and long-term existence would depend on broad political and citizen support. He was also an exuberant outdoorsman who climbed Mount Rainier and many other tall western peaks with comrades from the Sierra Club. One of Mather's more notorious promotions from his early days with the Pacific Borax Company was to pen letters from fictitious homemakers to fifteen women's magazines exclaiming the miraculous powers of borax and its myriad uses around the house. All were reportedly published. He went a step further in the elevation of the borax industry by coining the name "20 Mule Team Borax."[2]

Climbing the corporate ladder, Mather, who was graced with a bank president's distinguished silver hair and handsome face, eventually became co-owner of a large borax company and a millionaire. He embarked on philanthropy as a settlement house patron and rescuer of friends and colleagues in need. Above all, Mather was a people person who loved to mingle, make new friends, win new customers, and work a room. At times it seemed every hiker in Yosemite National Park and aficionado of the Sierra Nevada Mountains called him their close friend. He was at ease with wealthy, influential people, a skill that would serve him well when coaxing members of Congress to his side. Historians have described him as "almost pathologically fraternal."[3]

Albright was to Mather what John D. Rockefeller Jr. was to George Dorr. Albright was the detail-focused, plodding, even-keeled, behind-the-scenes, report-to-work-early-every-morning complement to Mather's big-picture and broad-strokes perspective and manic, glad-handing, extroverted personality. Mather once told Albright, "I like to do things in the field, not fiddle-faddle around an office pushing papers and digging around in details."[4]

The new National Park Service was born with Mather at the helm, yet within five months of its fruition, the intensity of the effort pushed Mather over the brink of sanity into a spectacular free fall into near total darkness. In early January of 1917, Mather's dinner companions at the Cosmos Club in Washington, DC, witnessed the charismatic director declare he was a failure with nothing left to live for. "He's raving, absolutely insane," reported Emerson Hough, one of Mather's friends.[5] The self-made millionaire, mountaineer, and father of the NPS babbled that he had accomplished nothing in his life and was quitting the Park Service. He rested his forehead on the table at the prestigious club and wept.

Mather's friends called Albright who raced down the four flights of stairs from his tiny apartment and then ran block after block through downtown DC to the club. The director had to be physically restrained and repeatedly stated his desire to end his life. Albright and some others hauled him to a private room and a physician administered a sedative. Although it was now midnight, Albright called Mather's wife, Jane, and fearfully explained the situation. Jane Mather didn't seem surprised. She explained at length his total collapse fourteen years earlier due to stress and overwork and how he spent four months in a sanitarium. Even after his release, it took over a year of rest and relaxing travel before he could return to work. Less severe incidents of stress-induced depression had reared their heads in 1906, 1912, and 1914, but Mather had successfully fended them off with strenuous adventures in the western wilds.

The next day, following Jane Mather's instructions, Albright and Southern Pacific Railroad executive E. O. McCormick quietly escorted Mather onto a train to Philadelphia. There they met Jane and had Mather admitted to a sanitarium. Two months later, his doctors described him as "very much depressed" with "no confidence in himself."[6]

Despite their odd-couple personalities, Albright and Mather were remarkably loyal to one another. Horace and Grace named their first child Robert Mather Albright. Mather subsidized Albright's puny government clerk's salary from his own pocket (until federal regulations made the practice illegal). Albright kept mum about the nature of his boss's illness, saying only that Mather "was worn out and needed a rest."[7] Albright then proceeded to cover Mather's as well as his own position for two extremely busy years.

While Albright was acting NPS director, new park superintendent Dorr headed back to Washington to lobby for appropriations to run Sieur de Monts. Throughout 1916 and 1917, America's notoriously stingy Congress had appropriated only $150 to operate the new national monument. Dorr consulted Albright and Lane on how to gain a more reasonable appropriation and it was suggested he "bombard the new chairman of the Appropriations Committee, Swager Sherley, with authoritative letters. Let him understand the history of the monument and its needs, show him appeals from important and influential people."[8] The men also agreed to push for an upgrade in status and title for the reserve from Sieur de Monts National Monument to Lafayette National Park (a tribute to the French ally of George Washington).

Pivotal to the effort was Dorr's invitation to Lane for a relaxing visit to his Oldfarm home in Bar Harbor and a tour of the new national monument with its unique combination of verdant mountains and Atlantic seashore. Because of his heart condition, Lane traveled the island on horseback while Dorr jogged beside him on foot, ready to catch the secretary if the horse lost its footing and he was thrown. As part of the late August 1917 sojourn,

Dorr and Lane decided to drop in on Junior at the Eyrie in Seal Harbor. The secretary indulged Rockefeller by allowing him to present "a map outlining his plans for an extensive system of carriage roads covering most of the eastern part of the island."[9] He also drove Lane around to show him the lay of the land. Lane was enthusiastic about Junior's development plan and gave his official approval. Lane also agreed that motor vehicles should be excluded from the carriage roads.

Apparently, the combination of ideal early autumn weather, gorgeous scenery, and the companionship of aristocrats far from the ceaseless bickering of Washington, DC, had quite an effect on Lane. In a subsequent letter to Dorr, Lane wrote, "You do not know what good you did my tired, politics-soaked soul by showing me . . . the beauties and possibilities of your island." Lane even suggested he would like someday to retire on Mount Desert Island "with you and your friends" and that Sieur de Monts should become "a demonstration school for the American to show how we can add to the beauty of Nature."[10]

Calling upon his enormous web of connections, in 1918 Dorr followed through with Lane and Albright's suggestions for obtaining operating funds from Congress. He conspired with former Harvard president Charles W. Eliot to meet with Congressman Frederick H. Gillet from Massachusetts, a heavyweight on the House Appropriations Committee. He also amassed a thick folio of letters from influential people like Theodore Roosevelt and George W. Wickersham; poet, psychologist, ornithologist, and millionaire Henry L. Eno; and Junior and Lane requesting a $50,000 annual appropriation. Dorr was elated when Congress approved $10,000 for Sieur de Monts, a respectable sum, especially for a national monument (which as a group was even more chronically starved of funding than the national parks), and considering there was an expensive war raging in Europe.

In the winter of 1919, dashing about Washington like a whirling dervish, hand-carrying bills between houses of Congress and the White House (there seemed to be no doors in government closed to Dorr), the sixty-six-year-old superintendent with the twelve-dollar annual salary, succeeded in winning both congressional and presidential approval for elevating his beloved monument a notch on the NPS hierarchy to national park. While Dorr had suggested the logical name—Mount Desert National Park—he was dissuaded by those who questioned the existence of deserts in Maine. Lafayette National Park was more familiar to the thick American tongue than Sieur de Monts and preserved the tribute to our faithful ally in Europe just as World War I was drawing to a close. Albright, in his biography, stated he preferred the name George B. Dorr National Park, a fitting tribute to the man's determination to preserve the Maine landscape.[11]

DRIVE-THROUGH NATIONAL PARKS

Acadia, 1919–1924

It is safe to say that one of Junior's two great pleasures in life was dancing—particularly with Abby—with whom he often tried out new steps in the privacy of the couple's spacious bedroom. Abby and Junior's six children occasionally complained that they had to compete with their father for their mother's attention, and dancing was undoubtedly one way for Junior to monopolize his wife's time and charms. Junior's Foxtrotting was also a rare instance in which he disregarded the stern edicts of his parents and Grandma Spelman. While he followed his elder's guidance in abstaining from alcohol, being a good Christian, saving money, not tolerating waste, and giving to charity, he rebelliously took up dancing while away at college and continued the defiance throughout his long life.

Junior's second passion was carriage roads. He seemed to revel in all stages of the process: pouring over the maps to discern the best route, heading into the field to test the options, contracting and supervising the work, and hitching up the horses for a ride. In the building of carriage roads, Junior and Senior believed in both quantity and quality. At their Forest Hill and Pocantico Hills estates, the Rockefellers had built the maximum number of carriage roads the landscape could contain, allowing them the luxury of variety during their years of pleasant walks and rides through the woods.

When Junior began building carriage roads around the Eyrie, all decisions were between him and the landscape. With the creation of the national monument, Junior became a collaborator with the National Park Service, his neighbors, the people of Bar Harbor, and many, many other interested parties.

The first formal objection to a proposed new carriage road was lodged in 1920 by George W. Pepper, a wealthy Philadelphia lawyer, Mount Desert Island cottager, and future U.S. senator from Pennsylvania. When Pepper's letter was delivered, the Rockefellers were at the Eyrie, as was their custom from the day after Senior's July birthday into September, with Junior up to his eyebrows in maps of proposed carriage roads and ongoing construction concerns. "The Amphitheater is as yet unbroken forest—a wilderness of tree-tops," Pepper declared in his handwritten letter to Junior: "Pierce this with a [carriage] road or roads and its character will vanish. The sense of remoteness which now gives it charm will be replaced by the realization of accessibility."[1]

As always, Junior responded to the criticism with grace. He self-imposed a cease-and-desist order on his own Amphitheater carriage road project and replied to Pepper, "Until the receipt of your letter of August fifteenth, I had assumed that the roads built . . . were favorably regarded and entirely approved by the people of the island generally."[2]

Junior then invited his neighbor Pepper for a parley. As a result, Junior, Pepper, Dorr, and other interested islanders laced up their hiking boots and scouted the route of the proposed Amphitheater carriage road, noting opportunities for splendid views as well as points of conflict where the new road might be visible from existing walking trails. At the conclusion of their 1920 rambles, Junior wrote a letter to the Northeast Harbor Village Improvement Society with a copy to Pepper.

"Because I was brought up in the woods I have always loved the trees, the rocks, the hills, and the valleys. For over five years I have been studying and preparing for the construction of the road under discussion, with every inch of which I am familiar." Junior then presented his carriage road mission statement: "To make available, views of unsurpassed beauty and sections otherwise inaccessible, to the many who could not reach them except with horses, as well as to that also large number of people who find walking on roads more comfortable than the rougher and steeper trails."[3]

In design, Junior was always striving to create a carriage road that offered the traveler a series of scenic views interspersed with more intimate landscape experiences. Thomas C. Vint, an NPS design and construction specialist, developed a special appreciation for Junior's carriage roads on Mount Desert Island. "He approached this [road] building as both an art and a science, studying every mile himself to give the maximum esthetic experience and the maximum comfort. He himself fitted the roads into the landscape to cause the least possible scar to the terrain."[4]

Sixteen boldly designed stone bridges punctuated the area's fifty-seven miles of carriage roads. Each bridge, whether spanning a stream, ravine, or other roadway, was crafted to delight the eye. Junior instructed the engineers to emulate the look of two of his favorite bridges in Central Park in New York City, famously created by Frederick Law Olmsted and Calvert Vaux. The senior Olmsted is widely regarded as America's first professional landscape architect, though as a planner he also made significant contributions to the establishment of Yosemite National Park and the NPS itself. In the early 1930s, his son, Frederick Law Olmsted Jr., assisted Junior and the Park Service with the Stanley Brook Road and bridges, mimicking the "'indigenous' structures of this sort to be found on many of the old roads of Mount Desert." Their Gothic style and old-world craftmanship helped create a feeling of distance from the modern world and arrival in a wild, timeless, charming realm.[5]

Renowned landscape architect and gardener Beatrix Farrand (also a student of Olmsted) advised Junior on adding the finishing green touches to many of his road projects. Farrand and her husband were longtime summer people on the island who designed several magnificent gardens in and around Eden, including Abby's famous Oriental garden. Both Farrand and Junior were in agreement that construction scars should be healed not with cultivars or ornamentals but by using mostly native and naturalized plants. Much of the flora was even cultivated from seeds collected on the island and propagated at the Rockefeller nursery.[6]

It is doubtful that any postconstruction restoration program ever received more careful and professional attention than the Farrand-Rockefeller collaboration. As Junior would discover in his 1924 journey to the western national parks, roadsides were treated as dumps in Glacier and Yellowstone, places of refuse rather than refuge. There, newly built roads were lined with stumps, limbs, rocks, and the entire carcasses of trees as if to highlight the difficulty of the route's construction. Conversely, Farrand and Junior reveled in the replanting of the carriage roadsides, niggling over details like:

> At the pool below the culvert winterberry and red berried elder, with possibly a few black berried elder would overhang the water and appear at home. On the far side of the Beaver Pool a plantation was to be tried of cardinal flower (*Lobelia cardinalis*) where this will be comparatively inaccessible. It would also be pleasant to try turtlehead (*Chelone glabra*), Joe Pye weed (*Eupatorium purpureum*) and meadow rue (*Thalicatrum polygamum*).[7]

Farrand and Junior's collaboration demonstrates that Junior learned from his father how to identify and recruit the best talent. He also emulated Senior's skill as a delegator. As a chronic perfectionist and worrier, Junior on his own—without the confidence and trust to delegate—might not have accomplished much at all during his life. But as a leader and a steady administrator—especially one who was willing to accept others' ideas and assume the role of second fiddle—Junior became extraordinarily productive.

The village improvement societies of Seal Harbor and Northeast Harbor met and held votes on the Amphitheater carriage road. In Seal Harbor, a resolution supporting the Amphitheater road passed by a narrow margin; in Northeast Harbor the vote was a landslide in favor of the new carriage road. The *Bar Harbor Times* panned those who opposed the road, labeling Pepper and his supporters as "certain summer residents" who acted against the wishes of "the majority of residents and summer residents of Mount Desert Island." The newspaper reminded its readers that "all of Mr. Rockefeller's roads are approved by the National Park Service and this particular road has the personal approval of Secretary [of the Interior] Lane and of Mr. George B. Dorr." The article concluded that "it sometimes takes an undesirable resident to bring out the appreciation that people have for a man and his work."[8]

Despite the favorable votes and subsequent petitions supporting the Amphitheater road, Junior decided to keep the project on the back burner for a time and continue extending other less controversial carriage roads. He had inherited from his father the traits of patience and perseverance, along with attention to detail—rare talents that served the two men well both in business and philanthropy.

*

In the postwar year of 1920, some 66,500 visitors made their way to Lafayette National Park. Every year more forsook the ferry and proudly drove across the causeway to the island in their shiny new automobiles. Inhabitants of Mount Desert Island were generally split on the question of allowing "motors" on the island. The year-round residents—the hardy Mainers who wrested a living from the sea, forests, fields, or tourists—mostly favored the cars and trucks which made earning a livelihood less arduous. One Mainer accused "city millionaires" of making Bar Harbor "a quiet, exclusive resort where their little clique can have full sway and where no state of Maine man is welcome."[9]

To the wealthy cottagers and other summer people, automobiles represented the urban clamor they came to the coast of Maine to escape. As Charles W. Eliot wrote as early as 1904, the summer people "do not wish to be reminded...of the scenes and noises amid which the greater part of their lives inevitably passes."[10] The seasonal inhabitants were influential enough to bar automobiles from Mount Desert Island until 1913, when locals finally prevailed to have the ban partially lifted, and motors were permitted on the Bar Harbor side of the island. In 1915, most of these restrictions were dropped and the suddenly ubiquitous contraptions were loosed to sputter and rattle along the island's few navigable roads.

Although Junior's father had made the bulk of his massive fortune refining oil into kerosene for use in lamps, the popularity of the automobile was another huge windfall for the wealthy family whose name was synonymous with Standard Oil. Its timing was fortuitous as well, since Thomas Edison's electric lights were gradually replacing kerosene lamps in the cities. Yet, despite his vested interests, Junior remained, contrarily, "a horse person." Part of his resistance was no doubt attached to sentimental memories of he and his father building carriage roads, riding tracks, and other horse amenities at their Forest Hill and Pocantico Hills estates.

When most executives were riding to work in machines manufactured in Detroit, Junior insisted on driving a two-horse carriage through Manhattan with the coachman idling in back. For Junior, horses were equated with spending time in the out-of-doors, removed from the stressors that aggravated his migraines, insomnia, anxieties, shingles, and sensitive stomach. Clip-clopping down tree-lined lanes with beloved wife Abby beside him in the surrey was as good as life ever got for Junior.

In a rare, extended interview with the local newspaper, the *Bar Harbor Times*, Junior recalled, "One of the things which attracted Mrs. Rockefeller and me most to Mount Desert island...was that there were no motors on the island."[11] In a 1915 letter to Dorr, Junior flatly stated, "I am very strongly opposed to automobiles on the Seal Harbor side of the Island; in fact, I deeply regret that they have been admitted to the other side."[12]

Yet Junior was generally able to see others' points of view and usually endeavored not to be cast as a villain in his philanthropic efforts. "There became evident a tendency to feel that motorists were being discriminated against in so far as road construction was concerned," he added.[13] Dorr in fact cautioned Junior at one point that carriage roads were seen by some as undemocratic and exclusive to the "leisure class."[14] In this spirit—and in the spirit of helping keep cars off his precious carriage roads—Junior volunteered to fund and help build the first motor road in Lafayette National Park, a 5.2-mile route from Bar Harbor to Jordan Pond known as the Mountain Road. The road was originally suggested by Dorr in response to a call for road construction proposals from his boss, Mather. Junior's monetary offer was slickly bundled with plans for several additional carriage roads in the national park and presented to Superintendent Dorr and NPS big wigs Mather and Arno B. Cammerer as a bipartisan package.

Junior offered to spend up to $150,000 ($2.5 million in 2025 dollars) to build the motor road, plus the monies needed to acquire necessary lands beyond the park boundaries.[15] Not surprisingly, the actual cost for construction turned out to be considerably more, over $400,000, due to rugged terrain, hardness of rock, and the extensive landscaping required to heal the construction scars. Junior took the overruns in stride. Although he was famously thrifty, he was also a perfectionist who never cut corners when it came to realizing one of his philanthropic visions. "When I thought a thing was worth doing," he told biographer Fosdick, "I made up my mind that the annoyances, the obstacles, the embarrassments had to be borne because the ultimate goal was worthwhile. Once embarked on a course I expected to see it through at whatever cost."[16]

The offer to sponsor an automobile road surprised many. However, for those who knew Junior well, they were aware that he often welcomed contrary points of view. As Dorr recalled, "Mr. Rockefeller says . . . that he gets along with him [eccentric island resident Richard Hale] splendidly, for they can not agree on anything and this gives them a good basis for friendly discussion."[17] The road offer, while conveniently helping to protect his carriage roads, was also an acknowledgment of the working class, full-time residents of the island who broadly supported good roads and auto access.

Of course, by supporting a motor road and aligning himself with the locals, Junior placed a wedge between his family and many of the wealthy summer people who were more of the Rockefellers' ilk. Knowing how much Junior loved his peaceful, horse-centric summers in Maine, this was no small sacrifice. Perhaps he recalled his time at Ludlow, rubbing elbows with the miners and dancing with the wives and daughters, when he made this decision. It is likely that, when he pondered whether or not building motor roads was the right thing to do, his pious, devoted nature and his mother's moral test ("Is it right, is it duty?") pressed him to make an unselfish decision.

Junior had advocated for a covenant that prohibited "motors" from his carriage roads for a minimum of twenty-five years. The whole point of these paths (used both by horse riders and hikers) was to encourage quiet enjoyment of the park's splendid scenery without disruption from automobiles and other modern aggravations. Junior unknotted this apparent contradiction in the 1928 *Bar Harbor Times* interview. "The preservation of the horse roads from the intrusion of motors has for all time been doubly assured in that the motor road affords as fine, as varied, as extensive, and as intricate views of the beauties of the park as do any of the horse roads." Junior was not being paranoid about such "intrusions," either.[18] He was well aware that many motorists had already succumbed to the temptation of steering their four wheels onto his beautifully designed and constructed carriage roads, in spite of legal prohibitions. Junior eventually built large, handsome gatehouses to help demarcate the start of carriage roads and to discourage automobile trespass.[19]

Inside the NPS, Albright, Mather, and Associate Director Cammerer were united in their push for better automobile access, a position that made distant parks available to a wider demographic swath of the American public.[20] "The great mass of visitors to a national park do not desire a walking trip over rugged terrain or strenuous climbs," Cammerer wrote in a 1922 report to Mather. "For the older, the less strong and active...who are the vast majority, means must be provided making reasonably accessible the features of special interest and beauty in the park."[21] First the railroads, then public roads, were Mather's way of "selling" national parks to the masses. Though counterintuitive today, historian David Louter points out that, in the early twentieth century, "automobiles provided Americans with the authentic experience they desired from the natural world." Car camping was viewed as a middle- and working-class way to get away from it all.[22] The old way of seeing the parks, by riding a train to Yellowstone or Yosemite and booking a five-day excursion with transportation, accommodations, and guides, was not something the average family could afford.

Other outdoor enthusiasts, however, especially the wealthier ones, maintained that building a road into a wild place was the quickest way to destroy the beauty they longed to protect. It was also certain to make a rich person's hideaway less exclusive. George Pepper roared again in the autumn of 1923 when he caught wind of construction on the Mountain Road, the park's first motor road. In an editorial in the *New York Herald Tribune*, Pepper (unselfconsciously) called the route a "rich man's folly,"[23] Now a U.S. senator, Pepper pounced on the opportunity to halt this much more disruptive type of road in his backyard wilderness. As a senator, Pepper now had ready access to the secretary of the interior, and he used this leverage to persuade Secretary Hubert Work to immediately halt all road building in Lafayette until a formal hearing and inspection tour could be conducted.

Pepper was not alone in his objection to automobiles in the park. Alexander Forbes, a noted professor of physiology at Harvard, logged his advocacy for wilderness preservation directly with Secretary of Interior Work. "I feel that the mountains are exceptionally adapted to enjoyment by pedestrians and climbers, and that the character which has rendered them so exceptionally attractive...will be seriously marred, if not lost altogether, if they become cut up by a network of motor highways."[24] The *New York Times* printed summer resident J. Gresham Machen's letter inquiring, "Are the national parks to be used to destroy the natural beauty or are they to conserve it for the benefit of generations yet unborn?"[25]

The Mountain Road at Lafayette was representative of Stephen Mather's grand plans for development of the national parks. Not only did Mather see roads as a way for tourists to experience nature, he knew they were key to developing a broad constituency for the parks. Political support, economic development in park gateway communities, and adequate federal funding, Mather recognized, would come only after Americans started experiencing and eventually loving their parks. In 1920 he wrote that our national parks had a "road problem" (namely, not enough roads), and that this deficiency was "one of the most important issues before the Service."[26]

Other conservationists of the early twentieth century likewise saw roads as a way of increasing visitation and thereby building a fan base for the national parks in the long run. Even John Muir came out in favor of roads in Yosemite to gain advocates and head off another controversy like the flooding of Yosemite's Hetch Hetchy Valley. Appalachian Mountain Club member Allen Chamberlain wrote that conservationists should "stimulate public interest in the national parks by talking more about their possibilities as vacation resorts." Like Mather, Chamberlain maintained that "if the public could be induced to visit these scenic treasure houses," they would "come to appreciate their value and stand firmly in their defense."[27]

Meanwhile, on Mount Desert Island, wealthy summer people sent letters of protest to the National Park Service and regional newspapers claiming the motor road would lead to a "constant rumble and roar of the automobile, disturbing to the ear." Others revealed their class prejudices and desire to keep Lafayette as their own semiprivate backyard idyll. Travel writer Herbert Gleason summed up the snobbish attitudes of some of the wealthier seasonal residents: "The proposed development would bring in a 'peanut crowd' of the Coney Island type, and that the park would speedily be littered with egg shells, banana peels, old tin cans."[28] "Peanut crowd" undoubtedly referred to middle- and working-class tourists.

Although Junior was stung by some of the personal attacks against him that stemmed from the roads controversy, he was remarkably open to feedback. "Would it seem desirable," he wrote, "to have given the opponents of the roads the fullest opportunity to study and discuss the question and express their views. Were a decision to be reached before the summer people have gone on island, they might have ground for feeling that they had not been given full opportunity to explore and consider the matter."[29]

While George Dorr was the quintessential hardy hiker and lover of wilderness, he believed the motor road was the appropriate response to the rapidly increasing number of visitors coming to the island in their automobiles. He also felt the roads were necessary not only for egalitarian public access, but also for his staff to efficiently move about the park and guard their domain from fire, poachers, illegal logging, and other perils. Therefore, Dorr diplomatically pushed back against the antiroad crowd and warned those in Washington of the negative publicity that would be generated from halting construction and laying off the substantial road crew employed by Junior and the NPS. Dorr's strategy was successful, and he and Junior were able to convince Secretary Work to keep their projects active until a parade of federal officials could leap at the opportunity to escape steamy

Washington the following summer for a look-see at Lafayette. The reprieve also gave Dorr plenty of time to rally park supporters and year-round residents who were wholeheartedly in favor of developing both motor and carriage roads in the nascent park.[30]

The stakes for Dorr and the future development of Lafayette National Park were high going into the March 1924 hearings in Washington. If Senator Pepper was successful, Dorr was likely to lose his position as park superintendent and Junior's ideal plan for park roads was probably at a dead end. Consequently, the ever-popular Dorr made certain his coalition of proroad Mainers dominated the hearing. The proceedings were so heavily stacked in favor of the roads that Pepper immediately retreated and left a disgruntled member of the Bar Harbor Village Improvement Association, Harold Peabody, to prosecute Dorr's and Rockefeller's plans. Peabody's off-the-cuff denunciations backfired when he took potshots at a road that Junior had built entirely on Rockefeller property.[31]

On the home front, the silent majority was finding its voice. The Board of Selectmen (Selectmen, Assessors and Overseers of the Poor) from the town of Mount Desert sent Junior a short letter stating they had been directed "unanimously" by the town to express their appreciation of the "development work which you are carrying out in Lafayette National Park."[32] John C. Clement, coproprietor of the Seaside Inn in Seal Harbor, wrote: "It is seldom that a region is so fortunate as to number among its residents a benefactor who is willing to so generously contribute to its well being."[33]

Philadelphia physician and Northeast Harbor summer resident Richard H. Harte's comments perhaps best support the position that Junior, Cammerer, Albright, and Mather were trying to articulate about Lafayette and other national parks.

> Before leaving Northeast [Harbor], I took the opportunity to get a carriage, and with Miss Wheelwright drove all over the roads which you have introduced; and I felt that the afternoon thus spent was one of the most interesting ones experienced during the summer: due entirely to the fact that you have made it possible for others besides mountain climbers to see and enjoy the natural beauties of the place.[34]

When Mather, Work, and elected officials from Maine made their inspection tour of the park, they were, again, impressed by the quality and careful alignments of the Rockefeller (NPS approved) roads. By late July, Work had given the green light to Dorr and Junior to proceed on all planned carriage and motor roads, including a controversial route to the summit of Cadillac Mountain. The one bone thrown to Senator Pepper was the stipulation that all future road plans had to be approved by the secretary of interior's office. And while this seemed like a minor stipulation at the time, consequential only in little Lafayette, it became one of several important stepping stones along a path to more comprehensive master planning for all national parks.[35]

Through all the automobile-induced brouhaha, it is important to recall the Mainer's earlier accusation that millionaires were making Bar Harbor "a quiet, exclusive resort." Wilderness historian Paul S. Sutter and others have cautioned that excluding cars and roads from a landscape "served the recreation and aesthetic interests of an urban leisure class," while acting against locals and marginalized populations.[36] Today, as the cost of vacationing in a crown jewel park spirals upward and beyond the means of many, the NPS's egalitarian goal of parks for the benefit and enjoyment of *all* the people is teetering. If the majority of our democratic society comes to regard taxpayer-funded national parks as exclusive resorts for an urban leisure class, our system of parks will find itself weighing on precariously thin ice. When demand outstrips supply in parks, it leads not only to overcrowding, but also to price inflation. While it is undesirable to decrease demand—since our increasingly urbanized populous needs the parks more than ever for their mental and physical health—it is past time to seize opportunities to expand existing parks and create new ones to serve a broader swath of lovers of the out-of-doors.

Future demand for protected natural areas can only grow as our population continues to increase and the widespread implementation of artificial intelligence potentially boosts the amount of leisure time we have for outdoor recreation. Junior's transition from builder of carriage roads to developer of motor roads was another step in his conversion from Old World lord of the manor to New World public lands benefactor. As the owner of a large estate on Mount Desert Island, Junior started building carriage roads as a hobby and for the enjoyment of his family and friends. His initial foray into motor road development, the Mountain Road, was partly motivated by his desire to keep cars off his carriage roads, but also to share the beauty of the place with the public. As he ventured further into the business of constructing motor roads, he became ever more closely aligned with the egalitarian goals of Albright, Cammerer, and Mather, to create and develop a system of national parks that would be enjoyed by all the people. From the beginning, Mather, as NPS director, felt strongly that the parks "belong to everybody" and that his agency had to "do what we can to see that nobody stays away because he can't afford it."[37]

The full extent of Junior's remarkable conversion is best captured in a later conversation recollected by Kenneth Chorley, one of Junior's chief conservation associates. The men were looking at a newly paved road in one of the parks, which struck Chorley as intrusive. "What are these parks for, Mr. Chorley?" Junior asked rhetorically. "The average American can't afford to go into secluded areas or to have private trips into the parks. He must travel on such a highway. That's the whole point of the national park system."[38]

*

It should not surprise us that Junior, Dorr, and the Eliots were not the last people to fall in love with the beauty of Maine and decide to fight to see some part of it protected. In 2016 President Barack Obama signed legislation creating Katahdin Woods and Waters National Monument in Maine. At the time, the monument consisted of 87,500 beautiful acres of rivers, lakes, boreal forest, moose, bear, fisher, and lynx. The land was purchased and donated by Roxanne Quimby, who, like John D. Rockefeller Sr., rose from rags to mind-boggling riches. Like Junior, Quimby became enamored with the Maine woods and began buying land to protect it.

Quimby was reportedly a fan of fellow Mainer George Dorr and Richard Louv, the latter an advocate for getting kids out of the house and into nature.[39] Quimby's park idea was taken to the mat by timber company interests and some sportsmen's groups. Even many of Maine's elected officials, including the governor, refused to support it.[40]

The contemporary philanthropist fought a long battle for NPS governance (rather than state or nonprofit) because she wanted to stimulate the local economy through tourism, and she believed the NPS was "a superior brand" that could attract people and facilitate the public's enjoyment of the Maine woods. Because of the lack of support from Maine's congressional delegation, President Obama went the national monument—rather than park—route.[41] Quimby also contributed $20 million to an endowment to help take care of the place once it was protected.[42] Such endowments, of course, make the addition of new parks much more palatable to the NPS and elected officials. In its press release announcing the monument's birth, the White House press secretary noted Acadia National Park had been created by a similar public-private collaboration exactly one century earlier.[43]

Because she was a businessperson who helped build the Burt's Bees® brand from a roadside honey stand to an international commercial empire, it should not be surprising that Quimby recognized the value of the NPS brand to the American public. The fact that we are loving parks to death certainly verifies their mass allure. For a century, visitors to national parks have been delighted by the pristine scenery, the beautifully landscaped roads, the strict protection of the plants and wildlife, the lack of commercial schlock, and the artfully designed and well-maintained trails. We have learned to love the visitor centers and museums, the rustic, tree-shaded campgrounds and picnic areas, and the rangers in their gray shirts and flat hats. Even the better lodges, restaurants, and shops have become treasured aspects of our vacation rituals. While other public lands have their own intrinsic purposes and values, nothing has captured the world's hearts like America's national parks.

LONGING FOR ENLIGHTENMENT

National Park Museums, 1920–1932

In his 1920 National Park Service annual report, NPS director Stephen Mather bemoaned the urgent need for "adequate museums" in all of his national parks. At the time of Mather's call to action, there were only three or four rinky-dink displays of specimens and artifacts in the national park system which might be called—with an elastic stretch of the imagination—museums. None of them, however, even approached Mather's standard of "adequate."[1]

Since 1915, staff in Yosemite National Park had been displaying some stuffed birds and mammals, pressed wildflowers, and watercolor sketches in a crowded room in the park headquarters building. Also in 1915, Mesa Verde superintendent Thomas Rickner appealed to then assistant secretary of the interior Mather for a little money for a museum. Rickner stated: "It has been a matter of wonder to tourists, and a disappointment to them, that there was no [artifact] collection for them to examine."[2] The following year, Mather reported to the secretary of the interior that rare artifacts were being plundered by tourists because the park had no place to store them. His request for $50,000 for a park headquarters and museum at Mesa Verde was denied. By 1917 Rickner had received an allocation of $22 and permission to build a six-foot-tall, five-foot-wide glass-fronted case in a log building previously used as a ranger quarters to display some pottery, tools, and human remains. The following year Rickner and his staff added five more cases and a small lounge with a fireplace for evening ranger talks.[3]

Responding to the needs highlighted by Mather, Yellowstone Park superintendent Horace Albright promoted Milton Skinner (a nature guide and road construction supervisor) to the role of park naturalist. Skinner immediately got busy cobbling together a makeshift museum and visitor center in Fort Yellowstone's old bachelor officers' quarters at Mammoth Hot Springs. The former bachelor pad featured around a hundred rocks, two mounted animal heads, a stuffed eagle, a "contorted tree," and eighty pressed wildflowers.[4]

The traveling public could not have been more delighted. By 1924, Albright and Skinner were recording 23,000 annual visits to their little shrine, an impressive number when one considers that only 144,000 people in total visited Yellowstone that year. Skinner quickly got busier, collecting specimens and expanding the exhibit space by another 1,500 square feet.[5]

In Yosemite in 1921, visionary Chief Naturalist Ansel F. Hall converted some spare rooms of the old Christian Jorgensen art studio in Yosemite Valley into a temporary museum. The log cabin studio was described by Sierra Club board member Francis P. Farquhar as "a shack that no one would care to live in much less keep their valuables in."[6] And even though Hall admitted that he initially "discouraged 'sightseers'" (people looking for amusement only), from visiting the museum, his collection of projectile points, baskets, dried wildflowers, and a ten-foot-long geology model built by Hall himself was viewed by 55,811 people in 1923 (out of 130,811 total Yosemite visitors).[7]

Mather expressed his personal commitment to Yosemite by recruiting, and in many cases digging into his own pockets for, subject matter experts and summering professors to present programs for tourists and field-tripping students.[8] During the mid-1920s, one-third of Yosemite visitors took advantage of the offerings of the nascent Yosemite Nature Guide Service. In Yellowstone, 75 percent of visitors attended a ranger program.[9] Still, Mather pressed hard for more and better interpretation in the parks. His 1925 memo to Secretary of the Interior Hubert Work declared: "One of the most important problems before the National Park Service today is to respond to the pressing demands for information, not only about roads, trails, and hotel accommodations . . . but also about the geology, trees, birds, flowers, mammals, Indians—in short, about everything that is preserved in its natural state in these great national playgrounds."[10]

As Hall's little collection of Native American baskets and other artifacts grew, he became increasingly concerned about their long-term preservation. Unfortunately, there was absolutely zero chance that Congress would fund construction of an "adequate" museum in Yosemite; they had in fact imposed a $1,500 spending cap on building construction in national parks, perhaps to emphasize their low ranking in the hierarchy of federal funding priorities. Regardless, Hall was discovering that the enthusiasm from locals and park visitors for his museum sometimes led to offers of contributions of money, artifacts, and even rodents to feed the snakes in his live reptile exhibit. Hall, therefore, got into the fundraising business with the object of someday building a large fireproof museum for Yosemite. The young ranger's passion for the natural and cultural history of the park was contagious; by 1923 he had over $6,000 ($113,000 in 2025 dollars) in the new museum kitty.[11]

Hall was also making important connections in the rapidly expanding world of American museums and museum management. During the summer of 1921, while on a backpacking trip in the Yosemite high country with Farquhar (a friend of both Hall and Mather), the party met up with Chauncey J. Hamlin—another of Mather's many, many chums—who would soon become president of the American Association of Museums. The fast friendship between Hall and Hamlin led to Hall chaperoning Hamlin's son, a recent prep school grad, on a tour of Europe and the Middle East. For Hall, the trip was a chance to visit the great museums of the world in preparation for his Yosemite project.[12]

Before Hall embarked, however, he made one of the most fateful decisions of his long, fruitful career: he created a nonprofit organization to look after his museum fund while he was abroad. The resulting Yosemite Museum Association was the seed that grew into a vast network of nonprofit park partner organizations that today exist in every national park and monument and provide over $200 million annually to support public lands education and science projects.[13]

While Hall was seeing the Old World sites, Hamlin was busy pushing the park museums idea to the American Association of Museums (AAM) and others. The Laura Spelman Rockefeller Memorial, which was created by John D. Rockefeller to honor his late wife, had begun providing the AAM with $10,000 per year in operating funds in the early 1920s, a show of support for public elucidation in the burgeoning world of science, natural history, art, and cultural history museums.[14] The memorial also supported the New York Public Library, the American Museum of

Natural History, and the Metropolitan Museum of Art. The social link between the AAM and the memorial was likely John D. Rockefeller Jr's. friendship with AAM officer Henry F. Osborn, director of the American Museum of Natural History in New York and one of Junior's top conservation mentors. Junior had also crossed paths with Hermon C. Bumpus, fellow Brown University grad, in 1897 when Bumpus was raising money for a Brown endowment. Not surprisingly, Junior rescued the campaign with a $500,000 (at least $20 million in 2025 dollars) contribution. Bumpus, who was the first president of the AAM, had served as director of the American Museum of Natural History and, up in Maine, later worked alongside George Dorr as a trustee of the nonprofit Mount Desert Island Biological Laboratory. Because the Laura Spelman Rockefeller Memorial (LSRM) was a major contributor to AAM operations, Hamlin was picking up documents at the memorial offices one day and happened to share with Director Beardsley Ruml (a young, creative, free-thinking psychologist) his interest in museums in national parks. When Ruml said he wanted to learn more, Hamlin hurriedly formed the AAM Committee on Museums in National Parks.[15]

By early 1924 Junior had been working with Mather and Assistant NPS Director Cammerer for several years on land acquisition and carriage road projects in Acadia. This relationship likely led Cammerer and Mather to make a big ask of $200,000 ($3.7 million in 2025 dollars) for the design and construction of new museums in Yosemite, Mesa Verde, Yellowstone, Acadia, and other NPS sites. Cammerer noted that if the AAM handled the monies instead of the federal government, there would be less federal red tape to navigate, and the parks could avoid competitive bidding requirements in the construction process.[16]

Abiding by the slow and cautious traditions of the Rockefeller philanthropy playbook, the memorial suggested that the NPS and AAM take baby steps into the national park museums business. Rather than take on half a dozen museums simultaneously, Ruml suggested they whittle their request down to one or two projects and proceed from there. Hamlin and the committee did so, and in July 1924, the LSRM trustees approved $50,000 for construction of a museum building in Yosemite Valley. The package included $10,000 for exhibits and furnishings, $10,500 for staff for the first three years of operation, and $5,000 for the expenses of the AAM museums in parks committee. Hall received the good news from Hamlin via radio while he and Hamlin's son were still aboard the ship headed home from their grand tour.[17]

Hall started work by hiring fellow University of California—Berkeley grad Herbert C. Maier to design the museum, a decision that turned out to be every bit as auspicious as his creation of the Yosemite Museum Association. Maier's variations on Park Service rustic architecture came to symbolize museums and education in the parks and contributed to an enduring style sometimes called "parkitecture."[18] Architectural historian Sarah Allaback described the style as "characterized by use of native materials, the desire for architectural simplicity reminiscent of pioneer craftsmen, and an implied association with the landscapes." Anyone who has admired an NPS building sporting native boulders, giant log beams, and roughhewn lumber and shingles that seems to have grown organically from the landscape, is a fan of parkitecture.[19]

Even before starting on the main museum in Yosemite Valley, Hall and Maier decided to warm up by knocking out a tiny satellite museum atop one of Yosemite's most scenic summits—Glacier Point. The now revered Glacier Point Lookout, aka "Geology Hut," was designed and constructed while the plans for the main Yosemite museum were under review. Built almost entirely of native stone, the lookout frames unforgettable views of Half Dome and other landmarks through its arched windows and interprets their geological story with small exhibits. The minimuseum was funded by various pots of money offered by the Yosemite Museum Association, park concessionaires, and the LSRM. Today it is recognized as the first trailside museum, an extremely effective style of open-air museum based on the premise that the *park* is the museum and facilities should complement, rather than detract from, the natural wonders. Constructed in just over a month by NPS staff, the satellite museum was open by late 1924 and was an instant hit. By 1927 it was recording fifteen thousand visitors annually.[20]

On November 6, 1924, Hall and Mather helped lay the cornerstone for the main museum in Yosemite Valley. Hall expressed regret about the absence of representation from the memorial, but he excitedly showed around the blueprints of the museum to the large crowd gathered on the chilly day. He and the AAM's vision was grand. Visitors would enter through a foyer appointed with large photographs for orientation to the park and be greeted by a park ranger. The adjacent room would hold the large-scale geology models and rock specimens, followed by a room of flora and fauna exhibits. Next was an open-air auditorium with daily programs, then the Native American exhibits, gold rush and pioneer-era exhibits, followed by outdoor gardens with trout pool. Upstairs one would find a library, offices for the park naturalist (Hall) and several seasonal nature guides, a press room for printing the park's famed journal—*Yosemite Nature Notes*—a darkroom (for processing photos), taxidermy room, 150-seat lecture hall, clubroom with fireplace for meetings, and caretaker's apartment.[21]

Thanks to "an unusual period of good weather," progress on the museum moved rapidly. In six short months, the fireproof museum—fortified with floors of native granite and concrete—was more or less completed.[22] In April 1925, Mather and Secretary of the Interior Hubert Work made the long pilgrimage to Yosemite to officially accept the donation of the museum from the AAM, Yosemite Museum Association, and the LSRM (representation from the latter not present). Hall made sure to escort Mather, Work, and their entourage through the old museum in the artist's studio "shack" before unveiling the beautiful new one. Mather and Work were emotional at the ceremony, both having long dreamed of the day when Park Service facilities would be designed and built commensurate with the grandeur of the landscapes they supported. According to Hall, Work "expressed himself as greatly impressed" and signed a memo, prepared by Mather, calling for the development of "a broad educational policy for the national parks and monuments with the guidance of the American Association of Museums" and oversight from Hall.[23]

In 1925, with the Yosemite museum projects largely in the rearview mirror, Hall's busy mind turned to the future of the Yosemite Museum Association. His vision for the new and improved YMA was, as you might expect, grandiose. It was tailored not simply to building museums, but to helping the Park Service meet the vast educational needs of the accelerating number of visitors traveling to Yosemite in automobiles. The YMA was born again as the Yosemite Natural History Association (YNHA), its mission: to gather and disseminate information on Yosemite's natural and cultural history, publish *Yosemite Nature Notes*, expand the park museum, develop the park's nature guide service, promote scientific research, study the remaining Native Americans in Yosemite and preserve their arts, customs, and legends, and maintain a park library. Within eight years, six other national parks, from Utah to Hawaii, boasted similar, self-sustaining nonprofit partner organizations.[24]

While still flush with the glow of the successful model museum projects in Yosemite, Mather, Work, Hall, and Bumpus, with the AAM, decided it was time to go back to the LSRM well. They made a big ask, $259,000 ($4.7 million in 2025 dollars) for museum/visitor centers in Yellowstone, Grand Canyon, Mesa Verde, and other parks. The proposal was inspired by Bumpus's recent tour of nearly all the existing national parks and his enthusiasm for the educational possibilities of each. However, the temperamental Ruml, who believed the memorial should be focusing more on social work, once again balked at the size of the proposal. Bumpus et al. scaled back to $105,000, which Ruml felt was still too large of a next step. The compromise was for the LSRM to fund planning, construction, and exhibits for two additional inexpensive "trailside" museums, one at Grand Canyon and one at Bear Mountain in Palisades Interstate Park—up the Hudson River from New York City—for $22,500. Once these demonstration projects were completed, and the principals had had a chance to gauge the public's reaction, further projects might be considered. The memorial's modest commitment reflected not only the cautious Rockefeller approach to philanthropy, as well as its preference for partnerships, but also the memorial's desire that eventually other entities (e.g., the federal government) would step in and fund museums in national parks.[25]

Bear Mountain was likely chosen because of its proximity to population centers as well as the Rockefellers' long-term involvement with the Palisades. Many of the AAM's officers also resided in the New York City area. Both museums were to be designed by Maier, and some architectural historians call the Bear Mountain structure the first true trailside museum since reaching it entails a half-mile hike. The trail to the Bear Mountain museum, is, in fact, part of the Appalachian Trail which runs through the middle of the facility.

The architect Maier's goal with trailsides was to minimize barriers between the museum patron and the park landscape by employing large open entryways, outdoor patios and verandas, and windows without glass. The Bear Mountain museum encouraged visitors to interact with the landscape by providing pails to fetch water from a nearby pond and microscopes to study the aquatic organisms they had collected. The Glacier Point Lookout featured a telescope purchased by the Yosemite Museum Association that allowed visitors close-up views of Half Dome and other peaks across the valley. While Maier experimented with rustic architecture in designing the main Yosemite Museum, he advanced parkitecture a step further with the trailsides.

By far the largest and most impressive of the first generation of trailsides was the Yavapai Point museum on the south rim of the Grand Canyon. In this design, Maier integrated rustic style with the look of already existing Grand Canyon structures created by the famous southwestern architect Mary Elizabeth Jane Colter. NPS landscape architect Ethan Carr recalled, "Like Colter, Maier also referred to Native American building traditions in architectural details."[26]

Bumpus, the Minnesota-born biologist turned museum director, also made a major contribution to the trailsides. Acting in his role as an AAM officer, Bumpus had become infatuated with the educational opportunities in national parks. His progressive ideas (so progressive he had been fired from the American Museum of Natural History) affected both the design of the museums and the contents of their exhibits. In the early days of museum development in America, Bumpus had gone out on a limb to emphasize education and the museum visitors' experience over the traditional attitude that museums were somber places stuffed with artifacts that some tenure-hungry scientist had recently hoarded.[27]

"To lead these people away from direct contact with Nature, to beguile them into a building where they are surrounded by artifacts . . . is contrary to the spirit of the enterprise," Bumpus believed. "The real museum is outside the walls of the building and the purpose of the museum work is to render the out-of-doors intelligible."[28]

The Yavapai Museum was situated to provide one of the broadest possible panoramic views of the Grand Canyon, one that even includes a narrow glimpse of the canyon-creating Colorado River far below. The exhibits attempt to relate the complicated story of how the canyon was formed, a task made easier by positioning museumgoers where they can view the exhibits and the landscape simultaneously. A variety of organizations made their top scientists available to Hall and the AAM to contribute to the exhibits' content, including the U.S. Geological Survey, the American Association for the Advancement of Sciences, and the National Academy of Sciences. John C. Merriam, an affiliate of the Carnegie Institution and collaborator with Junior on the campaign to save redwood trees, was also a key contributor.[29]

At three thousand square feet, the Grand Canyon museum is one of the largest trailsides. Yet its low profile, exterior of battered native stone, and covered terrace curved to match the undulations of the canyon rim, make it seem as much a part of the natural landscape as a juniper shrub. World War I surplus spotting scopes were mounted on the terrace to offer visitors enhanced views of the river, the top of Cedar Mountain, and the canyon's famous rock strata. A native plant garden with labeled specimens was added as the perfect complement to the trailside experience.

The Yavapai museum was formally opened on July 19, 1928, and was considered "an immediate success," according to an authoritative historic structures report. It attracted "much favorable comment" and "appears to have grown as a continuation of the canyon walls."[30] New Yorkers and New Jersians were also delighted with the Bear Mountain trailside, and NPS brass were eager to keep the museum ball rolling.

Up in Maine, Lafayette National Park Superintendent Dorr, Interior Secretary Work, and Dr. Robert Abbe—a famous physician and Mount Desert Island cottager who had recently purchased some Native American stone implements—were collaborating to create a trailside cultural museum in Lafayette. In the summer of 1924, Work had encouraged Dorr and Abbe by sharing with them a "telegram conveying a gift of $60,000 from the Laura Spelman Rockefeller Foundation [*sic*]" for a similar museum in Yosemite.[31] The grant from the foundation inspired Abbe and others to begin a grassroots fundraising campaign to build a museum, conduct research, and secure more artifacts for display. Junior pitched in $15,000 in honor of his neighbor Abbe, who was in poor health at the time.[32]

In the late summer of 1928, the compact Abbe Museum opened near a popular park trailhead and the famous Sieur de Monts Springhouse. Premier archaeologist Dr. Warren K. Moorehead was engaged by the museum to study the cultural history of Mount Desert Island and to produce literature for the museum and general distribution. Thereby the Abbe trailside museum became "the first institution in Maine to support archaeological research."[33]

Back in Washington, Albright, Cammerer, Hall, and Mather were collaborating with Interior Secretary Work to draft a letter directly to the LSRM for a big request. The secretary stated, "Since it has been proven that small museums, advantageously placed and equipped for the definite purpose of giving popular instruction, are of basic importance, I suggest that $118,000 ($2.1 million in 2025 dollars) be granted to the American Association of Museums." The funds would cover seven museums in Yellowstone and include $10,000 to study education programs in the national parks.[34]

Because of the success of the trial museums already completed, as well as the solid relationship between Albright and Junior, and because the request had come directly from the secretary of the interior, Ruml and the LSRM did not quibble; they approved the entire request. As with many of his philanthropic endeavors, Junior stipulated that "no public announcement would be made of the gift."[35]

Yellowstone's first LSRM-funded and Maier-designed project was a gorgeous trailside museum near famed Old Faithful geyser. In 1931, Bumpus reported, "The building at Old Faithful was but little more than completed when we found that we had underestimated the reaction of the public."[36] Consequently, the AAM and Park Service diverted some funds from the main museum at Mammoth (the expansion of Skinner's old museum had to be put on the back burner due to a larger reassessment of development of the entire Mammoth Hot Springs area) to add onto the jam-packed Old Faithful museum. The bigger, better Old Faithful facility was a museum/visitor center with information desk, open-air auditorium for ranger talks, exhibits on geyser mechanics and flora and fauna, as well as a wildflower garden. Albright praised the structure's fine rock and log work and declared the Old Faithful museum was "finer than the one at Yosemite."[37] Of all the museums described in this chapter, only Old Faithful is no longer in existence—it was replaced with a larger structure in 1971 because even it was too small to accommodate the eruption of curious visitors to Yellowstone.

The smaller trailside at Madison Junction, located between West Yellowstone and Yellowstone Lake, was sited where explorers in the Washburn Party of 1870 camped and discussed the future of Yellowstone country. Completed in 1929, the museum offered exhibits on park history and a tranquil view of the blending of the Madison and Gibbon Rivers. Architectural historian Albert Good called the museum "minor in size, but not in its contribution to park architecture. The spacious 'landscape' window serves to project the outdoors into the museum interior."[38]

Norris geyser basin boasted a true trailside in that the trail to the main array of gurgling, spouting, sulfuric wonders runs smack dab through the museum's central breezeway. Like most of the other Yellowstone museums, its construction suffered frequent delays attributable to the short season, minor squabbles, and other gremlins. Yet Bumpus and Albright never wavered in their commitment to the highest standards: "Our position is that this whole project, to be successful, must be carried on in an intimate and cooperative manner and that no step be taken that does not meet with the entire approval of the Service."[39] Norris was one of the larger trailsides, featuring an information desk, two exhibit areas, and quarters for visiting scientists. The exhibits on one side of the museum explained the functions of various thermal activities: fumeroles, mudpots, hot springs, and geysers; the other was devoted to flora and fauna. The carefully selected structural logs on the museum interior display a special rustic embellishment, natural burls and other deformities that add beauty and interest.

The adorable museum at Fishing Bridge featured a central breezeway which framed a tantalizing view of Yellowstone Lake. The exhibits highlight wildlife, especially waterfowl and other species commonly found around the freshwater ecosystem. A very large, standing taxidermized grizzly bear was the star attraction for many years. A park visitor in 1938, Flora S. McHarg, took the time to track down and write Maier and express her delight: "An inquiry at the desk did not reveal the names of those of genius who planned location, building interior, terraces and surrounding landscaping. There seemed nothing left to be desired. The selection of the tree trunks for the doorways, like tied-back draperies, was a stroke of genius."[40]

According to architectural historian Laura S. Harrison, Maier's museums are "the best structures of rustic design in the National Park System," serving "as models for hundreds of other buildings constructed throughout the nation in state, county, and local parks under auspices of the NPS during the work relief programs of the 1930s."[41] Other noted architectural historians, Kiki L. Rydell and Mary S. Culpin, lavish perhaps the highest possible praise on Maier's work, stating his "buildings best exhibit the notion that structures of any kind in a national park should harmonize with nature to the point of being unnoticeable."[42] Maier and his colleagues also contributed to the design of the NPS arrowhead logo worn on park employees' uniforms.[43]

The collapse of the American economy in 1929 and the advent of the Great Depression marked the transition in national park museum funding from the LSRM and the AAM to New Deal programs like the Civilian Conservation Corps (CCC) and the Works Progress Administration (WPA). Ironically, the financial catastrophe ushered in an era of ample federal funding for parks and even more ample low-cost labor. The New Deal not only fostered President Franklin D. Roosevelt's "Tree Army" (the CCC), it spawned an unparalleled boom in the development of roads, trails, museum/visitor centers, and other facilities on public lands. That the majority of new structures adhered to Maier's rustic template was not just the continuation of tradition, it was partly directed by Maier himself who was in charge of the CCC in the Southwest.[44]

The LSRM and AAM had succeeded in what they intended to do: jump start interpretation in national parks when there were no federal funds available to do so. As usual, Junior got a lot of bang for his buck with the museums. He and his associates had the heft to make certain their contributions were used prudently and effectively and were deployed at an early stage in a project so the ripple effects would be greatest. Secretary of the Interior Ray L. Wilbur wrote the Rockefeller Foundation (which had recently absorbed the LSRM) in 1931 to assure them "your practical interest in the park educational work has been of great influence to bringing additional support to the program, both governmental and private." He informed the foundation that the commitments of staffing and equipment made by the NPS had been fulfilled.[45]

The ever-gracious Bumpus proposed to the Rockefeller Foundation that each of the new Yellowstone museums include a "tablet" recognizing the efforts of the AAM, Rockefeller Foundation, and NPS "for the Benefit and Enjoyment of the People, 1932."[46] Replying on behalf of the Rockefeller Foundation, Thomas B. Appleget declined the recognition with an expression of humility that is almost incomprehensible today. "Realizing that our part in the accomplishment of certain projects has been relatively unimportant and the major credit should always rest elsewhere, we have attempted to discourage any acknowledgement."[47]

The museums are often-overlooked examples of Junior's and the Rockefeller institutions' aid to the national parks during the Mather–Albright–Cammerer years, when the NPS was just getting started and there was a dire need for public-private cooperation. The reason for the lack of acknowledgment regarding the museums and Junior's involvement in national park philanthropy in general has a lot to do with Junior himself.

The philanthropist's biographer, Raymond Fosdick, claimed Junior was "one of the most modest, unassuming, unpretentious men imaginable." Junior conducted much of his philanthropy anonymously for a number of reasons, some of which were his humble nature and attitude that his contributions were all made possible by his father's financial success, not his own. (A secondary reason was that news about a big Rockefeller donation always provoked a fresh tidal wave of requests for money from individuals and organizations, all of which had to be reviewed and often responded to.) Junior once turned down an acknowledgment for a large gift to the Harvard Business School by demurring "such recognition is most embarrassing to me."[48] When Fosdick broached the topic of writing his biography, Junior protested. "What on earth would you find to write about me?"[49]

Junior's and the foundation's rejection of plaques and other forms of recognition support the assertion that Rockefeller philanthropy was motivated more by a sense of religious duty and commitment to promoting "the well-being of mankind" than the need to redeem the family name.[50] The museum work was a continuation of support provided by Abby, Junior, and his father for educational institutions, including the American Museum of Natural History. Like landscape architecture and carriage-road building, museum development was yet another stepping stone in Junior's conversion to a monumental patron of the national parks.

COLORADO HIGH DESERT HIGH

Mesa Verde, 1924

The twin Packards carrying the Rockefellers and their entourage came out of the east, having crossed the Continental Divide at Wolf Creek Pass (10,856 ft.) before proceeding through the southern Colorado towns of Pagosa Springs and Durango to the village of Mancos. It was July 3, 1924, and the big touring cars kicked up rooster tails of dust. The party's destination, Mesa Verde, would be the third national park on the itinerary of this Rockefeller family summer vacation. Grand Canyon in Arizona and Bandelier in New Mexico (as well as the towns of Sante Fe and Taos) had been early highlights of the first two weeks of the group's sampling of the American West.

Junior rode in the lead car accompanied by Dr. Richard Corwin and Bert Mattison, two executives from the Colorado Fuel & Iron Company (CFI) headquartered in Pueblo. The trailing Packard, separated from the lead automobile by a roiling dust storm and several car lengths, was driven by another CFI employee and carried Junior and Abby's three eldest sons: John III, eighteen; Nelson, sixteen; and Laurance, fourteen. According to Junior, the trip had been undertaken "wholly for the sake of the boys."[1] Regardless, Junior was quick to seize the opportunity to separate himself for a bit from the three teens, a deft maneuver that would not have surprised anyone in the extended Rockefeller family.

Corwin and Mattison had, over the years, become cohorts and perhaps friends with Junior. They had worked together to improve relations between labor and management and uplift the miserable plights of miners and mill workers and their mostly immigrant families in the desolate company towns of southern Colorado. Corwin, who had been trained as a taxidermist at Cornell but went on to become a renowned surgeon in the region (he also dabbled in archaeology and eugenics), had become a "cherished friend" of Jesse Nusbaum, superintendent of Mesa Verde National Park.[2]

Superintendent Nusbaum (pronounced noose-baum) had planned to meet the V.I.P. visitors at the park entrance. However, Nusbaum, who had never been good at idling, quickly grew impatient and steered his Buick another eight miles to the picturesque cow town of Mancos, Colorado, where he might spend his wait time more productively. There, at the old park office ("old" because Nusbaum had recently relocated park headquarters from Mancos to the national park; one of his many management decisions that enraged locals as well as their representatives in Washington, DC), the superintendent could accomplish a bit of work while he watched out the front window for the Rockefellers' approach.[3]

Approach they did, their Packards barreling down main street and exiting the city limits before Nusbaum had time to lock the front door and fire up his ranger car. The superintendent was thereby forced to eat the dust of his would-be guests for five long miles before he resorted to using his "exhaust siren" to finally pass the trailing car containing the boys and their driver. His hard-won position in second place then allowed him to employ the siren again to persuade the lead car to pull to the side of the road.[4]

As the untethered dust clouds overtook the three automobiles and gradually dissipated, Mattison introduced Junior to the park superintendent and Nusbaum informed Junior and the CFI executives of his plans for the day. Junior then walked back to the third car and filled in the others on the itinerary. He agreed to abandon Corwin and the Packard to join Mattison and Nusbaum in the ranger car. Soon the three automobiles were slowly grinding up the steep, narrow, harrowing, and appropriately named Knife Edge Road toward Mesa Verde (Spanish for "green table"). While they ascended, Junior, who was naturally bashful and socially awkward, realized he had a pretty good story to tell. He related to Nusbaum that his siren had made an impression on the three boys. They (Nelson in particular) had in fact hoped the superintendent was the county sheriff intent on writing their father a ticket for speeding. Such a comeuppance, Nelson reasoned, might have humbled their straightlaced dad who had been hard on their older sister Abby (Babs) for garnering two speeding tickets already in 1924. The second offense made headlines when the judge issued only a fine, and no jail time, something the press perceived as favoritism toward the rich. The unfavorable news reports, with their hints at corruption, distressed Junior greatly.[5]

The boys' bristling at their father's puritanical bearing was a reaction shared by many in the Rockefeller clan. They were tortured every morning by their father's ritual of prayers and Bible verses before the first bite of breakfast, something only their beloved mother Abby managed to sidestep (she often took breakfast in her bedroom). As flappers, modernists, and the Roaring '20s convulsed around them in their New York City neighborhood, the Rockefeller children were painfully aware their father favored the company of his reserved church crowd to the celebrities, artists, and playboys the rest of the family gravitated toward. Only Junior's father, John D. Rockefeller Sr., and mother, Laura Spelman—all teetotaling, pro-abolition, pro-women's suffrage, and pro-prohibition northern Baptists—could match Junior's prudence.

Steam geysered from the radiators of the hulking Packards as the vehicles sputtered up the precipitous grade. Consequently, Nusbaum stopped the caravan frequently to point out sites in the Montezuma Valley and to wait for the gurgling to cease. Little in the way of guardrails or shoulder separated the roadside from a near vertical fifteen-hundred-foot descent to the valley floor.

Although Nusbaum was acting under strict orders from his boss, National Park Service Director Stephen Mather, not to solicit money or otherwise badger Junior on his much-needed vacation, once atop the mesa, the superintendent wasted little time in conducting a tour of the modest log building which had formerly served as a ranger station and recently been adapted as the park's museum. Meager as it was, the combustible structure housed a noteworthy collection of "pots, implements and skulls."[6] Nine years earlier, Mather had lobbied for funds for a real museum for the park, complaining that "many curios and rare objects ... were being carried away by tourists." His request to Congress for $50,000 for a museum and office facility received no serious attention.[7]

The view from the porch of the makeshift museum was magnificent. The vast cliff dwelling known as Spruce Tree House was strategically tucked beneath the rim of the adjacent canyon, only a stone's throw away. Standing on the porch was like being in the bleachers of a stadium where the field of play was a historic setting almost too well preserved and too picturesque to believe. Spruce Tree House's 114 stone rooms and fourteen meeting areas burrowed into the head of the box canyon invited all visitors to contemplate the breadth and richness of Native American life pre-European contact. Of course, for residents of New York City, the specter of so many families crammed into multistory apartments of stone and masonry was likely not as alien as it might appear to visitors from Nebraska.

Nusbaum was several inches taller than Junior, and the energetic superintendent was as confident as the millionaire was diffident. The son of a construction worker and a man with limited formal education in archaeology, Nusbaum had nevertheless risen in the ranks to become one of the foremost experts on Southwestern culture. Horace Albright called Nusbaum "one of the best superintendents we [the NPS] ever had."[8]

When Nusbaum had assumed the superintendency three years earlier, at age thirty-two, he inherited a park he labeled "an unholy mess."[9] Most of the superintendents since the park was established in 1906 had been temporary placeholders or underqualified political appointees who spent most of their time outside the park in Cortez or Mancos. Mesa Verde's first full-time superintendent, Hans Randolph, was a hard-drinking politician who got mixed up in some local banking rivalries and went about his park duties armed and wary. After narrowly escaping an assassination attempt, he was dismissed in 1911 for padding payrolls and misappropriating funds.

Local rancher Thomas Rickner followed Randolph. Although Rickner is credited with starting the little museum the party had just toured, he was also rumored to be illiterate.[10] Rickner fostered a legendary system of nepotism in the park by first hiring his daughter Oddie and son-in-law, Chief Ranger Fred Jeep, to run the campground concession. Jeep, though suffering both physically and mentally from the consumption of bad booze, was digging pots from the ruins and selling them, a cardinal sin in a historic park like Mesa Verde.[11] The ranger's relatives ran a high-pressure, low-quality tour guide service that resembled an East Coast extortion racket. Jeep's son, Friz, the self-proclaimed "best guide in Mesa Verde," got started in the lucrative tour business at age five.[12] Nusbaum, with some self-serving bias, described the content of the fledgling guides' interpretive talks as "atrocious."

Prior to Park Service involvement, conditions at Mesa Verde were much worse. The park's administrative history recounts that shortly after the area's widespread "discovery" in 1888, "one ruin after another was subjected to wholesale commercial looting by pot hunters."[13] Noted pathologist and amateur archaeologist T. Mitchell Pruden lamented in 1903 that, while earlier pot hunters tore "down the walls of the ruins to delve beneath the rooms," such destructive efforts had lately been diverted to the even more appalling looting of burial mounds.[14]

Superintendent Nusbaum mused: "If Congress could have seen the necessity for preserving American antiquities about 20 years earlier...our Museum would have been the most important on the Southwestern cultures in this country."[15] Fortunately for Mesa Verde, all was not lost; there remained a staggering number and variety of archaeological sites, some in remote and difficult-to-access canyons and others peacefully buried beneath the shifting sands.

After Rickner was canned for cronyism and nepotism, NPS brass Mather and Cammerer decided to overturn the tradition of allowing Colorado's elected officials to appoint superintendents as rewards for political support or as political "plums" to friends and family. This time the NPS took the initiative to discover a young, industrious, trained archaeologist who hailed from Colorado, possessed impeccable references, and had previously worked in the park.

Soon after reporting for work in June 1921, Nusbaum astutely observed that visitors to the park were running amok. They climbed through the windows of the ancient structures and posed for pictures standing atop the teetering walls. He summoned the courage to fire Jeep and the whole lot, an act that predictably estranged the Nusbaums from the families who occupied much of the land in the Mancos Valley below. He replaced the nepotistic syndicate with hard-working and skilled employees, most of whom were Navajo. Nusbaum then added even more fuel to the fire by announcing that park superintendents belonged in the park. He, wife Aileen, and stepson Deric moved into a tent at an elevation of 8,200 feet, on Mesa Verde.

*

First things first, the Rockefellers were covertly checked in at Spruce Tree Camp, where the group occupied three rustic cabins (and platform tents for the boys), then given a quick tour of the park superintendent's adobe home (designed, built, and furnished primarily by Jesse and Aileen). The Nusbaums were compulsively industrious workaholics with a passion for sharing their knowledge of Mesa Verde. Such qualities delighted Junior. The manic pace of the rest of the day was likely tailored to the needs of the boys who had been cooped up, either in cars or on public relations tours of CFI facilities, for the last few days. The frenzy of new sights, dizzying heights, hand-over-hand ascents and descents, and dramatic stories also suited Junior, whose insatiable curiosity helped compensate for his introversion.

The superintendent oriented his followers to mesa-canyon terrain and provided overviews to many greater and lesser cliff dwellings. Soon Nusbaum convinced the group to tighten their shoelaces and suppress their acrophobia for a hike along a narrow, rocky divide so the superintendent could interpret the progression of Mesa Verde occupants—from Late Basket Makers living in subterranean, mud-covered huts on the flat mesa top to the cliff-dwelling residents referred to in the day as Anasazi (Pueblo peoples today)—who thrived before abruptly departing seven centuries earlier.

By late afternoon the entourage had returned to the superintendent's comfortable quarters to enjoy "cold refreshing drinks (soft, not hard)."[16] Junior especially was impressed by the home's pueblo-style architecture, which had been inspired by the Hopi Indians, the dwellings of Sante Fe, and the works of Mary Colter. The two-thousand-square-foot home's interior featured eighteen-inch-thick stone walls that were beautifully plastered and whitewashed. Outside, the sandstone block walls, protruding vigas (rough-hewn roof rafters), and stone stairs made the structure the perfect complement to a historic site dedicated to preserving the ancient architecture of the Southwest. The home was so noteworthy, in fact, that tourists demanded guided tours. It also set the standard, not only for the score of Mesa Verde facilities that followed, but for all culturally appropriate national park architecture for decades to come.

Next up was the traditional Nusbaum welcome-to-Mesa-Verde steak fry. For the special occasion, Jesse had "procured the choicest T-bone steaks available in Durango" and proceeded to broil them "cowboy style" over an open fire. The superintendent was amused when Junior "attempted to handle his with a fork and knife" and finally came to his aid by demonstrating the more traditional southern Colorado carnivorous method of using one's fingers and incisors. "From then on, he [Junior] was perfectly at ease about the camp and grill. He was somewhat appalled when all were tendered a second steak—but also relished it, as I recall."[17]

The next day, July 4, was a glorious moment for Junior and the future of America's national parks. At Junior's request, Nusbaum conducted a private tour of Cliff Palace for the Rockefeller group. Cliff Palace was the park's most magnificent site, the largest cliff dwelling in North America that once sheltered around one hundred people and perhaps served as the center for ceremonies and administration. Junior spent hours ducking into the myriad rooms and kivas, noting the similarities in materials and construction to the superintendent's home. He relentlessly peppered Nusbaum with questions about the culture and lifestyles of the pueblo people. "Where did they farm? How did they farm? Where did they get their water? What were their ceremonies?[18]

Junior also quizzed Nusbaum on the National Park Service. He longed to know everything about the young agency, its interpretive operations, law enforcement strategies, and development plans.[19] There, in the perfect silence of Mesa Verde, far from his father's somber shadow, Junior was inspired. He vowed to Nusbaum that he would soon return to Mesa Verde with Abby and the couple's two youngest sons so he could share the experience with the rest of the family.

After lunch at the superintendent's house, Nusbaum showed Junior around the "headquarters area," such as it was. The superintendent shared his vision of how the headquarters would be laid out in relation to the museum, parking area, and view of Spruce Tree House. Junior, forever interested in landscape architecture, was impressed by Nusbaum's skills in blending facilities with the landscape and was simultaneously just as "amazed at the meagerness of the appropriations made by Congress and the amount of work that had been accomplished with these funds."[20] It's hard to imagine that Nusbaum could have resisted mentioning that he and Mather were trying to scrape together $350 from their barebones budget to replace "our pumping plant in the head of Spruce Tree canyon which is in the last stages of falling to pieces."[21] If the pumps failed, summer visitors to Mesa Verde would have no water.

Unsurprisingly, Junior also found the museum and the park's interpretive programs of particular interest. After little more than twenty-four hours at Mesa Verde, he had concluded that without the assistance of competent guides, exhibits, or other forms of explication, the average visitor to Mesa Verde would walk away with a very shallow appreciation for what they had seen. Nusbaum explained that his museum project had been accelerated with a $5,000 donation from Stella M. Leviston, a national park enthusiast from San Francisco.[22] His requests to Congress to fund construction of the museum had been repeatedly denied (and would continue to be well into the 1930s).[23] Just a month earlier, Nusbaum had had to reallocate a thousand dollars from the museum case fund to road work because "extreme drought has caused serious shale slide below Knife Edge Road."[24] According to Nusbaum, Junior was astonished that the superintendent had accomplished so much with only $5,000.

That afternoon, Junior chose to spend the hours in a more or less solitary fashion, sitting on the portico of the superintendent's house writing letters and gazing off at the ancient dwellings and tree-studded mesa. He dashed off a Western Union Cablegram to wife Abby at the Ritz in Paris: "mesaverda [sic] supremely beautiful. canyon supper last night. indian cliff pageant tonight. long for you." If one considers the austere prose Junior ordinarily employed in diaries, correspondence, and other forms of travelogue, the use of "supremely" is high praise indeed. (It also set him back an extra nickel or dime he didn't need to spend.)[25]

It was good that he rested, for the evening would be a momentous one. After dinner, the Nusbaums and Rockefellers seated themselves at the edge of the cliff overlooking Spruce Tree House, basically box seats for the big show that was being performed for their benefit. Several hundred park visitors also assembled to the Rockefellers' right and left. As darkness fell, fires were lit beneath the deep overhang and sulfurous railroad flares were ignited, bathing the scene in a smoky, reddish glow.

The presentation was "Eagle Woman," a pageant written and directed by Aileen Nusbaum. The eighteen actors were mostly park employees of Navajo descent and they wore colorful costumes designed and assembled by Aileen. Aileen based the drama on a Zuni story and related Navajo themes and developed it with the help of a Navajo medicine man and other park employees. The pageant included traditional Native American songs and dances and was narrated by the superintendent.[26]

Junior, an avid patron of the performing arts in New York, was beside himself with delight. After the show, he thanked each one of the performers profusely and offered to reimburse the superintendent for all expenses. Nusbaum had to explain that there were no costs to reimburse since the performers all volunteered their time (though a hat was passed to the audience after the show) and Aileen, like most spouses of park superintendents and other Park Service employees during the era, was an unofficial and unpaid government volunteer. Even the flares had been donated by the railroad.[27]

One of the actors, Sam Ahkeah, future chairman of the Navajo Tribal Council, endeared himself to Junior when he removed his turquoise ring and presented it to the millionaire.[28]

Following the show, the party retreated once again to the front porch of the superintendent's house. Gradually, the boys and others moseyed off to bed, leaving only Jesse and Junior to enjoy the cooling high-desert air. "It was a heavenly night—with the moonlight highlighting the canyon cliff walls," Nusbaum recalled. "As it [the moon] faded down, star light & milky way increased to maximum brilliance." After a bit, Junior pronounced that he "had never known such peace and quietude, and never before had the stars seemed so near."[29]

Junior must have considered Jesse a kindred spirit, even though the two men had been strangers forty-eight hours earlier. They stayed up until nearly midnight enjoying the beaming stars and each other's company. Junior again thanked Nusbaum for his hospitality, and complimented him on his good works, especially his efforts to interpret the complex history of Mesa Verde in a way that ordinary tourists on vacation might grasp. Then Junior did something he rarely even considered doing, he talked about himself. He confessed that he had not been ready to accept the burdens of the Rockefeller business empire and philanthropies when stress, failing health, and demonization by the press persuaded his father to retire before age sixty.

Although we can't know exactly what Junior revealed to Nusbaum under those Colorado stars, it could have been similar to a confession he made to his advisor and old fraternity brother Thomas M. Debevoise in 1932. "I did not seek or choose to be the recipient of great wealth.... It has not meant the greatest happiness. From my earliest years I have had but one thought and desire, namely, to be helpful to Father in every way in my power. I have gloried in the greatness of his unparalleled achievements in industry and his world wide services to humanity."[30] Likewise, longtime Rockefeller associate Frederick Gates once surmised, "He [Junior] would have preferred... to cut loose from his father's fortune and make for himself... a wholly independent career." But he was an only son, the heir to nearly unimaginable wealth.[31]

The massive inheritance and sudden change of life plans triggered a year of depression in Junior. In 1922 he had been a patient of Dr. John Kellogg at his sanitarium in Battle Creek, Michigan. After a battery of tests, his infirmities were diagnosed as "auto-intoxication brought on by strain."[32]

As Junior gradually reassembled himself, he decided he must turn over the business side of his father's far-flung empire to Senior's generally competent officers and devote his energies to philanthropy. His newfound goal was to be "as successful in the distribution of funds for public benefit as his father was in amassing these funds." Knowing that such enormous amounts of charitable giving were almost as likely to do great harm as great good, Junior was on the lookout for people and organizations that could "secure maximum benefit or accomplishment" from finite Rockefeller dollars. He was drawn to benefactors who treated philanthropic dollars the same way they handled their own hard-earned bucks.[33]

During the visit, Nusbaum also shared his hopes and fears with Junior. Topping the list of his anxieties was Aileen's failing health, diagnosed by doctors as a heart ailment, which resulted in several extended hospitalizations. According to Nusbaum biographer Kathy Fiero, "Jesse was obsessed with money," or at least with the astronomical medical bills (somewhere around $80,000 in 2025 dollars) that strained the couple's marriage.[34]

As superintendent, it would have been impossible for Jesse to hire his wife as a Park Service employee because nepotism rules forbade a federal employee from supervising their spouse. Aileen, however, did eventually become gainfully employed as a nurse at the much-needed Mesa Verde hospital she helped design and build. Eventually the hospital (funded by monies garnered outside the park's operating budget) was designated by Congress as the Aileen Nusbaum Hospital. (Ironically, the hospital, located within the boundaries of a national park created for the preservation of Native American culture, was segregated between white and Native patients.)

The Rockefeller party departed the next day, but not before Junior asked the Nusbaums for a list of all involved in the "Eagle Woman" pageant so that he could send them silver and turquoise rings and large, brightly colored silk handkerchiefs as thank you gifts. For Ahkeah and Albert Jones, the Medicine Man, Junior asked Nusbaum if he might purchase even more elaborate presents the next time he was down in Sante Fe, for which Junior would promptly reimburse.[35]

Junior also mentioned that he would like to "chip in" to finish the museum, with the understanding it was actually "a Government responsibility," and the government would take over and operate the museum once its virtues were acknowledged.[36]

*

It's interesting to contemplate how the trajectory of Junior's conservation philanthropy would have developed if Jesse and Aileen had not been so accommodating and such fine examples of productivity. As it turned out, the stop at Mesa Verde was a giant step in the expansion of Junior's national park philanthropy from Acadia to the wide world beyond. Junior mentioned multiple times that he was impressed at how hard the couple and their park service staff worked, and how much they accomplished with extremely limited resources.

And even though Junior insisted the summer trip of 1924 was mostly for the benefit of his eldest sons, he seems to have had a good time also. He found "peace and quietude" in the desert, was intrigued by the cultural resources being preserved, and even commenced a lifelong friendship with the Nusbuams, who, despite being professionals, would likely have been otherwise outside his New York social circles. As an idealistic Christian and perfectionist, Junior must have also been pleased with the rising standards for NPS facilities, research, and educational programs, as well as its earnest, altruistic, hyperproductive staff.

John D. Rockefeller and his son, John D. Rockefeller Jr., walking on the streets of New York City in 1915. Library of Congress.

Abby Rockefeller with son David in 1919. Photo by Arnold Genthe. Genthe photograph collection, Library of Congress.

John D. Rockefeller Jr. around the time of the creation of Sieur de Monts National Monument in Maine (1915–1917). Harris and Ewing photograph. Library of Congress.

Members of the Mount Desert Island village improvement association on Jordan Pond in 1923. George B. Dorr is on the right. H. W. Gleason photograph. National Park Service, Acadia National Park.

The Rockefellers' cottage, named the Eyrie, on Mount Desert Island in Maine, prior to the major remodel. Rockefeller Archive Center.

Harvard president Charles W. Eliot at Jordan Pond with the Bubbles in the background. Eliot, George B. Dorr, and John D. Rockefeller Jr. were primarily responsible for creating what would become Acadia National Park. H. W. Gleason photograph. National Park Service, Acadia National Park.

George Dorr, Arno Cammerer, and Stephen Mather on Beech Cliff in what would become Acadia National Park. H. W. Gleason photograph. National Park Service, Acadia National Park.

John D. Rockefeller Jr. testifying before a congressional committee investigating the incidents at Ludlow, Colorado. Bain News Service, Library of Congress.

NPS director Stephen Mather at the cornerstone ceremony for the new museum in Yosemite National Park. Yosemite Historic Photo Collections.

Recently completed Yosemite Museum. Photo taken for Director Mather. Ralph H. Anderson photograph. Yosemite Historic Photo Collections.

The trailside museum at Madison Junction in Yellowstone National Park. Historic American Buildings Survey, Library of Congress.

John D. Rockefeller Jr. (left) at Cliff Palace in Mesa Verde National Park, 1924. Denver Public Library.

Horace M. Albright in 1933, his last year as director of the National Park Service. National Park Service photograph.

The introduction of automobiles into national parks soon spawned a boom in camping. National Park Service photograph.

Old Faithful geyser in Yellowstone National Park. National Park Service photograph.

Lower Falls plunge into the Grand Canyon of the Yellowstone. National Park Service photograph.

Grassy Valley and snow-covered peaks. View across the valley of Jackson Hole to the Grand Teton mountains. Ansel Adams photograph. National Archives.

The Snake River and Grand Tetons peaks. The table flat valley of Jackson Hole contrasts with the jagged Tetons. Ansel Adams photograph. National Archives.

A 1930 group photo at the old Elbo Dude Ranch in Jackson Hole. John D. Rockefeller Jr. is back row, second from right; Abby Rockefeller is front row, fourth from right; Horace Albright is middle row, far left. Rockefeller Archive Center.

LONG TRIP THROUGH WONDERLAND

Yellowstone, 1924

For an anxious moment, Yellowstone National Park superintendent Horace Albright feared he had missed the Rockefellers. The massive Northern Pacific passenger train had arrived at its destination of Gardiner, Montana, only slightly tardy at 11:45 a.m., on July 12, 1924. The engine now sat restlessly at the end of the line in the one-horse town, the locomotive's pistons hissing steam and its broad stack seeping thick black smoke. Passengers peered from the windows and eagerly rose to their feet as graceful African American porters in navy blue caps scrambled to begin the deboarding process. *But where in the devil was the private rail car?*

Flustered, beginning to feel the July heat in his wool park ranger uniform, Albright emerged from his dusty seven-passenger White touring car and pulled a small notebook from the heavy jacket's breast pocket. Carefully checking his personal schedule, he concluded (and history would confirm) he was in exactly the right place at exactly the right time. Regardless, he no doubt felt the sweat gathering in the hatband of his Stetson "flat hat." Gardiner was perched at an elevation of just over a mile, but the noonday sun felt close and unfiltered. The entwined smells of coal smoke and creosote were stinging, and the heat rising from the locomotive's boiler would have caused Albright's view of the distant foothills to oscillate.[1]

As the porters helped the lines of finely dressed tourists step tentatively down from their passenger cars, Albright picked his way along the rustic post and beam platform toward the end of the train. An explanation eluded him. In 1924, traveling families did not alter their painstakingly planned schedules, especially to once-in-a-lifetime destinations like Yellowstone National Park. He had personally helped plot the family's unusual itinerary and there were few opportunities for diversion.

Now Albright's mood darkened. What if there had been some calamity that caused their plans to change? Although Albright's boss, Director Stephen Mather, had put a strict gag order on Albright—he was not, under any circumstances, to solicit donations—Albright couldn't help but get his hopes up. Of all the unfathomably rich families of the 1920s—the Vanderbilts, Morgans, Carnegies, Goulds, and Guggenheims—the Rockefellers were the wealthiest of them all. And they had elected to visit *his* painfully underfunded national park![2]

The year 1924 marked Albright's sixth summer in Yellowstone. Back in 1919, after Grace gave birth to Robert Mather Albright, she and Horace decided the time was right for him to leave the Park Service. Husband and wife had been apart for much of Horace's career, and he had "two excellent offers" from San Francisco law firms. He would be turning thirty soon. Spending more time with family back in California honestly seemed the right thing to do.[3]

Albright wrote Mather a nine-page letter of resignation. He gave his boss the highest praise, stating that everything the Park Service had accomplished was due to "your broad vision, energetic and enthusiastic work, your financial backing and your wonderful personality." Albright also mentioned that he had always wanted to be superintendent of Yellowstone. Within a few weeks, the two cofounders of the NPS had reached a compromise: Albright would serve as superintendent of Yellowstone during the busy summer season and return to Washington, DC, as NPS assistant director (field) during the winter. The latter part of his new title saddled him with responsibility for all the western national parks which, in 1919, was pretty much all the national parks, except for Lafayette.[4]

Back at the train station, Albright finally spotted an auspicious sign: an unusually tall train detective, replete with brown felt Western hat, looming above the crowd that was passing along the platform. Albright recognized the man as the armed employee of the Northern Pacific company who escorted special trains that carried VIPs or large amounts of cash. The Rockefellers had arrived.

"I know who you're looking for, he's in the car up next to the engine," the detective explained. Albright's mystery was solved; there was no private car, the Rockefellers were roughing it in coach. The detective and superintendent then made their way through the crowd until they were close enough for the Northern Pacific employee to point out John D. Rockefeller Jr.[5]

Albright thanked the detective and hurried to the first passenger car where a somewhat short, handsome, impeccably dressed gentleman stood critically regarding his three teenage sons as they emerged from the passenger car. Looking into Rockefeller's intelligent and bashful eyes, Albright nearly forgot his first instruction. "Mr. Roc—, Mr. Davison?"

Junior chuckled. "You know."[6]

The choice of Davison as an incognito traveling surname was not a random selection. It was the middle name of both Junior and Senior John D. Rockefellers and the name that represented puritanical goodness in the family lineage. Davison was the maiden name of Junior's saintly grandmother Eliza—Senior's mother—who lived a life of tolerance, responsibility, and necessary thrift. In contrast, her husband, family patriarch "Devil" Bill Rockefeller, was a notorious bigamist, con man, and snake-oil drummer.

Junior was not acting paranoid in his use of a cloaked identity. On an earlier Rockefeller family vacation, John D. Rockefeller III recorded that the "newspaper reporters were after poor father most of the time trying to talk to him. There were lots of very funny things put in the papers about us and lots of very untrue things."[7]

As the Rockefellers stood on the wooden platform and regained their land legs, things continued to deviate from Albright's craftily designed agenda. The park superintendent's intention was to swiftly corral the four Rockefellers, luggage, and traveling physician and load them into his old seven-passenger White motorcar and discreetly whisk them up the canyon road to park headquarters at Mammoth Hot Springs. Like the fake surname, the dusty jalopy was a bit of a ruse to detract attention from the Rockefellers, who were among America's most envied and loathed families.

Junior, however, had observed that "there were great piles of luggage being taken off the train and that the . . . porters were struggling."[8] To Nelson (future governor of New York and vice president of the United States), and Laurance (future conservationist and venture capitalist), Junior quietly stated, "There's an awful lot of baggage for those porters to handle, how would you boys like to help them?"[9] To which the boys replied, "Yes, father."[10] Like a good Rockefeller, John D. III (future philanthropist), age eighteen, continued recording in his ledger the trip's expenses, including tips to porters.

Handling the luggage took some time. As he waited, Junior removed a small ledger of his own from the inner pocket of his wool suit jacket and recorded some notes. Albright observed the millionaire's attention to detail.

The next itinerary improvisation involved Mr. Rockefeller climbing into Albright's open car while instructing his boys to ride up the canyon in the touring cars with the other dudes. "They need to mix with the crowd," their father explained to Albright.[11] Having grown up lonely and isolated on large estates, Junior no doubt hoped that his children would become better socialized.

First stop on the Albright-Rockefeller tour was the already famous Yellowstone (also known as Roosevelt) Arch. It was a forty-foot-tall, rustic stone monument dedicated two decades earlier by ardent conservationist and Rockefeller trust buster President Theodore Roosevelt himself. Albright slowed the car to point out the inscribed words: "Yellowstone National Park, Created by Act of Congress, March 1, 1872, for the Benefit and Enjoyment of the People." Junior would have approved of the use of native materials and skilled masonry work in the massive arch and likely noticed how it framed the distant Gardner River Canyon, the geographical entrance to the world's first national park. Junior was also interested when Albright pointed out the pronghorn antelope grazing in the flat leading to the winding canyon road. He peppered the superintendent with questions about the flora, fauna, and other park matters as they started up the steep, five-mile-long grade.[12]

Mammoth Hot Springs was quite a sight in the summer of 1924. It was a tourist hub, Park Service headquarters, and home of old Fort Yellowstone. It also showcased some remarkable hot springs' terraces displaying polychromatic, mineral-laden waters normally safely sealed deep beneath the earth's protective skin. Visitors were greeted with a wide array of noxious fumes and toxic gases (a smattering of dead birds reflected the latter), traditionally attributed to the bowels of hell. In the old days, before the National Park Service, clever souvenir vendor Ole Anderson placed everyday objects like corncobs, hats, and horseshoes on racks amongst the fuming terraces and observed as they became coated in travertine. The curios were then hawked to the growing number of travelers determined to prove to the folks back home that they had indeed visited Yellowstone and dutifully taken in the sites.[13]

Also occupying considerable space in "downtown" Mammoth were the prominent rows of fortress-like buildings known until recently as Fort Yellowstone. The army was called to Yellowstone's rescue after Congress created a three-thousand-square-mile national park but neglected to make any provisions for its management or protection. The ensuing fourteen years, 1872 to 1886, stand as one of the more notorious failures in the federal government's rookie years of managing public preserves.

Since there had never been a national park before and no one had the foggiest idea of what should be done with one (other than promote it to tourists to stimulate commerce), the first users were greedy market hunters, outlaws, unscrupulous miners, and archaeological site plunderers. Sportsman and author William Strong reported that in one winter "over four thousand [elk] were killed by professional hunters in the Mammoth Springs Basin alone." He also predicted that "few years will elapse before every elk, mountain-sheep, and deer will have been killed."[14] As for tourism, previewing the later sacking of Mesa Verde and most of the other early southwestern national monuments, "wagon tourists" came to the Mammoth terraces toting "shovel and axe, chopping and hacking and prying up great pieces of the most ornamental pieces they could find," reported Corps of Engineers captain William Ludlow. In the geyser basins, they carved their names into geyser cones and threw handkerchiefs, coins, rocks, bones, bullets, forks, watches, and other foreign objects into deep, blue Morning Glory Spring.[15]

Finally, following the publication of a number of outraged articles in the national press, and a call for wildlife protection from General Phillip H. Sheridan, the First U.S. Cavalry was called in to salvage what was left of "the best idea we [Americans] ever had."[16] The soldiers, who were also called to rescue Yosemite and Sequoia, were the perfect civilizing cure for the chaos in Yellowstone. According to nearly all accounts, their training, discipline, toughness, organizational structure, and esprit de corps made them extremely effective protectors of the resources for which the park was created. The Army Corps of Engineers also braved the climate and wilderness to lay down some of the park's formative infrastructure.

Although the cavalry was initially ordered to Yellowstone on a temporary basis, their successes led to a permanent assignment and construction of the formidable fort. The latter was designed not only for the purpose of housing the soldiers, but was fortified in case of attack by Agaideka, Bannock, Crow, Nez Perce, Salish, Shoshone, Tukudeka, or other indigenous peoples who mostly had treaty and inherent rights to continue hunting game and gathering plants in Yellowstone country. Because the sight of Yellowstone's human denizens sometimes frightened the ever-growing number of tourists, gun-toting soldiers accompanied the white visitors onto the Mammoth terraces and around the geyser basins. The soldiers were also present to protect Yellowstone's resources and the tourists from themselves. These responsibilities included advising visitors not to dump laundry soap into the geysers (to inspire premature eruptions) and to resist the urge to hand-feed sandwiches to wild seven-hundred-pound grizzly bears.[17]

*

The reassembled (and marginally better socialized) Rockefeller family joined Albright on the second floor of the former commanding officer's headquarters and enjoyed the views from Albright's narrow floor-to-ceiling fortified windows. Albright had already spread the maps, brochures, and most likely a revised edition of the indispensable *Haynes Guide* (in which he was prominently featured) across his big War Department desk. As previously mentioned, Albright's itinerary was an odd one, and as he presented it, he must have wondered if his motives were embarrassingly transparent. Although Yellowstone's two million acres were much too vast to see in barely four days, and this was the boys' first trip to the world's most celebrated national park, Albright was directing the drivers and guide to squander the entire second day driving the length of Yellowstone National Park, and to continue beyond the park's southern boundary to persevere for more miles on a wagon road to Jackson Lake for a quick picnic and a glimpse of the majestic Grand Teton mountains.[18]

It was no secret that Albright was more than a little infatuated with the idea of adding the Tetons to Yellowstone. The Tetons were as close as one could get to the perfect mountains, rising dramatically from the table-flat plateau of Jackson Hole. By combining the Tetons with Yellowstone, the Park Service would not only amalgamate an incredible scenic area and tourist magnet, they would also reserve adequate wide-open spaces for the survival of iconic and beleaguered western wildlife like elk, bison, antelope, wolverine, bighorn sheep, and grizzly bear.

Yet there was Mather's pesky gag order on Albright and all the superintendents of parks the "Davisons" were visiting that summer. No requests for donations. No hand wringing over crippling deficiencies. No mention of the extremely urgent need to save the Grand Tetons from a flood of hot dog stands, dance halls, and other forms of tacky commercial exploitation. They were to help the Rockefellers enjoy the parks and completely ignore the fact they were the most generous philanthropists in the history of civilization.

The motives for Mather's edict are unknown. Perhaps he just wanted to give the Rockefellers a break from the constant pleas for alms that such philanthropists endure. Perhaps he felt the family deserved the opportunity for unfettered recreation and relaxation in America's wonderlands. Possibly he wanted to direct the Rockefellers' philanthropy himself, rather than leaving it piecemeal to the near-sighted whims of park superintendents. Or, most likely, it was a brilliant move by the former advertising man and PR guy who understood there is no charitable need closer to a philanthropist's heart (and checkbook) than the one he discovers on his own.

It's interesting to note that today's park superintendents are prohibited from requesting donations from wealthy potential benefactors, just as Mather dictated. Modern federal regulations place much of the responsibility for philanthropic solicitations in the competent hands of nonprofit park Friends groups and the National Park Foundation. However, superintendents and other park officials may still usher philanthropists around their magnificent landscapes and highlight their area's natural beauty, significance, and challenges.

Unlike Jesse Nusbaum, Albright did not offer to accompany the Rockefellers on their exploration. Yellowstone was a much busier park than Mesa Verde during peak season, and Albright likely couldn't spare the time. Therefore, guided by the hired drivers of their touring cars, the Rockefellers proceeded fifty miles south to the Old Faithful area. Its namesake geyser erupted to heights of 150 feet every sixty-five to eighty minutes, allowing ample opportunity for the Rockefellers to appreciate its spectacle. Their accommodations at the Old Faithful Inn, described by the *Haynes Guide* as "the most extensive log structure yet devised by man," must have stoked Junior's appreciation for rustic national park architecture.[19] This admiration of architecture appropriate for wildland landscapes would, in fact, in a very short time, affect the design of a handsome and influential museum in Yosemite National Park. At the inn, the Rockefeller party bumped into

Senator Tasker L. Oddie of Nevada and discussed the forest pests that were killing park trees. The senator invited the party to join him the following day, but Junior had to decline, committed as he was to his group's relentless southward progress.[20]

The Rockefellers' touring cars were off for the Grand Teton Mountains early the next morning, another sixty-seven miles distant (with speed limits of twelve miles per hour on grades and sharp curves and twenty-five miles per hour on straightaways).[21] They arrived at the Amoretti Inn, on the shore of Jackson Lake—with Mount Moran and the north end of the glacier-studded Teton Range rising across the blue waters—in time for a late lunch.

"Magnificent view of the Teton Mountains," Junior noted in his travel diary. Considering the diary's spare prose (mainly a simple listing of miles traveled and lunch stops), this must be interpreted as an extraordinary level of enthusiasm.[22]

After lunch, the Rockefellers headed back toward Yellowstone where they ducked sporadic rainstorms at the Lake Hotel before continuing to Canyon Village. The next morning, they descended 1,200 vertical feet on the harrowing Uncle Toms Trail into the treacherous Grand Canyon of the Yellowstone. Still brimming with energy, the group proceeded to drive to the 10,100-foot summit of Mount Washburn ("Magnificent view, gorgeous wild flowers") and tick off Tower Falls and Roosevelt Camp before starting back to the hotel—forty-three miles, Junior recorded in his travel log.[23]

As planned, the Rockefeller party dropped by Albright's office at the end of their excursion. As they stood out front, the rain-washed touring cars idling, Junior took Albright aside, far enough from the drivers to be out of earshot, but within sonic range of his three boys. "If I may have a word with you, Superintendent Albright?"[24]

The superintendent's heart must have been racing like a hummingbird's. He could no doubt visualize the Rockefeller clan picnicking up on the hill overlooking Jackson Lake and the granite Tetons protruding beyond the placid lake like spear points. Albright must have felt certain he was about to hear Junior profess his undying allegiance to the Grand Teton cause. *They MUST be saved, for the benefit and enjoyment of the people!*

Junior handed a piece of paper to Albright. But rather than displaying a six-figure pledge, it was a bill from the Yellowstone Park Company. "I don't like to bring up points like this," Junior said. "I expect to pay full rates, but I dislike being overcharged just because of who I am."[25]

Though no doubt disappointed, Albright carefully examined the bill and explained that the charge was on the high side because the party had hired two cars and two drivers, and the fare was based on each car being fully occupied.

"Yes, I see," Junior said. "It's a fair rate and I am entirely satisfied." The millionaire then recorded the charge, and tips, in his breast pocket ledger.[26]

John D. Rockefeller Sr. had schooled his son in an almost identical fashion thirty-seven years earlier while on a family vacation in Europe. After Senior fired their tour director for swindling, the great industrialist assumed responsibility for auditing every bill the party accrued during the remainder of their holiday. In one famous exchange, someone questioned whether the Rockefellers had actually consumed two entire chickens in one meal. Senior settled the matter by questioning his family as to the number of chicken legs each had enjoyed. When the tally reached four, Senior, too, declared he was entirely satisfied.[27]

The three teenage boys were no doubt savvy that the performance was largely for their benefit. Another tedious, eye-rolling lesson on how to navigate life as a Rockefeller. They had suffered plenty of such lessons already. Like the miniscule weekly allowances each child received and was then called forth every Saturday evening to account for, in writing—down to the penny. The exercise was not only about money management, but to instill the importance of saving and charitable giving. Even the youngest Rockefellers were expected to tuck a few pennies away for a rainy day and to dedicate a third of their pittance to charity (usually the church). Junior and his three sisters had endured the same weekly ritual from Senior when they were tots.

The boys' mother, Abby, was also fearful of what wealth might do to her children. Too much money "makes life too easy: people become self-indulgent and selfish and cruel," she warned thirteen-year-old Laurance. To young Nelson, Abby advised, "I am sure too much money makes people stupid, dull, unseeing and uninteresting. Be careful."[28]

Although Junior never claimed to dislike money, he certainly disliked wasting it. As a duty-bound Christian philanthropist, his fate was to expend the family fortune responsibly, wisely, efficiently, and productively. It was neither a terrible nor an especially joyful fate for Junior, but it would be made happier by the company of interesting, virtuous people like Albright, Cammerer, Dorr, Mather, and the Nusbaums.

THE LANDSCAPE ARCHITECT

Yellowstone, 1924–1930

Yellowstone National Park was having a busy summer. During the previous year, tourist traffic had jumped 40 percent, hitting a record high of 138,352. So far, the 1924 season was busier still. The reason for the park's recent surge in popularity was no mystery: automobiles.

The year 1924 marked the tenth anniversary of an enormously successful automotive publicity stunt nicknamed "The Vagabonds." A group of carefree and footloose typical Americans, including Henry Ford, Thomas Edison, Harvey Firestone, and John Burroughs, would hop in a Ford automobile for a week or two each summer and hit the road with several dozen of their journalist best friends. As the newsreel cameras whirred, the elderly roustabouts would roam America's few decent roads, camping, fishing, fixing flat tires, and just bumming around. They were also inventing a custom entirely new to many middle-class Americans: the summer vacation.

The nation's increasingly prosperous citizenry took notice. The flivver was not just a tool for utilitarian purposes like work and visiting Aunt Ida, it was God's way of showing he loved us and wanted us to go camping.

The motoring craze synched perfectly with NPS director Mather's support for the "See America First" crusade (even though the slogan originated with the railroads, who were definitely not benefiting from the automobile). Why should Americans be so hell-bent on visiting dirty old

Europe when they hadn't even seen the Grand Canyon, Half Dome, or Old Faithful geyser yet? As a former businessman, Mather grasped the importance of attracting people (customers) to the parks. Theoretically, the visitors and their commerce would then attract funding from Congress, especially those representatives with parks in their districts. In truth, Congress didn't care much about preserving pine trees and buffalo, but they recognized a growing revenue stream when auto camps, lodges, service stations, and roadside diners began springing up in their hinterland domains.[1]

The rising tide of self-propelled visitors also increased demands on the eight-year-old National Park Service to improve roads and add facilities such as picnic areas, campgrounds, and lodgings for its also increasing number of parks and monuments. One imagines Superintendent Horace Albright had this on his mind as he sorted through his daily mail on August 23, 1924, five weeks after the Rockefellers' departure from Yellowstone. He expected most of the letters to be complaints, and they were. Each required a personal reply from the superintendent or his delegates. One of the envelopes, however, no doubt quickened Albright's pulse. It had a Seal Harbor, Maine, postmark and the words "John D. Rockefeller Jr." handsomely set in the upper-left corner.

If the superintendent quickly scanned the two-page letter anxious to see the words "Tetons," "Grand," "Jackson Hole," or even "mountains," he was sorely disappointed. Rockefeller's letter was also a complaint, albeit an especially polite one. Dated August 15, it had been penned from the Rockefellers' summer cottage in Maine. Rockefeller first thanked Albright for making the arrangements for the brief family trip to Yellowstone and complimented the courtesy of Albright's ranger staff and the skills of the chauffeur.[2]

Junior then mentioned that he did have a problem with the esthetics of Yellowstone's roads. "I refer to the vast quantities of down timber and stumps which line the roadsides."[3] After all, at the superintendent's direction, the Rockefellers had just spent three days looking at alternately dusty and muddy roads. His assessment was not unjustified. In their rush to cut roads through the Yellowstone wilderness during the park's exceedingly short construction season, the road builders had conveniently piled the cut trees, boulders, and multitude of gnarly stumps onto the roadsides. Where road construction damaged the roots of standing trees, dying and denuded skeletons now lined the routes. Adding insult to injury, utility crews added not one but two sets of obtrusive powerlines (one on each side of the road). The cumulative effect was not unlike heaping broken picture frames, empty tubes of paint, and ripped canvases—along with a couple strings of barbed wire—in front of the Mona Lisa.

Rockefeller stated he *might* be able to fund a project (without mentioning any specific figures) that would address the incongruence.[4]

On the same mid-August day, Junior had written Superintendent Charles J. Kraebel of Glacier National Park to compliment his park and to also point out all the unsightly stumps beside the Many Glacier Hotel, and the likewise disagreeable "group of shacks" beyond the hotel, built originally to house sawmill workers.[5] As in Yellowstone, Junior noted the dead timber along the roadsides and wondered if it might be cleared. He asked Kraebel if "a little help from me" might facilitate the cleanup.[6] Unlike Albright and Nusbaum, Kraebel had not met the Rockefellers on their extended Glacier trip which included a lengthy backcountry adventure on horseback.

Director Mather was entirely sympatico with Junior on Yellowstone's messy roadsides and even more so with Many Glacier's clutter of decrepit shacks. On the very evening, in fact, that the Rockefellers had departed Many Glacier in 1924, Mather had given a talk to a group of some two hundred in the hotel lounge. Mather lambasted the hotel owners and Great Northern Railway by asserting "the old saw mill, warehouse, and many dilapidated buildings were a tremendous eyesore."[7]

Kraebel responded to Junior over a month later, largely with excuses as to why he hadn't done most of the suggested improvements. He bemoaned that the hotel company was reluctant to spend money on improvements that "gave no promise of revenue" and pointed out he had given the ugly roadsides "much study." He concluded that he was unable to provide Junior with any cost estimates for the work Junior had suggested.[8] Consequently, with his bureaucratic blinders firmly in place, the superintendent left Junior's offer of funding for cleanup on the table.

Incidentally, almost exactly one year after Mather gave his Many Glacier speech, the NPS director again visited Glacier and this time instructed the park trail crew to plant thirteen charges of TNT in the sawmill. Mather then invited the hotel guests outside to view the spectacle. The impromptu demolition caused Albright (and Mather's many other friends) to fear the director's mental health might again be precarious.[9]

Junior also wrote an acquaintance, Secretary of Agriculture Henry Wallace, about the condition of several trails on the national forest lands he had visited on horseback in July, and the impacts of sheep grazing on wildflowers. According to Junior, some trails were in "excellent condition." Others were "almost impassable" and still others "could not be traversed." Junior ended the letter by emphasizing the family's deep interest in public lands and the major gifts they had made to promote their development.[10]

Wallace's reply was entirely defensive, as if he were responding to any old complaint from yet another disappointed national forest visitor, entirely missing the point that Junior might be the answer to the agency's myriad problems. Wallace stated that maintaining the trails Junior described as impassable was "an immense task" and it was not "possible to do all that could be desired."[11] He also defended the livestock grazers, making the dubious claim that sheep didn't harm the wildflower populations. Junior responded with his usual courteous reply, thanking Wallace for his "full and interesting letter." One can only imagine how the trails in our national forests might look today if Wallace had only accepted Junior's offer.[12]

Albright replied as soon as he had finished the 1925 National Park Service budget request to Congress and Yellowstone's 1924 annual report. To Junior he wisely exclaimed: "The condition of the roadside is unsightly and there can be no question but that it detracts a great deal from the pleasure of the Yellowstone trip in the case of many, many visitors."[13] Although Albright was no landscape architect, he and Mather shared a high-brow vision for the esthetics of their growing system of national parks; that they should appear rustic, yet classy. Nothing cheesy, tacky, chintzy, sloppy, or second rate. Mather had in fact made note of the unkempt Yellowstone roadsides way back in 1915 and had been advocating a cleanup ever since. He had lamented, however, that the work would probably not be taken up by the government for many years "if ever."[14]

Despite Junior's failure to mention a single word about the Grand Tetons, Albright gushed his praise for the millionaire's insight and generosity. [Your letter is] "one of the most inspiring communications that I have received since I entered the National Park Service over 11 years ago. Interest such as you have shown in Yellowstone Park spurs one on to greater effort and makes one feel more than ever that we are in a public work that is worth while."[15]

Albright then went on to blame the nasty roadsides on a lack of finesse by the army engineers and a funding drought during the recent world war. Next, Albright launched into a carefully considered and extremely detailed plan for spending the Rockefeller money. He emphasized that he had requested federal appropriations for the purpose of roadside improvements on several occasions, but in the spirit of former House Speaker Joseph Cannon's adage "not one cent for scenery," the parsimonious ones in Congress had not considered it a priority.[16]

The superintendent bolstered his argument with copies of recent tirades from park visitors who shared Junior's displeasure with the shoddy road work. All in all, Albright's enthusiastic reply was over three pages, plus numerous enclosures. It was the antithesis of the weary dismissals penned by Superintendent Kraebel and Secretary of Agriculture Wallace.

Albright promised that he could immediately assemble special crews to embark on a series of roadside improvement test strips, variously located on sections of road that ranged in esthetic quality from marginally bad to ungodly. By carefully documenting the progress and labor hours on each foot of test strip, Albright would be able to extrapolate accurate estimates for grooming long sections of scenic road.[17]

It is not surprising that Junior's mind was on roadsides during the summer of 1924. Back along the rocky coast of Maine, representatives from the National Park Service and Department of Interior, including Secretary of Interior Hubert Work and NPS director Mather, were dropping by to investigate Senator Pepper's complaint regarding a network of carriage roads being built by Junior and the NPS in and near the new Lafayette National Park. The site visit followed a congressional hearing in March also focused on Junior and Superintendent Dorr's road construction plans.[18] It's noteworthy and frankly astonishing that the high-level skepticism of Junior's philanthropic work in Maine did not deter him from offering to help with projects in Mesa Verde and Yellowstone.

Junior agreed to pay up to $2,000 ($36,000 in 2025 dollars) for the test strips in Yellowstone, and by September 13, less than a month after the Easterner wrote his initial gripe, Albright had a hand-picked crew of laborers cutting dead trees, hauling stumps, rolling boulders, and burning the piles of timber. As an added benefit, the seasonal workers thus employed would otherwise have been laid off with little more to do than loiter about the bars of Gardiner. According to Albright, the "hardworking fellows who want to save a little extra money" were grateful for the work and hopeful of extending their productive season as far as possible.[19] Likewise, in Acadia, Junior's exuberant trail and road building was providing much-appreciated work for scores of local tradespeople.

Albright's September 15 letter, which marked the one-month anniversary of Junior's offer to finance roadside improvements, included numerous attached reports as well as engineering blueprints detailing "the exact condition of each foot of road between my office at park headquarters and Norris junction," a twenty-one-mile section of road. The superintendent also gushed to Junior that his letter "gave me great happiness and inspiration."[20]

Rockefeller dropped a check for $1,000 in the mail. He also dispatched a telegram from his Manhattan office with suggestions to make the work more efficient and visually appealing: "Am wondering whether dead trees stumps etc cannot be burned in piles where they lie or in the road thus avoiding loading on trucks and hauling them away stop ... ought not dead standing trees be cut to a greater distance from the road side than ten or fifteen feet stop"[21]

Albright agreed "entirely, and in fact, I am working along these lines at the present time." He changed his procedures to remove all timber within fifty or 100 feet of the roads, depending on terrain. He added a second crew and set them to work in the chilly autumn air on the road to Norris. Along one segment, where a wildfire had charred the lodgepole pine forest, a single crew pulled thousands of stumps and burned the debris *in situ*.[22]

Albright's letter of November 4, 1924, was another humdinger. Weighing in at five pages, the length of a decent magazine article, it bulged with detailed cost estimates and optimistic timelines. It also unleashed a flood of enthusiasm and affection for the two men's collaboration (and requested another $2,500). "Truly, no more important work has ever been undertaken in this park than the landscape improvement that you have authorized and I find that I personally am getting more pleasure out of supervising this work than almost anything else I have undertaken."[23]

"Almost anything else I have undertaken" is weighty when one tallies Albright's accomplishments. During Mather's illness, Albright had drafted the "Statement of National Park Policy" (a.k.a. the NPS creed), an enduring outline of park management philosophy.[24] Together, Albright and Mather had made the parks somewhat safer for tourists by throwing out many of the bad-apple concessionaires who had been making a living serving ptomaine-laden meals to park visitors and dumping raw sewage into the pristine mountain waters. One aristocratic guest of the National Hotel at Mammoth Hot Springs (where unacquainted lodgers were sometimes forced to share the same bed) described the facility as having an "air of discomfort."[25] As acting director of the Park Service, Albright had wheedled from Congress the first NPS appropriations bills and appointed trustworthy superintendents and key staff to run the parks and the agency that managed them. As Yellowstone's superintendent, Albright fought off persistent corporate efforts to dam the major waterways in Yellowstone and the Grand Tetons—short-sighted boondoggles to secure more water for irrigation, mainly requested by the Utah-Idaho Sugar Company.

During the fall of 1925, Albright sadly apologized because work had bogged down in a swamp: "I did so want to make an extraordinary showing that could encourage you to let us go on with another project next year."[26] The crew had fallen four miles short of their goal for the year because of difficulty clearing debris which had been "thrown in [the swamp] at the time the road was constructed."[27] Worse still, Albright had underestimated the cost of the clearing and therefore had to hit Junior up for an additional $4,000 to finish the twenty-mile stretch. Junior was amenable and continued to write checks and compliment Albright on the quality of the work and detail provided in the reports.

In January of 1926, when Albright learned that the Rockefeller family would be revisiting Yellowstone and other western parks over the summer, his enthusiasm for roadside cleanup overflowed its cup. Albright commended Junior, asserting the Yellowstone project "has had a profound influence not only in Yellowstone but in all of the other national parks."[28]

Lead Park Service landscape architect-engineer Daniel R. Hull in fact considered "the cleanup of roadsides and other park areas...to be the most important improvements [in national parks] in the mid 1920s." By the end of the 1920s, postconstruction roadside cleanup "became an integral part of park service work and was funded under annual appropriations."[29]

*

Although progress on Yellowstone's roadside beautification project was confined to the shoulder seasons, when the tourist traffic was thin and crews were available, remarkable headway was made over the next four years. Albright wrote Rockefeller with regular progress reports, photographs, and revised cost estimates and Rockefeller responded with checks and encouraging words. For example, in October 1926, Albright proudly reported his crews had "cleaned up the roadsides of thirty-one miles of the Yellowstone road system."[30]

Also in the fall of 1926, Assistant NPS Director Arno Cammerer provided Junior with a sneak peek at Yellowstone's annual report. The praise for the roadsides was profuse: "The effects obtained are almost unbelievable.... one can hardly realize that the highway has not been removed to a new location. The grass and flowers ... present a truly park atmosphere that did not exist before this work was undertaken."[31]

Road work in the summer of 1927 had some special admirers. President and Mrs. Coolidge were touring Yellowstone and Albright took the time to point out the roadside improvements and mention Junior's role in the work. Mrs. Coolidge was especially impressed "and paid high tribute to both your father and yourself for your wonderful work for humanity."[32]

"The park has never been more beautiful where the roadsides have had two to three years to get entirely back to nature," Albright wrote in 1928.[33] In his year-end report, Albright added, "The Yellowstone roadside clean-up has stimulated interest in this sort of work by State and county authorities. So much work is now being carried on along Western roads that it can no longer be observed and recorded by National Park officials."[34]

By 1930, nearly every blighted roadside, including those with extensive fire damage, had been groomed. The cost to Junior was modest, approximately $50,000 ($923,000 in 2025 dollars), yet the impacts were enduring. The Yellowstone roadside improvements coincided with the first substantial congressional appropriations for building and improving roads in national parks. By incorporating the new standards, refined by years of trial and error in Yellowstone, the engineers and construction crews began creating roads that passed esthetic muster right from the start. Even into the 1930s, the beautified roadside campaign continued to spread. "During the year 1936, approximately $7,000,000 will be expended in *road-side* beautification" nationally, Albright reported to Junior.[35]

The fortuitous meeting of Albright and Junior in Yellowstone in 1924 led to a lifelong friendship, unparalleled conservation accomplishments, and a remarkable correspondence that included over thirteen hundred letters and telegrams.[36] For persons interested in fundraising for conservation projects, there will likely never be a better outline for donor-beneficiary relations than the Albright-Rockefeller model. Following Mather's orders, Albright never directly requested contributions from Junior. Rather, Albright helped the family enjoy the park and made himself open to their inquiries. When Junior showed interest in funding roadside cleanup (rather than annexing the Grand Tetons), Albright didn't pout, he quickly provided detailed cost estimates and specifications on how the work would be accomplished. He kept Junior fully apprised (providing maps, photos, and spreadsheets) of progress and actual costs versus estimates. One can only imagine that things would have turned out very differently if Albright had been a lackadaisical administrator who tolerated unnecessary delays, fuzzy math, and mediocre work.

Of all of Albright's many, many accomplishments during his nineteen-year career with the NPS, none was more impactful than his response to the Rockefeller offer in 1924. Albright also showed his gratitude to the philanthropist often and always emphasized the long-term, positive impacts of their collaboration.

Undoubtedly, just as Junior's love for the beauty of Mount Desert Island and his interest in landscape architecture led him to conservation work in Maine, his correlated appreciation of well-groomed scenic roads sparked the prodigious collaboration between Albright and Rockefeller. And while Albright's work with Junior on Yellowstone's roadsides had been fruitful, it was not the project foremost on the superintendent's mind when the two men had met at the train station in Gardiner in July of 1924.

COUNTRY MICE, CITY MICE

Mesa Verde, 1925

Aileen and Jesse Nusbaum did not hesitate to leave Mesa Verde National Park in southern Colorado and take up Junior on his invitation to visit the family in New York City. The couple arrived at 10 West 54th Street on a Thursday in late February 1925 for an early supper followed by an 8:30 p.m. performance by the New York Philharmonic Orchestra at Carnegie Hall. It was the Nusbaums' first opportunity to meet Abby Aldrich Rockefeller, Junior's charming, witty, extroverted wife, and the rest of Abby and Junior's brood: Babs, the chronic speeder (age twenty-one); Winthrop, academic underachiever and future governor of Arkansas (age twelve); and David, the amateur entomologist and future president of Chase Bank (age nine).[1]

Junior reciprocated the Nusbaums' hospitality at Mesa Verde by conducting a thorough tour of the family's nine-story, 102-foot-tall mansion, the tallest private residence in New York City. The first floor housed a somewhat plainish formal marble reception area and Junior's home office. Above that was one of the public floors, a Roman-columned expanse segregated into a dining room, music room (with electropneumatic Aeoline organ), and dancing room. A large portrait of Junior's mother, Laura Spelman, loomed over the dining room. The third floor supported Abby and John's bedroom, plus a library where "books spilled over from their shelves."[2] It also contained Abby's office with its floor-to-ceiling windows and cheerful, floral-patterned drapes. The fourth floor encompassed bedrooms for all six children as well as their tutors, nurses, and personal assistants. The entire fifth and sixth floors were reserved for visiting family and guests (though the Nusbaums, for some reason, did not sleep over). Another flight up was the gymnasium with squash courts and various steam rooms and baths. The rooftop hosted a playground, sunroom, and summer sleeping area.

When Abby and Junior ran out of space to display their porcelains, tapestries, and Impressionist paintings and prints, they converted a residence next door into an art annex and connected it to the main house with corridors on three floors. The Nusbaums were especially impressed with Junior's collection of Chinese porcelains and one of the famous circa 1500 Hunt of the Unicorn tapestries (eventually donated to the Cloisters). The visitors noted that Junior fast-walked through Abby's "room of rooms," containing her modern art collection, as if even an accidental glance might turn him into a pillar of salt.[3]

Junior's lack of enthusiasm for modernism is also evident in many of his philanthropic projects and lifestyle choices. The resurrection of Colonial Williamsburg, a salute to America's founders and drive for independence, was one of Junior's most personally rewarding and expensive projects. He was a very generous donor for several historic restoration projects in France and the Middle East. His office was a veritable museum of antique furnishings with all its modern gadgetry well camouflaged.

Jesse reported that he and Aileen enjoyed a "very fine but modest supper" with the Rockefeller family and two additional guests and that the atmosphere was informal with the youngest children climbing into their mother's lap, hugging her and mussing her up. In Jesse's perplexing words, Junior and Abby "live just a modest comfortable, unassuming home-life [in their nine-story manse!]" and that they are focused on their children "who go about with them all over the house."[4]

Although it had only been seven months since Junior and the three older boys had visited Mesa Verde, the park superintendent had plenty of progress to discuss regarding the Rockefeller-supported museum construction project and archaeological expeditions. "We finally got enough water to plaster the museum, to run cement floors and build firewalls," Jesse had reported in an earlier letter.[5] Nusbaum had also assembled a crew of mostly Navajo workers to excavate the rear of Spruce Tree house, a dark, dusty realm that earlier "pot hunters" had understandably bypassed. Using Rockefeller funds, the superintendent had purchased special lights and respirators, which made the hunched-over toil in claustrophobic conditions almost bearable. They unearthed the mummy of a three-year-old child wrapped in a woven mat of rushes, as well as corn, beans, yucca seeds, and evidence of domesticated turkeys. All in all, enough materials were recovered to fill a museum case and "illustrate features of the early culture" not covered in current collections.[6]

As generous as Junior was toward the museum project and Jesse's archaeological research, he was even more supportive in his personal gifts to the Nusbaums. A month after his first visit to Mesa Verde, Junior tendered a check for $5,000 ($92,000 in 2025 dollars) to the couple: "That both you and Mrs. Nusbaum may be free from anxious thought" and free to devote themselves to public service. Since Junior had gotten an earful from Jesse regarding his and Aileen's medical debts during the 1924 visit, Junior was able to calculate that $4,000 would take care of the past-due bills, leaving $1,000 for Aileen to spend to spruce up the home and to relieve her "from unnecessary care and burden." Junior concluded with: "I am sure that you will accept this expression of my friendship and goodwill in the spirit in which it is sent and as a mark of appreciation of what you both are, what you stand for and what you are doing."[7]

For the Nusbaums, opening the letter with its enclosed check was a lottery-winning moment. Jesse "yelled for Aileen. Together we read it, time and again, and wept for sheer joy in each other's arms."[8]

After dinner, Jesse and most of the adults took a limousine ten blocks to Carnegie Hall, while Junior insisted on squiring Aileen in his old Detroit Electric "car." Junior, whose father had successfully cornered the international oil market four decades earlier, received a fair amount of ridicule from his family over his choice of transport—not because of the vehicle's nonpetroleum fuel, but because the diminutive auto had garnered the nickname "Grandma Duck" (after the Disney comic) and to some, it cast its driver and passenger in an unflattering light. Junior bragged the electric convertible started instantly and quietly, ran silently, and spared the riders the insult of unpleasant exhaust fumes produced from the combustion of Rockefeller-refined gasoline. Truth be known, even in 1925, consistent with his reverence for the past, Junior preferred a horse-drawn carriage to any form of modern conveyance.[9]

*

On the following morning in New York City, the Nusbaums visited Junior at his recently renovated office on the twentieth floor of Standard Oil Headquarters at 26 Broadway. Jesse noted the hazy view of the lower bay and Bowling Green and described the office's interior as being clad in oak paneling from an English castle. He later mentioned built-in bookshelves protected by leaded glass and a carefully curated array of three-hundred-year-old chairs. Hidden doors allowed Junior's assistants to discreetly enter and exit his office and, adding to the sense of magical subterfuge, the telephone and intercom system were hidden within the desk.[10]

Aileen and Jesse took a special interest in a heavy wooden table dating back to 1608 that was apparently handicapped by a cumbersome leaf-extension system. When Junior confessed it took three strong assistants to extend or retract the leaf, the Nusbaums began to probe, pry, and systematically investigate. They eventually discovered a concealed lever under the leaf that allowed one person to easily adjust the table length. According to Jesse, "I never saw a man [Junior] more tickled."[11]

The office tour was followed by lunch at the Bankers Club on the fortieth floor of the Dupont building. Aileen gushed about how much good Junior's funds were accomplishing in both Mesa Verde and Yellowstone and painted with broad strokes the Park Service's ambitions for adding new parks and improving facilities. Jesse reported to Albright that Junior "is keenly alive to the whole proposition, and particularly so in regard to the museums." When Junior offered additional funding for Jesse's off-season excavations, the ever-efficient Nusbaums declared they still had plenty of cash left from his initial $1,000 contribution and inquired whether they should return the balance. (Yellowstone superintendent Albright had returned some of Junior's Yellowstone roadside improvement money when work came in under the estimate. This display of honesty and accountability no doubt paid huge dividends for Albright down the road.) Junior replied that he preferred the Mesa Verde overages go toward further archaeological work.[12]

The next day the Nusbaums were stopping by the Great Northern Hotel to retrieve their laundry when they were whipsawed by a calamity at the front desk. A panic-stricken clerk explained that Junior had called for Mr. Nusbaum and that she, assuming the call was a prank, had given a sassy reply and mocked Junior's voice. The hysterical clerk then "prostrated herself literally at my feet" begging forgiveness and imploring Jesse to apologize to the world's wealthiest man on her behalf. Junior had never batted an eye at the clerk's skepticism and had simply repeated his request until she finally came to the mortifying conclusion that she was indeed conversing with John D. Rockefeller Jr. When Junior eventually learned the Nusbaums had left the hotel, he proceeded to leave messages all over town and even dispatched his personal secretary to comb the streets of the city and track them down. Junior's message, when finally delivered, was that he had something "dreadfully important to tell us and could we not meet him for lunch at his home at noon."[13]

When the Nusbaums hurriedly complied, Junior confessed that there was in fact nothing dreadfully important to convey, other than he was lonely and couldn't bear eating in the large formal dining room by himself. He asked the Nusbaums to "please excuse his selfishness in wanting us to be with him." Furthermore, Junior declared that "he had so few real friends in New York" and that he and Abby rarely shared the same friendships, but the Nusbaums were the exception. Jesse interpreted Junior's words to indicate that he "is besieged on every social occasion . . . for money for this and that. His real pleasure comes from seeing a need and giving voluntarily I presume and not from request."[14] Jesse's inference, however, presumes that his own incessant moaning to Junior during the Mesa Verde visit about the family's crushing medical debts, did not qualify as a direct appeal.

Over the next few days the Nusbaums themselves "besieged" an array of wealthy New Yorkers who had expressed enthusiasm for Mesa Verde or who were friends of friends who might be willing to provide support for various projects, including a small hospital in the park. Jesse reluctantly bowed to pressure from William T. Grant to update his tuxedo for an upcoming fundraiser for which Aileen was likewise urged to don a big "ermine collared" evening coat borrowed from a Mrs. Oastler. Thusly attired, the Nusbaums accompanied the Grants of the W. T. Grant Department Stores to the theater and a post-performance gathering of some one hundred friends at the Grants' home in Pelham Manor. Both Aileen and Jesse spoke to the well-heeled attendees, and despite the fact that William "was worth about $20 million," Grant "flivvered" without contributing a dime to Mesa Verde.[15]

The remainder of the Nusbaums' New York campaign included more outings to theaters and orchestra halls and presentations to additional groups of upper-crust New York swells. Aileen and Jesse even made their pitch over the airwaves during a special noon-hour WAHG radio broadcast. They were invited by friends to Adolf Lewisohn's Fifth Avenue home for a reception for Willem Mingelberg, conductor of the New York Philharmonic. Lewisohn, a Jewish immigrant from Germany who had quickly risen to prominence as a copper magnate and philanthropist, was said to be interested in the parks. According to reliable sources, Lewisohn "was making very great bequests for various causes" (though most of his projects were confined to the New York City area). The big bash featured an elaborate supper, music, dancing, and—despite the restrictions of Prohibition—booze. Lots of booze. "They mixed cocktails for two hundred and fifty at a shake," Jesse reported.[16]

Yet when the music stopped and the last martini had been drunk, no significant pledges or contributions were made. This deficit illustrates the point that anyone in the fundraising business knows too well: recruiting donations, even from the ultrawealthy, is harder than it looks. This is particularly true if the prospective donors have no personal or emotional ties to the place needing protection. Or the people trying to protect it.

*

During the summer of 1926, Junior returned to Mesa Verde, this time with Abby and the younger offspring—David, eleven, and Winthrop, fourteen—in tow. As a middle child, Laurance, sixteen, was fortunate enough to be included on both Rockefeller summer vacations in the West. His two national park tours in three years may well have influenced his future as a conservationist and park benefactor. Either for the pleasure of the first-timers or because of Junior's love for routine, the Nusbaums repeated many of the features of the 1924 tour, including the cowboy steak fry, Cliff Palace excursion, and an outdoor pageant, this time "Fire," recently created by Aileen.[17]

For many years thereafter, Junior continued to support the Mesa Verde museum, which grew into a true gem among all NPS cultural history institutions. He also supported the Nusbaums, who suffered from overwork and an extraordinary number of serious ailments and extended hospitalizations. Throughout their lifelong friendship, Junior provided the couple with at least $14,000 ($250,000 in 2025 dollars) in personal gifts, mostly to cover medical bills.[18] He also gave enough to Horace Albright to help the family buy a home near Washington, DC, and covered some of the Cammerers' medical expenses. Junior also helped out Superintendent C. G. Thomson at Crater Lake, Oregon. Stephen Mather, too, supplemented the puny government salaries of Cammerer and Albright (as long as it was legal), with checks, gifts of stock, and finally inheritance. The contributions by Junior and Mather to overworked, underpaid NPS staff are further evidence of the depth of the two donors' dedication to excellence in the NPS.[19]

THE VIEW FROM LUNCH TREE HILL

Grand Tetons, 1926

The 1926 Rockefeller family vacation to Yellowstone National Park was considerably less hurried than their 1924 motor marathon. This time the Rockefellers—including Abby and the three youngest boys—had twelve long summer days to enjoy Yellowstone. The train trip would be Junior's third visit to America's flagship park. His first was made at age twelve, accompanied by parents John D. Rockefeller Sr. and Laura Spelman Rockefeller, and coincided with the U.S. Cavalry galloping into Yellowstone to chase away poachers and vandals.

Rather than riding coach as they did in 1924, this time the Rockefellers traveled first class. Abby's participation in this trip no doubt weighed in Junior's decision to upgrade. According to David Rockefeller, the family rode in the richly upholstered private Pullman car named "The Boston." Their entourage included a lovesick French tutor who corresponded incessantly with his sweetheart in France and a youthful doctor on loan from the Rockefeller Institute hospital.[1] Because Abby and Junior were a sickly lot—the former suffered from frequent stubborn colds, flu, insomnia, hypertension, and exhaustion; the latter prone to insomnia, myriad digestive disorders, bronchitis, sinus infections, persistent flu, migraines, shingles, exhaustion, and various severe nervous conditions—the onboard physician should hardly be viewed as superfluous.

Before arriving in Yellowstone, the family again visited the southern Colorado coalfields, site of the Ludlow massacre a dozen years earlier. Junior had toured the mines and steel mills on several occasions before and Abby had also visited previously. On this trip, the Rockefellers met with workers and union officials and toured the facilities. The stopover was likely intended as an educational opportunity for the boys, with lessons in mining, manufacturing, labor relations, corporate responsibility, and perhaps even mistakes and redemption.

Horace Albright again greeted the Rockefellers upon arrival in Yellowstone. This time the family was traveling undisguised, and Albright was free of National Park Service director Stephen Mather's gag order. By now, Albright and Junior were no longer strangers; they had been corresponding feverishly since the 1924 visit, exchanging hundreds of pages of letters and documents (mostly from Albright), and making legendary progress on reforming Yellowstone's ugly roadsides. Not surprisingly, Albright was eager to show off the esthetic improvements and insisted on personally escorting the family on the first half of their tour. Though an absolute perfectionist when it came to landscape architecture, Junior expressed delight at the work that Albright and his crews had accomplished. According to Junior, Abby and the kids also had a favorable impression of Yellowstone. Abby rated Yellowstone as the trip's "high-water mark."[2]

Like an English Pointer, Albright directed the Rockefellers due south. They tarried for barely half a day at Old Faithful and the sulfur-scented Upper and Lower geyser basins. Their haste allowed Laurance little time to snap photographs with the camera he had purchased with money earned from household chores. David was already engaging in what would become his lifetime infatuation: flipping rocks and logs to observe the beetles that scurried out from the damp darkness. They spent the night at Old Faithful Inn and departed early the next morning for another half day of bouncing along motor and wagon roads until ... off to the right! . . . above the pines! . . . their heads in the clouds! . . . the jagged, snow-clad peaks of the Grand Teton Mountains![3]

They arrived at Jackson Lake Lodge around noon and by early evening Albright was ushering the Rockefellers up a small hill, boxed suppers in tow, and positioning them for the most advantageous view of the long, deep alpine lake and the north end of the towering mountain range. As was the custom for sightseers of the day, the family was dressed in semiformal attire—suit jackets for the males, an ankle-length dress and stylish hat for the matronly Abby. Albright recalled the party spent over two hours on the hill where they watched for moose in the willows as the sun set behind the Tetons. The group sighted at least five surprisingly large moose along the lakeshore.[4]

For all except Albright, the next day's journey was the travelers' maiden trip into the heart of Jackson Hole. At an elevation of nearly seven thousand feet, the morning was crisp and bright, the kind of July day in the Grand Tetons that made your heart race and your skin tingle. They crossed the Jackson Lake dam and continued through stands of fragrant lodgepole pine and occasional glimpses of Jackson Lake until they topped a hill and the broad, flat valley of Jackson Hole opened before them. Unfortunately, the euphoria of the thin air and postcard-perfect scenery was short-lived. While the morning rays spotlighted every spire and crevice of the formidable mountains, it also revealed a lot of nauseating schlock. "Why are those telephone lines on the west side of the road, where they mar the view of the mountains?" Junior inquired. Albright blamed the U.S. Forest Service. The always calm, jovial, and witty Abby became visibly agitated as they passed what Albright later described as "a woebegone-looking old dance hall, some dilapidated cabins, a burned-out gas station, a few big billboards, and here and there a sign advertising cattle range land for sale." Finally, she asked Albright if something couldn't be done to halt the desecration?[5]

This was Albright's opportunity to formally unfurl his dream of adding the Grand Teton Mountains and the surrounding valley and foothills to Yellowstone National Park. He summarized the meeting he'd had three years earlier with a number of conservation-minded locals at Maud Noble's little cabin in the hamlet of Moose. The hostess was a single woman from a wealthy Philadelphia family who had come to Jackson Hole for a respite at a dude ranch and never left. Also present were a cattle rancher, a newspaperman, a businessman, and two dude ranchers. All were concerned with the unregulated, tacky, Coney Island–style development that was proliferating to service the stampede of hungry travelers arriving in their automobiles, known locally as "tin can tourists."[6]

Although Albright had been present at the conclave, the group's focus was not toward annexing Jackson Hole into Yellowstone. Rather, their goal was to gain federal protection without all the restrictions (especially regarding hunting) that accompany national park status. Dude rancher and writer

Maxwell Struthers Burt articulated this vision best when he described it as a "museum on the hoof,"[7] a sanctuary where wildlife, dude ranches, and the flavor of the Old West were all perpetuated. At the end of the meeting, the action plan was to recruit "a wealthy 'angel'" back east to purchase private property in Jackson Hole and donate it to the federal government as a national recreation area, or some such thing.[8]

In the three years between 1923 and 1926, the group's earnest crusade grossed a measly $2,300 (mostly donated by Albright, Burt, and Mather). The direct mail campaign to leading conservationists was a dud. Rancher Jack Eynon and newspaper publisher Dick Winger's trip back east to recruit the Morgans, Vanderbilts, Whitneys, and other Gilded Age plutocrats was, according to historian Robert W. Righter, "an abysmal failure."[9] The idealism of that lamp-lit meeting on the banks of the Snake River and in the shadow of the south end of the Grand Teton Mountains seemed doomed to evaporate in the razor-sharp light of day.

*

The next stops on Albright's tour included visits to two of the area's most famous dude ranches, the Bar BC and the JY. At the Bar BC, Albright and the Rockefellers visited with Struthers Burt and wife Katharine Newlin Burt. Struthers was no ordinary cowpoke, but rather a Princeton grad who devoted part of his waking hours to writing poetry, novels, and works of nonfiction about the American West. Katharine Burt was an even more prolific and popular writer than her husband, publishing better than two dozen novels (mostly Westerns) and helping adapt several of her books to the movie screen. The Burts pioneered the JY Ranch, the first dude ranch in Jackson Hole, but after a falling out with a partner, they moved on to build the more upscale Bar BC.[10]

Although in earlier years Struthers had been vehemently opposed to federal protection of Jackson Hole, in 1924 he fretted in one of his most famous books, *The Diary of a Dude-Wrangler*, "I am afraid for my own country unless some help is given it—some wise direction. It is too beautiful and now too famous. Sometimes I dream of it unhappily."[11]

Two years later Struthers penned an influential article for *The Nation* calling the Grand Tetons "the finest mountains in America" and cautioning, "I saw my country attacked by every known variety of water-power man, land scheme man, and ruthless get-rich-quick man. I saw . . . the increasing hordes of automobile tourists sweep the country like locusts. In short, I saw at work the old American passion for killing the goose, both spiritual and physical, that lays the golden egg."[12] Struthers had also met with Junior's chief conservation associate Chorley on at least two occasions prior to the meeting with Junior and Abby in Jackson Hole. Considering the years Struthers had spent in futile fundraising efforts, he must have looked upon his meeting with Junior the way a Snake River angler watches a trophy cutthroat trout cautiously rise to his dry fly.

Both the Bar BC and the JY were extraordinary properties acquired for next to nothing via homesteader land grants and bargain-priced purchases. The Bar BC was situated near the Snake River for a straight-on view of the Grand Teton, Teewinot, and the rest of the Cathedral Group in the very center of the spiny mountain range. The JY stood on the shores of Phelps Lake, a crystal-clear body of water that was spread at the feet of the Grand Teton mountains like a natural reflecting pool. After a pleasant lunch, standing on the lake's cobbly shoreline, Abby and Junior must have wondered if the sublimity of the scenery in Wyoming would forever diminish their enjoyment of the country estates they inhabited back east.

After the ranch visits, Albright set the Rockefellers up for the kill. He chose an overlook that offered a scene similar to the one captured by Ansel Adams sixteen years later in one of the photographer's most famous black-and-white masterpieces: "Tetons and the Snake River." One's eye was led across the vast flat expanse of sage and willows to the imposing, serrate, nearly vertical mountains. Their spiky summits seemed to promise the mountaineer a traversable route to heaven.

Out of the car, the Rockefellers sat on fallen pine logs while Albright waved his arms and pointed his finger to passionately describe the particulars of his and his cohorts' vision. He summarized all the failed and ongoing initiatives to protect the land they so pleasantly viewed. Albright wanted not just the mountains protected, but the valley of Jackson Hole, too, including the elk, bison, deer, coyote, bear, moose, antelope, and bald eagles. Acquiring the land from the cowboys, dude ranchers, and vacation homeowners would be difficult, but not impossible. At an elevation of seven thousand feet, in a climate where snowfall was generous from October to May, in a locale over one hundred miles from the nearest city, any agricultural enterprise was marginal and had persisted more from the proprietor's love of the landscape and lifestyle than economics.

When Albright's soliloquy was concluded, he looked into Junior and Abby's eyes for a reaction, but they sat silently and watched the sun set. "There was no word or comment," Albright recalled. In fact, the topic never resurfaced during the rest of their camping trip together.[13]

THAT ISN'T WHAT I HAD IN MIND AT ALL

Grand Tetons, 1926

Although John D. Rockefeller Jr. was fabulously wealthy in the autumn of 1926, he would never have been confused with the ranks of the idle rich. In the realm of education, beginning in 1924, Junior committed some $28 million ($494 million in 2025 dollars) to fifty-seven universities and research centers in thirty-nine countries, mostly through the General Education Board. Much of this support came in the form of grants to promising students, teachers, and institutions in the fields of physics, astronomy, math, agriculture, and archaeology. Notable results in those few years included construction of the two-hundred-inch telescope at Mount Palomar in California, an instrument that allowed astronomers to peer three times farther into space than previously possible.[1]

Thanks in part to the abolitionist traditions that ran through the Spelman (Junior's mother's) side of the family, Junior and multiple generations of Rockefellers provided support for African American education programs, including Spelman College for Girls in Atlanta (now Spelman College). The institution is America's oldest private, historically Black, liberal arts college for women. Among its attendees were Alberta W. King, mother of Martin Luther King Jr.; Alice Walker, author of *The Color Purple*; Stacey Abrams, former minority leader in the Georgia House of Representatives; and Evelyn M. Hammonds, dean of Harvard College. Junior likewise was boosting Tuskegee Institute (now University) in Alabama, the Hampton Institute (now Hampton University) in Virginia, and Fisk University in Nashville. He also generously financed the United Negro College Fund, an organization he chaired for ten years, steadfastly guiding its support toward thirty-two institutions.[2]

During this era of Prohibition, Junior, who wholeheartedly supported the alcohol ban, whose grandmother entered saloons to pray with the customers for the strength to give up drink, was forced to accept that the widespread lawlessness and corruption spawned by Prohibition was worse than drunkenness alone. His pivot away from the Anti-Saloon League toward a research-based compromise solution on alcohol made headlines and was one of the larger nails in the Eighteenth Amendment's coffin.[3]

Starting in 1923, Junior poured considerable resources into the Interchurch World Movement and the Institute of Social and Religious Research. The latter organization conducted seventy-seven in-depth studies on the significance of religion on societies worldwide. Although Junior's original goal was to unite all denominations of the Christian faith, when that aim proved untenable, he continued to pursue the general improvement of humanity through religion. The institute's work produced dozens of important publications on the impacts of religion on society and advocated for a significant change in the way churches approached overseas missions, guiding them toward a greater emphasis on health care, agriculture, and general education and less proselytizing and meddling with the traditions of other cultures.[4]

Concurrent with projects in Acadia and Yellowstone National Parks was Junior's involvement in restoring several French monuments, including the Rheims Cathedral and the sprawling palace at Versailles. The once magnificent cathedral, where kings, Joan of Arc, and revolutionaries had stirred, was bombarded during World War I so that only a shell remained when Junior and Abby visited in 1923. Versailles and its acres of gardens, statues, and other art were rapidly decaying, not a direct casualty of the hostilities, but a victim of French austerity spawned by the costly war and reconstruction. Leaky roofs at the palace were causing water damage to large swaths of the interior; mushrooms were sprouting on structural beams, the roof of Marie Antoinette's theater had caved in, and the overgrown gardens were filled with derelict statues and fountains.[5]

Aware that French pride could interfere with his plan to help, Junior approached the situation gingerly. He first worked through his contacts in France to assess the extent of damage to the landmarks and to research the French government's capabilities for their restoration. When he was satisfied with the reams of information compiled, Junior wrote a fawning letter to the premier of France, first complimenting the nation on its "not only national but international [art] treasures" and then stating his "feelings of deep regret" toward the current conditions. "I should count it a privilege to be allowed to help toward that end," Junior said, "and shall be happy to contribute one million dollars [$18.5 million in 2025 dollars], its expenditure to be entrusted to a small committee composed of Frenchmen and Americans."[6]

Pleased with the early results from his contributions, Junior added an additional $2 million to the restoration projects. The work continued until 1936. Although French ambassador Jean Jules Jusserand thanked Junior profusely ("The whole of France now and later will bless the name of one who, unasked and simply for having seen the danger, came to the rescue"), it was a later incident at the palace that endeared Junior to the French people.[7]

During a 1927 visit to Versailles to observe the ongoing restoration, Junior rode out to the palace in late afternoon, just as the gates were closing. After he was unceremoniously shooed away by a guard, Junior returned to Paris with plans to visit the following morning. However, fate favored the humble on this day. Someone at the palace recognized Junior as he was departing, and a rescue party was dispatched to track him to his hotel and apologize profusely. The next day French newspapers headlined the embarrassing incident and the citizens of France fell in love with the unpretentious Yankee. Years later, when asked why he didn't play his Rockefeller card at the gates of Versailles, Junior replied, "I can't imagine myself doing such a thing!"[8]

Other personal crusades originated from reading a newspaper article or a conversation with an altruistic acquaintance. Wife Abby once confessed to Junior's attorney and advisor, Raymond B. Fosdick, "I never know where John is anymore, but he is out saving the world somewhere!"[9]

Junior was also engaged with developing some of the old family property at Forest Hill near Cleveland. The "Rockefeller Homes" planned community was considered innovative for the time, a "harmonious village in an open, park-like setting." It included over eighty stately, two-story French Norman–style homes featuring terra-cotta roofs, copper gutters, and backyards that combined into a common greenspace. Junior donated the remainder of the Forest Hill property for a hospital, school, and public park.[10]

The fall of 1926 was an important turning point in Junior's vision for developing the site of the old Billings Mansion in upper Manhattan into Fort Tryon Park, site of an important Revolutionary War victory for the Colonial Army. He was on the cusp of hiring the Olmsted Brothers, headed by the sons of one of the designers of Central Park, to landscape the dramatic site. Another of his crusades of the era was a collaboration with European scholars to establish world peace.

The season was also an especially prosperous one for Junior, Abby, and their six children. Nine years earlier, Junior had begun receiving a series of very brief notes from his father, including this one: "I am giving you 166,072 shares of the stock of the Standard Oil Company of California."[11] Over a short period, as the probability of a major increase in federal estate taxes (up to 25 percent in 1917) loomed, the senior Rockefeller transferred over $450 million ($13.6 billion in 2025 dollars) to his only son.[12] Senior's and Laura's four daughters and their families were bestowed a much smaller share of the Rockefeller fortune, although they did receive sizable allowances and frequent gifts. Importantly, Senior's benevolence to his children came after the elder Rockefeller had endowed his institutes and foundations with some $447 million ($14 billion in 2025 dollars).[13]

The transfer of wealth placed Junior at the top of the hierarchy of the world's richest persons and made him America's top payer of federal income taxes in 1923 and 1924 (the last years that newspapers had access to such information).[14] In 1935, Junior was the sole person in America's top 63 percent income tax bracket.[15] Over Junior's lifetime, author Ron Chernow calculated the philanthropist paid $317 million in federal, state, and local taxes (in non-inflation-adjusted dollars).[16] Despite the hefty bite, Junior never publicly complained about high taxes on the wealthy. "The rich haven't always hated taxes," reported contemporary sociologist and Princeton professor Shamus Khan, and they even took pride in being included in the top brackets. "No editorials were written by John Jr. to suggest class warfare, or that the rich were being unfairly singled out."[17]

In response to his father's short notes and gargantuan gifts of dividend-paying stocks, Junior acknowledged, "The opportunities for doing good which you have made possible to me are unlimited and wonderful."[18]

The Rockefeller fortune and institutions continued to boom throughout the Roaring '20s with the U.S. stock market surging by nearly 500 percent between 1921 and 1929. A domestic manufacturing boom delivered automobiles, washing machines, radios, and other consumer goods to families at home and abroad. Commercial air travel was shrinking the nation. All these new cars, trucks, tractors, earth movers, and airplanes needed refined petroleum to run, and the "baby" Standard Oils (Esso, Standard Oil of Kentucky, Ohio Oil Company, etc.), spun off from the breakup of Standard Oil by the "Trust Busters" in 1911, were perfectly positioned to fulfill their thirst.

While the Rockefellers remained thrifty in many ways, even as their fortune reached the stratosphere, they did splurge on some significant upgrades to their domiciles. In 1911, Junior convinced his father to purchase a small piece of property near the family's large and comfortable New York City brownstone so that Abby and Junior could construct a "very modest but roomy house."[19] With the assistance of architect William W. Bosworth, the initial vision for a humble abode proliferated into the nine-story mansion the Nusbaums had toured.

Summers were largely spent at the Eyrie on Mount Desert Island in Maine, while holidays, birthdays, and long weekends were enjoyed at Pocantico Hills. The latter estate began when Senior followed his brother William's migration from the city to the bluffs overlooking the Hudson River in Westchester County, New York. Back then, in the late nineteenth century, the landscape was mostly rural and teeming with forests and wildlife. It was a suitable country place, within thirty-five miles of the city yet far removed in terms of the pace of life and with plenty of fresh air and room to spread. Senior and Laura moved into a large but plain old house and began acquiring vast swaths of land. By the end of the 1920s, Pocantico Hills had grown to approximately six thousand acres.[20]

After his parents' home, the Parsons-Wentworth House, burned to the ground in 1902, Junior unwisely insisted his parents build a real mansion, suitable for the world's first near billionaire.[21] Senior was more interested in landscaping, golf, and philanthropy than building a statement house, but Junior was highly motivated, at least in part by a jab made by Senior's nemesis, Ida Tarbell, in *McClure's Magazine*: "No one of the three houses he [Senior] occupies has any claim to rank among the notable homes of the country."[22]

The task of building a noble home for his disinterested parents proved to be one of the most daunting tasks ever undertaken by Junior and Abby.

Senior and Laura were indecisive on style and parsimonious to the extent of vetoing anything that smelled remotely of extravagance. The committee of architects that was eventually hired bungled several key elements of the project. When Kykuit (meaning "watchtower" in Dutch) was finally completed in 1909, Laura, by then seventy years old and suffering from emphysema and other respiratory ailments, hated almost everything about the pretentious palace. For example, the overly tall chimneys could never be coaxed into drawing properly and frequently filled whole rooms with smoke. Mysteriously, a few of the fireplaces somehow belched smoke even when not in use. Adding insult to injury, the house was more than $200,000 over budget, a fact that irritated Senior to no end and shamed his son.

Junior and Senior kept a close and cordial relationship throughout Senior's long life, partly because the elder Rockefeller seemed to instinctively reject the temptation to become the overly domineering patriarch to his insecure son. Perhaps Senior sensed that the wayward desires of his own father, the fraudster and bigamist "Devil Bill" Rockefeller, still lingered in his blood and he wished the strain to terminate with his generation. Senior rarely criticized his only son and was full of praise and encouragement whenever Junior pulled through a rough patch.

The Rockefellers finally waved the white flag of surrender after two years of tinkering with and failing to solve any of the money pit's gross deficiencies. In 1911 the family conceded to gut the place and rebuild. The resulting four-story, forty-room Georgian villa with views of the Hudson River was handsome and functional and would eventually become Junior's and Abby's home, as well as home to their children and grandchildren in decades to come. Senior was satisfied that it lacked the scale and ostentatiousness of a Vanderbilt home or even that of his brother and business partner, William. Junior recovered from the humiliating failure without suffering a nervous collapse and, through the ill-fated project, gained valuable experience in managing a complex design and construction process while navigating the treacherous waters between family and business. These skills, including tenacity, would prove invaluable in his conservation efforts as well as his other gargantuan philanthropic endeavors.

*

At the end of September, 1926, Horace Albright wrote Junior to provide some answers to the questions he and Abby had raised concerning the rescue of Jackson Hole from the scourge of "dance halls, lunch stands, and other objectionable structures."[23] Albright had relied heavily on Richard Winger—one of the participants in the Maud Noble cabin meeting and Jackson, Wyoming, newspaper man—to compile some estimates as to the cost of purchasing most of the land between the Snake River and the Grand Teton Mountains. It's obvious that Albright was chastened by the sum when it was presented to him. In a letter, he humbly confessed to Junior, "I am naturally very disappointed that his figures, both as to acreage and values, should have run so high and I am afraid the project may strike you as one that is impossible of further consideration." In recent years, Jackson Hole land values had stretched as high as $100 per acre and were rising relentlessly as more and more automobile-enabled tourists discovered the remote alpine haven.[24]

There was no reply from Junior for over a month, which must have been a long, anxious period for Albright. In early November he wrote Junior another letter updating him on how very well the roadside grooming was going in Yellowstone and reminiscing on all the joy and satisfaction the project had brought to both men. He also mentioned that he would be in New York later that month and would be happy to pop by 26 Broadway if that were convenient. Junior replied that he would like to see a map of the proposed land purchases as "the boundaries we talked of last summer are no longer fresh in my mind."[25]

November 20, 1926, was a fair and cool day in New York City when Albright arrived for his appointment. Even though the superintendent was no longer a small-town rube, his walk from the train station through the financial district and down Broadway to the Standard Oil building must have been intimidating for a man who spent his summers amongst the bears and bison in Yellowstone National Park.

Although Junior would have been delighted to see Albright, a kindred spirit in his love for beautiful landscapes and the quest to do good, Rockefeller was also likely overwrought, since that was more or less his permanent condition. Charles W. Eliot, the Harvard president and one-third of the Mount Desert Island triumvirate of conservationists, described Junior in 1925 as "always anxious and troubled."[26] While he could have been enjoying the winter sun on a Greek island or raising racehorses or sailing yachts, he chose instead enslavement to a Puritanical sense of obligation. He was also one of the world's most renowned idealists and perfectionists, traits that no doubt caused him frequent aggravation and disappointment as he plodded through the messiness and disorder of the world.

Junior's Christian faith was more than a motivator to do good deeds for humankind in general. It also related directly to his conservation projects, as he often remarked on seeing links between God and the beauty of the natural world. At a speech he gave to a large group in Chicago in 1946, he stated, "How can anyone who has witnessed the miracle of the spring doubt the existence of an eternal creative force in the world, call it nature, or God, or what you will."[27]

Despite his anxious disposition, Junior would have welcomed Albright with humility and graciousness to his twentieth-floor office. The two men were by then friends. The Albrights had even hosted the Rockefellers for supper at the Albrights' superintendent's quarters in Mammoth at the end of the Rockefellers' last Yellowstone trip. Grace had cooked a nice roast and helped churn homemade ice cream with the boys. Abby, who was open and comfortable with most everyone, talked in depth about her family and the progress on her father's biography. As a token of their appreciation, Abby sent the Albrights several thoughtful gifts: a beautiful handbag for Grace and three biographies for Horace. In a subsequent thank you note, Horace stated, "I do not know when Mrs. Albright has received anything that has made her happier."[28]

Once they had finished with their greetings, Albright presented Junior with a large map which the two men unrolled on Junior's antique table. Albright also apprehensively presented a sheet of cost estimates for purchasing fourteen thousand acres of private lands in Jackson Hole, pretty much everything on the west side of the Snake River. Albright's voice no doubt waivered when he announced the cost estimate of nearly $400,000 ($7.1 million in 2025 dollars), an amount he considered odious and which, he feared, might cause Junior to faint upon the Oriental rug.

As Albright later recalled in his many writings and interviews, Junior studied the map carefully, then removed his glasses and frowned.

"Mr. Albright, is this all there is?"

Albright was taken aback. "This is the property we have been working on," he replied.

Junior shook his head. "No, no. That isn't what I had in mind at all. You took us that afternoon to a hill where we looked out over the mountains and the whole valley. I remember you used the word 'dream.' *That's* the area for which I wanted you to get cost estimates. The family is only interested in an ideal project. And it would please me very much if you could get me data on that."[29]

Albright later recalled he was "completely flabbergasted." Maybe he was a rube after all. Yet how could he know precisely what Junior and Abby had in mind when they sat silent as statues after his oration on the promontory overlooking the vastness of Jackson Hole and the spectacle of the Grand Tetons?

"Why it might cost you as much as a million and a half or two million dollars [$26 to $44 million in 2025 dollars] to buy all that land!" Albright blurted, perhaps defensively.[30]

"Well of course, you don't know yet what it would actually cost," Junior replied. As a man who had the world's best accountants, advisors, real estate assessors, and attorneys at his beck and call, Junior didn't like to waste time with conjecture. "But that is what I'm interested in. I'm only interested in the ideal proposition. I would like to know about the land on both sides of the [Snake] river, and to the north."[31]

Junior then directed Albright to a meeting across the street with two of his most trusted associates, Colonel Arthur Woods and Kenneth Chorley, the people who would handle the day-to-day challenges of Junior's conservation and restoration projects. Woods, a former New York police commissioner, described Albright's condition as "completely overwhelmed" and liked to recall how the park superintendent "staggered" into his office that prodigious day.[32]

Albright certainly can't be faulted for thinking small when Junior wanted to go big. The men had been collaborating for over two years on the Yellowstone roadside improvement project, and Albright had been writing long letters with maps, photos, and documentation justifying every request for another $500 or $1,000 to keep the ball rolling. This may have been the first the Park Service employee had heard about "ideal projects." Yet that was the Rockefellers in a nutshell: they could quibble at length over a seventy-five-cent overcharge on a hotel bill and in the next breath unleash practically unlimited resources to eliminate the scourge of hookworm or the boll weevil in the American South.

As Albright left the Rockefeller offices that day, he described his condition as "near a state of shock."[33] His disbelief must have only increased as he spent the evening with Junior, Abby, and the kids at the Rockefeller mansion. For a humble civil servant who began his federal career sleeping at the YMCA, the evening must have stretched his perceptions of the known world. Later, on his way back to the train station for his return to DC, Albright recalled he was "walking on air."[34]

From that evening on, correspondence between Junior and Albright took on an even warmer tone. In the spirit of the 1926 Christmas season, Junior slipped a check for $1,000 (over $18,000 in 2025 dollars) into his card to Albright, for "rendering service to the people of this country utterly out of proportion to the financial recognition which the government makes."[35] Albright's thank you note was both grateful and jubilant. "Any mere words that Mrs. Albright and I might write or speak to you could never really express the happiness that your letter and gift gave us."[36] He pledged to invest the money in "safe bonds" and put the funds toward their children's education.[37]

Over the next few months, Albright diligently obtained estimates for the cost of an "ideal project" in the Tetons. Albright described the sanctuary "as the greatest scenic and wild life preserve ... on the face of the Earth." Including Yellowstone, the parkland would encompass ten thousand square miles.[38]

In early February Albright was back in New York City. The superintendent had furtively requested a few minutes to discuss "a matter that you [Junior] had asked him to investigate."[39] Albright described the consultation as short. Junior looked over the cost estimates—up to $2 million, and the land ownership map of Jackson Hole, then declared, "We'll do it."[40]

CLOUDS OF DUST, HONKS OF HORNS . . . AND SMELLS OF GASOLINE

Acadia, 1924–1932

As time went on, the Rockefeller family became increasingly fond of their summer home in Maine. Though Junior remained a landlubber, Abby, Nelson, David, and others enjoyed sailing and even purchased a modest boat and land on nearby islands. Once they started families of their own, Nelson and David came to own smaller "cottages" at Seal Harbor, sparing the family the discomfort of being shoehorned into the 107-room Eyrie. Junior partnered with friend and Mount Desert Island neighbor Edsel Ford to establish a family-oriented country club in Seal Harbor which was notable for its prohibition on alcohol.

Junior seemingly contributed to every fundraiser ever held on Mount Desert Island. He supported most churches on the island, and every school, museum, and community center. He gave to the fire departments, the camps, missionary societies, monuments, and the Red Cross. He purchased and razed the fire-damaged Glen Cove Hotel and donated its six acres to the town of Seal Harbor for a village green. In fact, an annotated listing of his philanthropic gifts during the 1920s, on Mount Desert Island and beyond, would fill a volume of several hundred pages.[1]

Also during the decade, Dorr, Rockefeller, and others collaborated to foster a world-class marine biology laboratory on Mount Desert Island at

Salisbury Cove. One vision for the lab, as articulated by an early proponent, was a place where "biologists would mingle with the owners of great estates at fund raising teas and receptions."[2] The facility drew scientists from major research institutions throughout the East and, by the late twentieth century, became one of the island's largest year-round employers. While the early focus was on marine biology, the lab soon broadened its scope to include human health issues like kidney functions and threats to the environment from petroleum and pesticides. Junior, who enjoyed intelligent conversation with scientists, was an early and frequent contributor to the facility.[3]

In August of 1926, while summering at Northeast Harbor, the ninety-two-year-old Charles W. Eliot announced to his son Samuel that he planned to die the following Saturday. That particular day of the week was thoughtfully chosen to facilitate travel of loved ones to his funeral. "The Sunday train was more comfortable than the week-day train," Eliot stated.[4] Eliot's estimated termination, however, missed the mark by a day and he passed painlessly on Sunday, August 22. The triumvirate was now a duo, and Park Superintendent Dorr, then in his seventies, continued to suffer from deteriorating vision. Dorr, however, remained an exceptionally vigorous person, thanks, perhaps, to his almost daily swims in the frigid waters of the Atlantic.[5]

A major controversy erupted in 1927 when Dorr and Junior released plans for new carriage roads in the vicinities of Eagle Lake and Bubble Pond. Some of the proposed carriage roads tracked close to the properties of prominent summer people and traipsed into the undisturbed wilds of Bubble Pond Valley. Members of the Bar Harbor Village Improvement Association (BHVI) were aggravated and formed a committee to squelch the plan.[6] Their strategy was to hire Charles W. Eliot II, grandson of the recently deceased triumvirate member, to lead a study and produce a report that would serve as a master plan for the island.[7]

The impressive report called for doubling the area of the park and establishing ten wilderness zones within its new boundaries. From the beginning, Junior, Dorr, and the higher-ups in Washington suspected the project's surreptitious purpose was to halt all new road construction. The actions of the citizens' group also countered an NPS initiative to create its own master plans for Acadia and other parks. "I am confirmed in my feeling that this committee was organized for the sole purpose of preventing or at least delaying as long as possible the further construction of any roads,"

Junior fumed.[8] Though widely distributed in 1928, the BHVI plan failed to find much support from the NPS or islanders. Eliot's vision was ahead of its time, however, in calling for a major expansion of Acadia National Park based on protecting whole ecosystems! Regardless, its influence gradually faded.[9]

There was a new sensation in the summer of 1930: plans were announced for Junior to fund and help manage construction of Ocean Drive, a fourteen-mile-long motor road from the Cadillac Mountain Road to Hunters Head on the coast of the Atlantic Ocean. The route, chosen after nearly a year of research by legendary landscape architect Frederick Law Olmsted Jr., would give the national park its first scenic coastal road and provide access to ocean beaches, tidal pools, and views of cresting waves smashing against rocky cliffs.[10] The proposed road, which Dorr called "a magnificent coastal drive, of inestimable value to the future," was also a key segment in a future park loop road.[11] In addition, it would employ some five hundred workers for three years during the early, dismal days of the Great Depression.[12] Its estimated cost was a breathtaking $4 million ($76 million in 2025 dollars). The "public-spirited offer" to construct the motor road "without expense to the United States" was praised and approved by acting secretary of the interior John H. Edwards in September.[13]

The ambitious project faced more than the usual number of barriers. Because of the hard rock, mountainous terrain, and esthetic requirements, the carefully planned drive would cost five times more per mile than the average Maine roadway.[14] There was highly charged opposition from wealthy summer people and wilderness advocates on Mount Desert Island. The town of Bar Harbor had to be convinced to donate an earlier, primitive version of Ocean Drive. And then there was the matter of Otter Cliffs U.S. Naval Station, an impediment that might have been circumnavigated by a less obsessive road planner, but one that, from Junior's perfectionist perspective, necessitated an extremely contentious relocation.[15]

At the *Bar Harbor Times*, a respectable weekly paper which dedicated most of its column inches to local politics, civic groups, and the social lives of society people, a steady trickle of letters to the editor protesting Junior's and the park's Ocean Drive and other road projects was dutifully presented. The letters came from summer people like Mrs. David Ogden, who refused to sell her land in the path of the new road until she was convinced that "a solid majority of both summer and winter residents favor construction."

Others feared their scenic refuge would be "invaded by clouds of dust, honks of horns, remains of picnics from flivver tourists and smells of gasoline."[16] Their objections were representative of leisure-class elites nationwide who opposed new roads in public wild areas while more working-class outdoor enthusiasts favored greater automobile access.[17]

The letters expressed both a desire to preserve the summer residents' backyard wilderness and a general suspicion of the Rockefeller plans based on a lack of knowledge of their entire scope and breadth. Summer residents are living "under a benevolent despotism...of one or two-man rule," complained cottager Richard Hale. On the other side of the argument, a summer resident countered that those opposed to public access simply feared a disruption of "the privacy of their picnics."[18] Many of the same letters, especially those from the wealthier, more genteel summer people who opposed the road, also landed on the desks of Superintendent Dorr and Interior Secretary Ray L. Wilbur.

The debate over motor roads in Acadia roughly coincided with the beginnings of a national discussion over automobiles, roads, and wilderness. At the 1926 National Conference on Outdoor Recreation, Aldo Leopold—considered to be the father of modern conservation—represented the minority at the conclave who believed that automobiles and other machines constituted the most obvious threat to wilderness on public lands. Pioneer wilderness advocate Bob Marshall published an article in early 1930 in *Scientific Monthly* calling citizens to rally around wilderness protection for the incomparable beauty of wild places and the benefits to Americans' physical and mental health.[19] The Wilderness Society was formed in 1934 and the first issue of its magazine, *Living Wilderness*, clearly drew the battle lines between roads and wilderness preservation.[20]

In response to the loud opposition, Junior announced in January of 1931 that he was pulling the plug on the Ocean Drive project.[21] He explained his decision as a reluctance to impose "upon even a small minority of the people who frequent Mount Desert Island something they do not want." He added, in a letter to the editor of the *Bar Harbor Times*, "nor do I care to be the continuing cause of the regrettably bitter criticism and comment which have been expressed...[in letters and] in other ways as well." [22] *Bar Harbor Times* editor and publisher Franklin S. Albion responded with general praise for Junior's many contributions to the park but criticized him for a lack of transparency regarding the specifics of his road plans.

Albion did admit, however, that he had failed to visit Acadia National Park Headquarters in Bar Harbor to view a map of the proposed Ocean Drive route that had been on display in the superintendent's office for at least six weeks.[23] And, according to Cammerer, there had been "other local publicity given projects but no criticism or objections have come to our attention" prior to the *Bar Harbor Times* letters.[24]

Regardless, Rockefeller and Dorr were likely negligent in not sufficiently communicating their road development plans much beyond Park Service and Department of Interior staff. Albright, however, privately responded to the accusations of secrecy with defiance: "They were too busy enjoying themselves in their mansions and broad estates and too little interested in the affairs of the island and the park.... Everybody but the rich and exclusive knew all about the program."[25]

The upwelling of support from the no-longer-silent majority who favored Ocean Drive was rapid and unprecedented. "All Maine Joins in Expressing Approval of Park Road Program" was the bolded headline on the front page of the February 11, 1931, issue of the *Bar Harbor Times*. Although "All Maine" appears to be an exaggeration, if one reads the letters and endorsements that fill numerous pages of the February 4, 11, and 18 editions, one might conclude the proclamation was understated. Endorsements by elected officials included the Maine governor, legislature, and numerous municipal governments. Pro-road civic groups included the Elisworth Chamber of Commerce, the Selectmen from Surry, the Bar Harbor Odd Fellows Club, Citizens of Bluehill, the Unison Rebekah Lodge No. 107 Sewing Circle, the Woman's Study Club of Bar Harbor, the Business and Professional Women's Club of Bar Harbor, the Grange of Salisbury Cove, the George E. Kirk Post of the American Legion, the Maine Automobile Association, and the Bar Harbor and Southwest Harbor Boards of Trade. Individuals' letters were published not only from Maine, but from Florida, New York, Pennsylvania, and Massachusetts as well. Excerpts from editorials supporting Junior's plans were reprinted, including those from the *New York Sun*, *Bangor Daily Commercial*, *Elisworth American*, and *Portland (ME) Sunday Telegram*.[26] The *London Daily Mirror*, bemoaning the growing number of unemployed in America, published two contrasting paragraphs describing how the rich were deploying their riches. The first was a note from the social pages stating that Mrs. Virginia Graham Fair Vanderbilt had thrown a party "at her palatial country estate at Manhasset [Long Island]" costing $200,000. The second was a dispatch that Junior would spend $500,000 to build a public highway "in order to provide work for unemployed men."[27]

No doubt flattered by the sudden international influence of his small-town weekly, the *Bar Harbor Times* editor conceded: "There is neither time nor space today for the publication of more than a mere fraction of the letters we have received urging Mr. Rockefeller to go on with his road building program." His paper also rightly concluded that the park's road issue was not simply local, but rather a question for the whole nation.[28]

In the early 1930s, despite the nascent national debate over wilderness, there was little opposition to roads in national parks (other than by people who would be displaced by their construction). Roads like Skyline Drive in Shenandoah National Park and Trail Ridge Road in Rocky Mountain would soon become as iconic as the parks themselves.[29] Mather had been pushing for a "Park-to-Park Highway" that would tie together western parks from the Grand Canyon to Mount Rainier. In many areas, the push for good roads was a major selling point for new parks. And the Depression-era New Deal programs were about to unleash a massive effort to build new park roads and major federal parkways. In this context, the sides in the roads controversy on Mount Desert Island might be characterized as a few privacy-seeking, wealthy elites who were anti-road versus a large group of islanders who were less affluent and more public spirited, who favored auto access for the benefit of themselves and others.

There were of course numerous exceptions to this lumping. For example, Northeast Harbor summer resident Dr. Francis G. Peabody's letter to the editor reads:

> There are probably few summer visitors who can report, as I can, a residence on the Island for sixty years, or who can claim a more thorough acquaintance than I can with its beloved hills and bays....No one can care more for the preservation of natural beauty on the island than I. We must approach the proposal of Mr. Rockefeller by recognizing that he has the same intense interest in safeguarding the picturesqueness and charm of Mt. Desert....He has rescued great areas from the logger's axe and from squalid settlements...by the establishment of Acadia Park, and by the munificent generosity of one neighbor, our hills and forests are made reasonably secure from devastation, and that the rugged beauty of the Island may be made more accessible to all.[30]

Bar Harbor Times editor and publisher A. F. Sherman remarked that Peabody's letter "said something that many of us would like to say and said it in a way that perhaps only he could have said."[31] The February 11 edition of the paper also featured a map of the proposed road adapted from a photograph taken of the notice posted in Superintendent Dorr's office.

Despite the strong encouragement from Mount Desert Island's year-round residents and the public at large, a handful of summer people continued to express their vehement opposition to additional motor roads in the park. This only highlighted how much Junior, the ultimate wealthy summer resident, had evolved. Back in 1915, Junior sounded much more like the elites who now opposed his roads projects. In a letter to Eliot regarding the national monument project, Junior fretted, "Do you not feel, that the establishment of the [national] monument will bring an undesirable class of tourists to Bar Harbor in their automobiles who, if automobiles are admitted to the south side of the Island, will be a real nuisance to the residents there?"[32]

As a person who disliked motor vehicles and revered the privacy and quietude of the woods, Junior was now aligning himself with the locals in a battle to build, of all things, a very expensive road that could attract the multitudes in automobiles. And while his initial foray into motor-road construction likely had the main purpose of protecting his precious carriage roads, the Junior of the 1930s, through his experiences with Yellowstone, Mesa Verde, and other national parks, had since evolved toward becoming an authentic conservationist, progressively joining a high-minded club whose members included Albright, Cammerer, Mather, Eliot, and Dorr.

And while his promotion of automobile roads appears to be a contradiction, Junior had long been in the same camp as John Muir and his Sierra Club members in expounding that urbanization and industrialization fouled the human spirit and only time in nature could heal these wounds and foster a closer relationship to the divine. Even his massive investments in historical projects like Colonial Williamsburg and Middle East archaeology served to turn back the clock to periods in history with less mechanization and more nature. One might conclude, therefore, that Junior, though he personally favored nonmotorized travel, decided it was more important to take an egalitarian approach to enjoyment of scenic areas than to impose his preference on the masses. Likewise, Albright, Cammerer, Dorr, and Mather—all hardy horsemen, hikers, and adventurers—promoted park roads that would encourage the less capable general public to experience the natural world.

Junior had joined the NPS leaders who believed that national parks would only be protected (and funded) if lots more people visited them. Mather's marketing brilliance had even convinced the public that visiting American national parks was patriotic. In 1923, Mather wrote that national parks "do more toward stimulation of national pride and contentment" than do the activities of any other federal agency.[33]

Junior's decision to renew his support for Ocean Drive was no doubt driven by the passion of Maine's year-round residents and a decisive vote by the town of Bar Harbor to abandon to the NPS their old Ocean Drive right of way. The fact that Maine and the entire nation remained in a state of shock from the Great Depression also tipped the scales toward employing hundreds of workers to build the road. With help from the Park Service in Washington, Dorr and Junior gradually restarted the Ocean Drive planning process. In 1932, Junior reflected in a letter to Cammerer, "If they know I am indifferent [to road building], then the proposal begins to be considered on its own merits; and when, as in this case [Ocean Drive], the merits are found to be very considerable, like all other blessings they brighten as they take their flight."[34]

*

The U.S. Navy, however, proved to be an even more obstinate adversary than the wealthy cottagers. The navy's Atlantic-facing radio station at Otter Cliffs was crucial for communicating with ships, planes, and submarines, and the department had no inclination to surrender it to a pretty road for tourists.

Yet Junior and Dorr had a potential, though longshot, solution for the naval station problem. Back in 1922, Dorr had met a wealthy widow, Mrs. Warner M. Leeds, while dining at the Jordan Pond Tea House. The charming Mr. Dorr began talking up his effort to create a public park with donated lands, and Leeds wondered if the sprawling preserve she and her two daughters had inherited from deceased husband John G. Moore on the Schoodic Peninsula might be of interest. Dorr affirmed that it would, indeed. He considered the view from the rugged, isolated peninsula "one of the greatest in the world." Unfortunately, Leeds died unexpectedly not long after her conversation with Dorr and before she had a chance to discuss the matter with her daughters. Despite this setback, Dorr was still able to convince the two daughters, Faith Moore and Lady Lee of Fareham, both of whom resided in England, to donate their two-thirds of the property inherited from their railroad and telegraph mogul father to the park. Mr. Moore had acquired the land as a private preserve for hunting and fishing with his friends and had been a good steward, citizen, and philanthropist. His conservation legacy, combined with the vexing reality that property taxes on the two thousand acres of idle land had recently doubled, were no doubt motivators to his daughters.[35]

There was one catch, however. The anglophile siblings could not stomach the park's French moniker, Lafayette. Their contribution was contingent on the name of the park being changed to something more acceptable to British snootiness.[36]

The peninsula became crucial to the park's future when Junior and Dorr proposed that the naval station and its looming radio listening tower be relocated from Otter Cliffs to Schoodic. Otherwise, the road was destined to cut through the station and detract from the scenery.[37] Dorr had no problem charming Rear Admiral Philip Andrews and convincing him that Schoodic was an ideal site for the new station, especially since the Otter Cliffs base had been slapped together during World War I by an amateur radio operator who leveraged the facility for a naval commission and commanding officer position.[38] Quarters on the windswept cliffs had never been intended for permanent occupancy and had therefore become dilapidated. Despite the economic chaos rattling the United States, Junior offered to bankroll half the cost of the navy's relocation.[39]

Unfortunately, Officer Andrews retired before the deal could be closed and his responsibilities were transferred to Captain S. C. Hooper, the director of naval communications, who viewed the exchange with much greater skepticism.[40] After all, the Otter Cliffs location was ideal in that it was relatively free from interference from other radio signals and offered a straight shot to Europe. During the war it had proved invaluable for intercontinental communications, so much so that ninety soldiers were assigned to its operation and protection. Even during peacetime, on a foggy day, Hooper claimed "often forty ships are assisted in their navigation" by the station. Otter Cliffs was also less isolated than Schoodic, making it more economical for the navy to operate and more appealing to its staff.[41]

Dorr called upon one of Maine's congressmen, John E. Nelson, to pressure Hooper to move toward a compromise. Hooper refused at first but finally acquiesced to a site visit. After two days of mucking about in the cold rain, the captain suggested the radio station be upgraded with new offices and staff housing—but remain on its current site. Hooper believed he could convince Junior of the efficacy of this approach if a meeting could be arranged; consequently, the principals set up a half-hour phone call between Junior in New York and the naval commander. Junior did not budge.[42] Once he had made up his mind to do something, he was as hard to redirect as a snapping turtle. "From the outset my interest in the motor road project has related to the whole plan and not to the construction of any part of it. It was conceived on an ideal basis, and on that basis I believe it should be carried through, if at all. I cannot think that in the long run the public interest will be served by unhappy compromises."[43]

Frustrated and humiliated, though reluctant to return to Washington without some sort of progress to report, Hooper agreed to a second visit to Schoodic. Auspiciously, the driver reached the tip of the peninsula in record time, making the site seem less remote, and the sun finally burned through the fog, illuminating some of the most gorgeous scenery on the Atlantic seaboard.[44] By October of 1932, Hooper had agreed to relocate the Otter Cliffs Radio Station to Big Moose Island on the Schoodic Peninsula. He stipulated only that the Interior Department build the road and causeway to the radio station and install the utility lines.[45] Fortuitously, the new service road would be scenic as well as utilitarian and therefore serve both park visitors and the personnel of the new naval station. Junior helped acquire land needed for the road and paid for a well to provide fresh water for the new base.[46]

Other significant obstacles persisted, however. Lafayette National Park was not authorized to acquire lands beyond Mount Desert Island. The Schoodic Peninsula, across Frenchmans Bay from the island, had never been envisioned for inclusion in the park. Dorr solved both deal busters by hosting Michigan congressman Louis C. Cramton, member of the House Appropriations Committee—and stalwart supporter of the national parks—and his wife for three days in Bar Harbor. Once again, Oldfarm and its resident gentleman worked their magic. After the tour of the park and relaxing fall evenings around the big fireplace, Cramton agreed to push through legislation that would not only expand the preserve's boundaries but would change the name of the park.[47]

Dorr claimed to have been the one to choose Acadia as the site's latest name.[48] "Acadia," though not as difficult to pronounce as Sieur de Monts, is, unfortunately, often confused with "Arcadia." Both words have similar pedigrees and mean "pastoral" or "place of rural peace." However, while Arcadia can be traced to an agrarian region of southern Europe, *l'Acadie* was the French name for their colony in eastern Canada and portions of Maine. How this escaped the prejudiced eye of Lady Lee is not known, though she may have been appeased by Albright's assertion that the moniker "is of native origin, coming from an Indian word apparently describing the region."[49] More recent scholarship traces the words *akadie* or *cadie* to the language of the indigenous Mi'kmaq people.[50]

As the nation and federal government slipped further into the quagmire of the Great Depression, Mainers lobbied hard for the relocation of the naval radio station and subsequent road construction. They celebrated the news in late 1932 that tests by U.S. naval radio engineers at Schoodic Point had proved the new site "entirely adequate for naval broadcasting and receiving purposes."[51] The *Bangor Daily News* reported on the effort by "virtually every civic organization in Eastern Maine" to push forward on Ocean Drive.[52] In the fall of 1933, the *Portland Press Herald* crowed "Over 225 Men Employed" in building the new naval station and scenic highway.[53] When the federal funds for moving the station were finally allocated, one of the National Park Service's chief engineers sent a note to associate director Arthur E. Demaray reminding him that the justification for the funding was to create employment during the Great Depression.[54]

By 1934, even Junior's bank account had shrunk alarmingly (contracting from almost $1 billion to less than $500 million). During the Depression, his philanthropic commitments and massive investment in Rockefeller Center far outpaced his income, forcing him to liquidate stocks at greatly diminished prices and actually borrow millions of dollars to make good on his obligations.[55] The philanthropist wondered if it might be time for the federal government—with its constellation of New Deal programs—to step in and foot the bill for some of Acadia's unfinished roads. "I am just wondering whether there is any possibility of getting from the Administration of Public Works Fund, or whatever its proper name may be, funds for the construction of the continuation of the motor road in Acadia." Junior assumed the federal government was looking for just such projects to put the unemployed to work.[56]

Cammerer seemed rather shocked at the suggestion. "Frankly, as you had offered to build this without expense to the National Park Service, the thought [of using federal funds] had not occurred to me." He went on to explain that the limited monies available for road building in parks were already spoken for. "Any possibility of Federal participation in the completion of the road you have in mind is therefore now entirely out of the picture."[57] Junior was let off the hook, however, for covering half the cost of relocating the radio station. Apparently, the secretary of interior decided Junior had already given more than his share for Acadia and that taxpayers could cover the move.[58]

Yet, as a show of appreciation to the navy, Junior did decide to bankroll the new staff quarters at Schoodic. Characteristically, Junior refused to make any compromises on the quality of the housing. He helped secure the assistance of famed architect Grosvenor Atterbury, designer of the Acadia carriage road gatehouses, to create a three-story brick-and-beam Acadian lodge on par with the millionaire's "cottages" around Bar Harbor. Even art-centric Abby Rockefeller pitched in by reviewing the structure's early drawings and warning against blending Georgian and French architectural styles. The result was "Rockefeller Hall," a massive (80-foot-by-133-foot) truly magnificent French Norman Revival-style apartment complex that must have surprised and delighted Navy personnel reporting for duty at the new radio listening station. And while new technologies eventually made the navy radio operations obsolete, the high-class housing continues to be well utilized by the nonprofit Schoodic Institute for research and education purposes.[59]

*

As work on Ocean Drive and some long-delayed segments of carriage roads restarted, another main park thoroughfare was completed: the Cadillac Mountain Road. Junior's role in the creation of the 3.9-mile road was actually minor compared to other Acadia road projects. He had been an early advocate, arguing it would prevent roads from intruding on other Acadia summits. He personally acquired most of the land upon which the road would run and built a short starter road for the route. He monitored the planning and advised Dorr on the placement and construction of the scenic byway (all of which were ignored by the Park Service engineers).[60] But perhaps his greatest contribution to the entire national park system was the standards he established for all previously constructed roads in Acadia and elsewhere. The bar set by Rockefeller carriage and motor roads was indeed much higher than those employed by most public works projects.[61]

The federal government elected to fund the new Cadillac Mountain Road as part of its massive effort to stimulate employment through infrastructure development. Unfortunately, by awarding the construction contract to the lowest bidder, as per federal guidelines, the builders made a mess of the project. Assistant landscape architect for the NPS Charles E. Peterson groaned that the grading work caused "the worst piece of landscape damage which any road has done in a national park."[62] Rock, dead trees, truck tires, rubbish, construction equipment, and other debris had been strewn all over the route. The parking area, which was not even close to being sufficient for even 1930s levels of visitation, was "devastated" by poorly executed blasting. Yet, because the contractor had seriously underbid the job, the project managers felt powerless to enforce the need for further work.

As plans for a grand opening for the Cadillac Mountain Road began to ferment, the landscape architects realized that everybody from the president to the secretary of interior might well attend. They could visualize all too well how "the road would be judged against Rockefeller's excellent work." Roadsides piled with blasted rock and the corpses of trees would be compared to shoulders designed by Beatrix Farrand and skillfully planted with native shrubs and wildflowers. For the federal engineers and landscape architects involved, it was going to be like entering a frozen grocery store pie in the Maine State Fair baking contest.[63]

Fortunately, the ribbon cutting originally planned for the fall of 1931 was pushed back to July 1932. Both Peterson and Junior made strong statements to Albright, now director of the NPS, of how badly the road needed additional work. Albright found $11,000 for a special cleanup project and promised Junior "two of our very best men" to study and rectify the situation.[64] Peterson reported that much of the scarring was necessarily "concealed with blueberry sod." He further recommended to the Bureau of Public Roads that more stringent restrictions be placed on all future specifications for road work in Acadia. Even after the hoopla was over, twenty men were needed to complete additional cleanup tasks along the drive.[65]

Nothing better demarcated the end of three decades of squabbles over automobiles and roads on the island and ushered in the beginning of Acadia's new life as a full-fledged democratic American national park than the over-the-top ribbon-cutting ceremony on the summit of Cadillac Mountain on July 23, 1932. According to the *Bar Harbor Times*, hundreds attended the ceremony, including the secretary of the navy, Charles F. Adams, thirty-seven naval officers from three destroyers anchored in the harbor, Maine's governor Gardiner, Senator Fred Hale, and Congressman John E. Nelson; acting secretary of interior Joseph Dixon, and of course park superintendent George Dorr. President Herbert Hoover sent a message congratulating "the people of Maine upon the completion of the road."[66]

Characteristically absent was Junior, claiming a bad case of shingles. Still, Junior must have been feeling some comfort, as much as a compulsively humble person can, as he had just received a letter from Secretary of Interior Ray L. Wilbur. The secretary had toured Acadia and been pleased with what he saw. "Rarely have I enjoyed anything so much," Wilbur gushed of the park and Junior's projects. "Your roads are admirable, your bridges beautiful, and your plans complete. I want to thank you on behalf of the Department of the Interior and the Nation."[67]

Representing Junior at the July ceremony was his son, John D. Rockefeller III, age twenty-six, who stood shoulder to shoulder with Dorr as they cut the silk ribbon officially opening the Cadillac Summit Road. Local newspapers reported the Bar Harbor Band played "spirited marches" as microphones relayed the live event to radio stations and newsreel cameras panned the sublime landscape. Shortly afterward, Mother Nature stole the show with a summertime deluge of rain. An orderly relocation of the hundreds on hand to the Malvern Hotel in Bar Harbor was quickly accomplished and proficient orators filled the ballroom with copious praise and appreciation. They paid homage to park giants Charles W. Eliot (1834–1926) and Stephen Mather (1867–1930), both by then deceased. The crowd at the Malvern then toasted Dorr and Junior for their dedication and sacrifice. They proclaimed that the new road democratized the park's scenery, making "available to all men the matchless beauty of the spot."[68] As a thunderous finale, the guns of three United States destroyers, anchored far below in Frenchman's Bay, fired salutes.[69]

For many of those assembled that afternoon, and for millions following the story elsewhere, it was not the establishment of the national park per se that made the preserve on Mount Desert Island enjoyable for the people; it was the road. Only motor roads could transform Acadia from Old

World–style millionaire's private playground to public park. In his book *Windshield Wilderness*, historian David Louter reminds us that "knowing nature through machines" was once accepted as the perfect way to experience national parks.[70] Although in decades to come, continued road building in national parks would become much more controversial and often be perceived as an assault on natural resources or wilderness, in 1932, roads like Cadillac Mountain were largely seen as an opening of the gates for all the ordinary folk who wished to live like nobles of yore by freely experiencing the beautiful mountains, lakes, and seashores of Acadia.

In many ways it is remarkable that the bitter controversies in Acadia didn't sour Junior on public lands philanthropy, especially when one considers how his road projects conjured vehement criticism against him in the press and alienated him from some of his friends and neighbors. The peace of his idyllic retreat was frequently interrupted by formal reviews of his roads by elected officials and Department of Interior brass and a full-fledged congressional investigation, not to mention the more mundane challenges and expenses of building perfectly landscaped mountain roads in the rugged northern climes.

Junior's involvement with Acadia was unlike his philanthropic commitments in other national parks. In the case of Acadia, he was already emotionally invested in the beauty of the place (Mount Desert Island) before he was called upon to become a conservationist. He had spent halcyon days there, and he and his growing family were tied to the area's people and the island's environmental wellbeing. The bridge between investing $3.5 million between 1915 and 1960 (over $60 million in 2025 dollars) in creating and developing Acadia National Park and playing a leading role in so many other park conservation projects (most of which he had no personal connection with) must have been the NPS superintendents and directors he befriended.[71] Even though the Rockefellers rubbed elbows with presidents, members of Congress, kings, queens, princes, princesses, scientists, the world's greatest artists, wealthiest businesspeople, and most influential innovators, there must have been something about the humble public servants in green and gray that opened his eyes, as well as his wallet, to the incalculable worth of protecting nature and history. Not only did he come to like the NPS leaders, he appreciated that they always worked hard to assure excellent value for his philanthropic dollars. Through this collaboration, Junior came to realize how national parks fit into the larger scope of Rockefeller philanthropy, as another way to "serve the public good and improve the conditions of humankind."[72]

Still, it can be said that all of Junior's future conservation projects began with his falling in love with the natural beauty of Maine. Love of nature's beauty often brings out the best in us.

BOTH SORTS OF GREEN

Great Smoky Mountains, 1923–1926

John D. Rockefeller Jr. was not the only private citizen interested in America's national parks in the first decades of the twentieth century. People all over the country were making pilgrimages to Glacier, Yosemite, Yellowstone, Mesa Verde, or Acadia and coming away impressed by America's abundance of readily accessible scenic grandeur administered by the fledgling National Park Service. And, as post–World War I visitation to parks surged by 10 percent per year or more, enterprising citizens began forming ideas about the potential for additional national parks in their own vicinities. Along with its jazz, flappers, and increasing urbanization, the 1920s were about to become an era noted for the flourishing of America's national park system.[1]

In June of 1923, a prominent Knoxville, Tennessee, couple, Anne and Willis Davis, took an adventurous vacation to Yellowstone National Park. On the long train ride home, Anne had an epiphany: "We have seen some beautiful country and grand mountains, but none more beautiful and majestic than our own Great Smokies. Why, then, should there be national parks in the West and only one small national park [Lafayette] in the eastern half of the United States?" Anne inquired of Willis. "Why shouldn't our own Great Smokies be made a national park?"

After a moment, her husband replied, most considerately, "If that is the way you feel about it, I will see what I can do."[2]

Anne was an attractive, educated, wealthy society woman with plenty of connections to upper-crust citizens throughout the region. Willis, a captain of the coal and iron industries in Appalachia, watched the Great Plains roll by through the passenger car window as he contemplated his wife's revelation. He too had been stimulated by their national park experiences and had witnessed firsthand how a majestic American landscape could draw throngs of tourists and subsequent commerce to a site that was otherwise Nowheresville, USA.

A few months after their Yellowstone trip, Willis nearly went into shock when Anne announced that she would pursue a seat in the Tennessee state legislature. Her principal purpose for running was to create Great Smoky Mountains National Park and thereby right the geographical injustice between East and West (and, perhaps, in some Southerners' minds, North and South). Even the Davis children were amazed to see their mother's sudden metamorphosis from parent, homemaker, and socialite to politician.

But transform she did. Anne and Willis began talking up a Smoky Mountain national park with such evangelical fervor that friends and business associates likely resorted to jay walking across Knoxville's Gay Street just to avoid crossing the Davises' path. The couple's proselytizing, however, filled with the boundless optimism of 1920s small-city America, had only begun to foment. Within months, Anne's train-ride spark had ignited a bonfire of national park mania in the scruffy little textile and marble-mining town of Knoxville. The chamber of commerce and local automobile club (Willis served on the boards of both) jumped on the bandwagon and helped forge the Great Smoky Mountains Conservation Association (GSMCA). True to his pledge, Willis capably assumed the helm of the new, highly influential organization. The *Knoxville News* quickly became an important park advocate, trumpeting in August of 1923, "The beauty of the mountains will not preserve itself. Man must save his world from defacement by his own restless activity." The editors advocated the creation of a national park, "or at least a national recreation area."[3]

Rising to the occasion and to Willis's side was a bulldog of a man named Colonel David Chapman. A two-time war veteran, wholesale druggist, and George F. Babbitt–style civic leader, Chapman was a powerful addition to the GSMCA. His indomitable energy, business acumen, fiery boosterism, and determination to win at any cost would prove pivotal in the formidable battles that lay ahead. His willingness to throw and take a punch in a contentious, rib-cracking commission meeting was also considered an asset.

Willis was so excited by the snowballing momentum of the early park crusade that he personally made a pitch to John D. Rockefeller Jr.'s attorney, W. S. Richardson. Willis introduced himself out of the blue and immediately requested a large sum for the park project. His one paragraph letter suggesting that Junior purchase "this property to present to the Government as a National Park" was painfully premature, brief, and blunt.[4] Davis even went so far as to sic Dr. Arthur Kendall of Washington University (a professor of medicine) on Junior's people to warn them the region's dwindling virgin forest was "ripe for the axe," and if logging was not interrupted the "splendid primeval forest will be a myth."[5]

Though woefully naïve, Willis's request actually did open the door for Major W. A. Welch, who was known to the Rockefellers from the family's efforts to expand the Palisades park along the Hudson River in New York, to describe the Smokies project to key Rockefeller advisors. The official response from 26 Broadway was not encouraging, yet it did not completely slam the door in the Knoxvillian's face, either. Key associate Richardson wrote, "We are looking into the request presented to Mr. Rockefeller Jr. but fear that it is not a matter that lies within the scope of his plans in giving."[6] And, "There is little or no likelihood that Mr. Rockefeller would care to provide the land."[7]

Like Acadia, the movement to create a national park in the Great Smoky Mountains was a grassroots effort. Both were incited by civilians who had to start completely from scratch without a lot of outside support (and plenty of opposition). Unlike Acadia, which had much-revered conservationists like Charles W. Eliot and George B. Dorr to shepherd its birth as a national park, the Great Smoky Mountains campaign was spearheaded by businesspeople with tourist dollar signs in their eyes. There were exceptions to the profit motive, of course. Eastern hikers and conservationists like Horace Kephart, George Masa, Paul Fink, and others wished to stop the logging and protect the mountains they loved to explore. Their articles, photographs, and orations were important assets in the fight.

Yet the majority of Smoky Mountain park proponents were motivated by the prospect of better roads and increased trade between Knoxville, Atlanta, and Asheville, North Carolina. Scenic roads would attract auto

tourists who would rely on Knoxville for their accommodations, gasoline, curios, sundries, and supplies. Knoxville would become synonymous with towering trees, sparkling waterfalls, magnificent wildlife, and mountain splendor, and vice-versa.

What could stop them?

*

Obstacle number one: Upon approaching U.S. Secretary of the Interior Hubert Work about their big idea, Willis learned that a plan to create a second national park in the East had already left the station and was high-balling down the tracks. In fact, the post–World War I popularity of national parks in the West was fueling an epidemic of park envy up and down the Appalachian mountain range. Newspaper columnists, civic boosters, good-roads advocates, and the automobile clubs in thirty other cities were all clamoring for a national park to bring tourists and prosperity to their local elevated topography. And the seven-year-old National Park Service—led by Mather, Albright, and Cammerer—amplified their call.[8]

All three of the above founding NPS leaders were aware their impoverished agency was not getting any support from members of Congress hailing from the South. And why should they? The citizens they represented, with the exception of a few wealthy families like the Davises, could hardly imagine having the time and money to abandon their farms, factory jobs, or small businesses for a guided three-week sojourn to Yellowstone or the Grand Canyon.

Being dyed-in-the-wool Californians, Albright and Mather were naturally inclined to view national parks as largely western bonanzas. Yet politics were politics, and if they were ever to get their agency's miniscule budget expanded to properly meet the needs of the caravans of motorists now blazing their way toward park entrance stations, they needed Congress to become more sensible regarding appropriations.

Mather formally advocated for more parks east of the Mississippi River in his 1923 annual report to the secretary of the interior. "The only practical way national park areas can be acquired would be by donation of lands or acquisition of such lands from funds privately donated, as in the case of the Lafayette [Acadia] National Park."[9] The concept of philanthropy for parks was not at all foreign to Mather. Being a Californian, he was well aware that Muir Woods had become a national monument through donation of land, and Mather himself had contributed better than $200,000 (over $4.5 million in 2025 dollars) to Yosemite, Glacier, and other parks. The exact amount was difficult to assess, however, since, as Cammerer pointed out, Mather "rarely speaks of the things he does."[10]

To facilitate the expansion, Mather and Secretary Work pieced together the Southern Appalachian National Park Committee (SANPC) to scout the hills and hollers for the perfect southeastern mountain park. The committee, consisting of five men, all of them Yankees, set to work in the spring of 1924. Mather was adamant that the NPS shouldn't lower its high standards for national parks as it considered scenic areas in the East. Any new park "must continue to constitute areas containing scenery of supreme and distinctive quality," he stipulated.[11]

Director Mather wrote Junior requesting a contribution to cover the travel expenses of the SANPC (all of whom were volunteers and had gone "into their own pockets" for train fare and lodgings). Mather stated he was kicking in $250 toward the stipend and that perhaps Junior would also like to help.[12] Characteristically, Junior did not simply drop a check in the mail, although doing so would hardly have dented a minute's worth of dividends rolling in from his Standard Oil stocks. Instead, Junior responded that he "would be glad to know what the total expenses of the committee are estimated to be and whether others besides yourself and Mr. Gregg have made contributions."[13] One of Senior's guiding lights on philanthropy was "we wish to root the institution [beneficiary] in the affections of as many people as possible."[14]

Mather replied that SANPC member William Gregg had pledged $500. Therefore, they needed only $750 more to cover the estimated $1,500 in total costs. Junior, ever the parsimonious millionaire, eventually sent a check for $500, perhaps suggesting that borax millionaire and NPS director Mather might pony up another $250 himself, thereby making the three men equal benefactors in the worthy cause.[15]

The SANPC's scouting trips into the potential national park areas quickly drew a crowd. Politicians and reporters joined the parade, creating an All American 1920s media circus. Governors and members of the U.S. Congress welcomed committee members to town and treated them to flattery and dinners. Mayors and chamber of commerce leaders, clutching maps and photographs, elbowed their way into the vanguard, insisting the committee visit the sublime scenery in places like Tallulah Gorge, Georgia; Canaan

Valley, West Virginia; Clay City, Kentucky; Bakersville, North Carolina; Grandfather Mountain–Linville Gorge, North Carolina; Big Knob, Virginia; Roane Mountain, Tennessee, and twenty or so other sites.[16] Unfortunately, as the committee's tour progressed, it gradually became clear the Great Smoky Mountains had not even made the SANPC's relatively long short list.

When the Davises, Chapman, and the others heard of this slight, they were understandably roused. The GSMCA hurriedly organized a "fighting delegation" and stormed off to Asheville to ambush the committee in their lair—the tony Grove Park Inn with its Arts and Crafts décor and mountain-view golf course.[17]

The pro-Smokies group's ammunition included a fat portfolio of eight-by-ten-inch glossy black-and-white prints by Knoxville photographer Jim Thompson. The exquisite photos illustrated iconic Smoky Mountain scenes like the Chimney Tops, Rainbow Falls, Mount Le Conte, and Cades Cove. Because they had no chance of evading the sneak attack, the SANPC granted the Knoxvillians three hours to make their case. The Smokies boosters emphasized the heights of the mountains, the wilderness character, the variety of plant life, and the stands of virgin forest. At the end of the presentation, SANPC acquiesced to include the Smokies in their ongoing tour de parks.[18]

Just five days later, on August 4, 1924, SANPC members William Gregg and Harlan Kelsey arrived in Knoxville and were instantly commandeered by several members of the Smoky Mountain Hiking Club. Somewhere in the vicinity of two dozen hale and hearty men—mountain guides, newspaper reporters, and a string of park promoters that included Paul Fink, Willis Davis, Russell Hanlon, Johnny Morrell, and Colonel Chapman—volunteered to escort the committee members to the lofty summit of Mount Le Conte.[19] Apparently, all of those enlisted considered themselves to be in tip-top physical shape, because in their day, the trek up Le Conte was no walk in the park.

After a meaty breakfast at the Mountain View Hotel in Gatlinburg, Tennessee, the party proceeded by automobile and horseback to the trailhead. From there they forged ahead on foot for a couple of steep miles on a passable trail to elegant Rainbow Falls, one of the sites the committee members had admired in the Thompson portfolio. Lunch was ham-and-egg sandwiches washed down with coffee, surrounded by old-growth hemlock and basswood trees and the seventy-five-foot-tall waterfall.[20]

Then the going got rough. Conquering the waterfall's brink required climbing the "leaning tree," a tall evergreen with dense branches. Some in the party no doubt required a boost to reach the tree's lowest footholds. From the top of the falls the trail continued "right up the creek." Dense shrublands known locally as "hells" impinged the route. The twenty or so city slickers and wiry mountaineers wearily plodded in a lengthening line for another three miles to a soggy, boulder-strewn camping spot below the looming rock face called Cliff Top.[21]

As darkness fell, the men swaddled themselves in wool blankets against the drizzle and settled amongst the rhododendron for a damp, fitful sleep possibly interrupted by thoughts of black bears, rattlesnakes, and timber wolves. During the seemingly endless night, the boosters brooded that if they could only show the committee members a grand sunrise from Myrtle Point, one of Le Conte's three knuckle-like summits, all their hardships would be forgotten. With the fog and mist weighing ever more heavily upon the mountaineers' manes and bedding, there finally appeared in the east only the slightest hint that the suffocating gloom might someday end—possibly even within their own lifetimes. The party elected to forego breakfast and began a slow, miserable slog toward the rocky summit. The fretful Chapman worried that "the Point" might be socked in. Mountain guide and GSMCA member Paul Adams described the procession thusly: "We were a motley crew as we headed up across Cliff Top. Most of us had blankets around our shoulders to protect us from the rain-wet bushes."[22]

But then, as if ordained, the stifling, vaporous ceiling began to unravel. Within the gaps between the clouds, a night sky filled with dim stars faded to indigo. Fog and thunderhead clouds roiled in the valleys, but the highest peaks—Guyot, the Sawteeth, and Anakeesta Ridge—seemed to catch fire with the sun's primal rays. From their precipice, crowded by sand myrtle, the party gazed across a verdant wilderness unlike anything else in eastern America—forested ridge after forested ridge as far as the eye could see. "We were small spectators, awe-struck by the vast, primitive beauty of an extra-special Myrtle Point sunrise,"[23] Adams recalled. "For grandeur I know of no other view in the Southern Appalachians that quite equals Le Conte," committee member Kelsey later reported.[24]

Yet, Le Conte was just a warmup for Chapman and Willis Davis. Chapman had convinced Tennessee Governor Austin Peay to come to Knoxville where he spent two days lobbying Gregg and Kelsey. Three research scientists from prestigious universities were ushered in to lecture the committee on the hidden value of the Smokies—a temperate rainforest unsurpassed for its varieties of flora and fauna. The ancient mountain range boasted more different species of native trees than all of northern Europe and better than a thousand known varieties of wildflowers.[25]

Then there was the virgin forest—more than one hundred thousand acres of magnificent hemlock, tuliptree, American chestnut, red spruce, Fraser fir, maple, black cherry, basswood, red oak, black walnut, and silverbell, all of which reached record or near-record size in the sheltered folds of the rumpled mountains. The great arbors boasted circumferences up to twenty-five feet and heights comparable to a twenty-story building.

But the forest's days were numbered. Over the last two decades, industrialized logging companies had been punching railways deep into the mountains and were using steam-powered skidders and legions of men to shave the slopes to a stubble of forlorn stumps. They had conquered the steepest terrain and were hoisting their equipment onto the flanks of Clingmans Dome—the top of ol' Smoky. Hiker and park advocate Paul Fink reported on the "utter devastation," during a backpacking trip in 1919. "The whole basin of Big Creek...had been cut over. Fire had gotten out into the slash and completed the devastation, burning everything, even the organic matter in the forest floor."[26]

During the 1920s, the timber and wood pulp companies were cutting sixty acres of virgin forest each day in the Great Smokies. As word of a possible national park spread, they conspired to expedite their operations.[27]

*

On December 12, 1924, Secretary of Interior Hubert Work approved SANPC's report. SANPC reminded the secretary their comprehensive search had included the southern mountain states of Alabama, Georgia, Kentucky, North Carolina, Tennessee, Virginia, and West Virginia. They reiterated their goal of creating a second national park in the East where two-thirds of the population resided. After lavishing praise on the Smokies, declaring they "easily stand first because of the height of the mountains, depth of the valleys, ruggedness of the area, and unexampled variety of trees, shrubs, and plants," the judges placed the crown on the head of Shenandoah.[28] Not because Shenandoah's single ridge of diminutive mountains was superior to the dozens of other sites, but because its more modest topography was easier for building roads and its location was conveniently close to Washington, DC.[29] Congressman Henry W. Temple of Pennsylvania, chair of the SANPC, was already busy introducing a resolution to establish the new park's boundaries.[30]

The report, which Director Mather called "very fine," did, however, leave the door slightly ajar for subsequent efforts at some unspecified date to make the Smokies the second southern Appalachian national park.[31] Predictably, Chapman, the Davises, and all the other hard-charging Smokies boosters were, once again, incensed. The *Knoxville News Sentinel* reported the GSMCA had received the bad news with "astonishment and amazement."[32] By this time Anne Davis was an active member of the Tennessee state legislature and was relentlessly pressing her fellow lawmakers to commit funds toward the park. The GSMCA was lobbying their representatives in Washington, too. Seeing that Shenandoah was about to run away with the prize, park proponents in western North Carolina finally joined the Great Smoky Mountains crusade in earnest, thereby doubling the boosters' powers of advocacy.

The combined wails of protest from Smokies fans on both sides of the mountains did not fall upon deaf ears. Secretary Work, Mather, and others heard the clamor and recognized the thwarted boosters in North Carolina and Tennessee were plotting to rally representatives from other Southern states to block passage of the Shenandoah bill. They also calculated that adding the Smokies to the Shenandoah proposal might net them favor from members of Congress in at least three states.[33]

On January 11, 1925, Secretary Work announced in a news release that he was pressing Congress for the creation of at least two new national parks in the East: Great Smoky Mountains and Shenandoah. A month later, Congress responded with a bill confirming the Smokies and Shenandoah and adding Mammoth Cave to the list of new parks in the East.[34]

*

The second, seemingly insurmountable challenge facing the Great Smoky Mountain effort was the thorny reality that every acre of land within the proposed park boundaries was in private ownership. By comparison, the creation of other national parks, including Yosemite, Yellowstone, Glacier, and Mesa Verde (as well as President Theodore Roosevelt's myriad national forests), had been easy. Their high plateaus and pine-studded mountains, though beautiful to behold, offered little in the way of agricultural or mineral-extraction opportunities. These economically stunted lands had therefore lingered in the public domain. To create a new national park or forest in the West, members of Congress could simply draw a line around a map of some rugged terrain without ever rising from their leather chairs or extinguishing their cigars. In fact, Yellowstone became a national park in 1872, just three years after the first formal exploratory party returned from its harrowing expedition. As Secretary of the Interior Ray Lyman Wilbur would one day summarize, "It is comparatively simple to carve out an area from the public domain. It is a very complicated matter to displace established settlers and business organizations."[35]

To avoid casting the foreboding shadow of the federal government over the process, the NPS and park proponents delegated the contentious task of acquiring lands in the Great Smoky Mountains to local representatives from the states of North Carolina and Tennessee. Their job was a distressing one: Approximately 1,100 farms, some of them prosperous, idyllic, beloved properties in family ownership for three or four generations, had to be assessed by a committee and dickered over until both sides agreed upon a "fair" price. Sometimes the question of what constituted a fair price led to billable hours for lawyers and the buyers had to flex their eminent-domain muscle. Records show the North Carolina Park Commission paid an average of 1.6 times the tax valuation of the land they purchased, and the Tennessee Commission paid an average of $16 per acre, in the same ballpark as the Carolina buyers.[36]

Of course, well-to-do residents of Asheville and Knoxville cherished the idea of an eight-hundred-square-mile mountain preserve in their backyards, as well as the good roads and economic boom they believed were certain to follow. But the hardscrabble Smokies farmer was simply bewildered. Most could not imagine how their rural hamlets, replete with whitewashed churches, schools, stores, cemeteries, sawmills, and gristmills could be transformed into a wilderness park for tourists from Ohio. Gatlinburg resident Herb Trentham recalled his Aunt Mary Jean's reaction when the land assessors visited her remote cabin. "You keep talking about a park, a national park, what do you mean by that?" she inquired. When the men explained, she blurted, "You mean you're going to turn our land back over to them varmints [wildlife]?"[37]

Prominent Cades Cove resident John Oliver, who operated a lodge for tourists in the beautiful valley, complained that he and his rural neighbors were being forced to make sacrifices for the benefit of city folk. He advocated for better compensation for landowners in Cades Cove because farmers were being "dispossessed of their homes in order that city people might have a playground."[38]

Frank Maloney, a Knoxville civil engineer who had helped Cammerer survey the proposed park boundaries, optimistically reported to Cammerer on his land scouting in East Tennessee, particularly around Sugarlands, Greenbrier, and Cosby. "I find a large number of small land owners desirous of selling and feel we can count up several thousand acres of this class of property."[39] Though some of the younger farm families might have viewed the cash they were offered for their marginal farmlands as a welcome opportunity for a fresh start in Texas, Detroit, or rural areas outside the "Cammerer Line" (proposed park boundary), older folks and the wealthier landowners had no desire to cooperate. The Palmers of Cataloochee, for example, owned over a thousand acres and lived in a stately two-story home with gas lighting and a trout stream just beyond their doorstep. They managed apple orchards, livestock operations, and fields of corn and wheat, and operated a community bank and post office. In later years the Palmers even rented cabins and provided meals to tourists.

Some of the displaced fought for and received larger sums for their land and improvements while others negotiated lifetime leases that allowed them to stay on through their golden years, albeit as folksy, incidental tourist attractions. Lessees were highly restricted in their activities: "No hunting, no cutting or use of [standing] timber. No digging of medicinal roots and herbs," and no cultivation of new fields.[40]

The Smokies were not unique, however, as an American scenic wonder that had to be reclaimed from private ownership. Decades earlier, the famous conservationist and landscape architect Frederick Law Olmsted and others had rescued Niagara Falls from the hucksters and carnival barkers who had lined the shores of the river with hot dog stands, fences, and sideshows and then charged tourists a fee for a quick glimpse of the falls.[41] As tourists and journalists from around the world made their requisite visit to the wonderland, their critical reviews of the lowbrow exploitation of the beauty spot cast Americans as hopeless hicks destined to defile whatever they touched. "The lesson of which Niagara Falls may be said to be a pioneer teacher," stated Andrew Green, a commissioner for the new Niagara Reservation, "is the State's right of eminent domain over objects of great scientific interest and natural beauty—the inherent right of the people to free enjoyment of the wonders of nature."[42]

Almost exactly one century prior to the establishment of Great Smoky Mountains National Park, the federal government signed treaties with the Cherokee and other groups of indigenous people in the Southeast (Indian Removal Act of 1830 and Treaty of New Echota in 1835) which facilitated the removal of Natives from their homelands in the vicinity of the Great Smoky Mountains to Indian Territory in Oklahoma. The treaties were motivated by white settlers' desire for more agricultural land and rumors of gold in Georgia. The second treaty was soon revealed to be fraudulent because only a fraction of the Cherokee people had endorsed it. The forced removal was an infamous action known today as the Trail of Tears. In 1838, some 16,000 people were marched at gunpoint from Tennessee, North Carolina, and other southeastern states to Oklahoma. Thousands died and families were forced to try to adapt to a new environment with which they had no cultural connection.[43] Ironically, the creation of the national park by European Americans would attempt to mostly erase 125 years of white settlement and resource extraction by creating a federal reserve that, ideally, more closely resembled the Smokies when they were exclusive to the Cherokee.

Fortunately, some Cherokee were able to avoid the 1838 death march. Many of their descendants now live in the Smokies, on the Qualla Boundary, as neighbors of Great Smoky Mountains National Park and the Blue Ridge Parkway.

*

While the twentieth-century process of buying the farms and giving the 5,500 mostly white residents of the Great Smoky Mountains the old heave-ho was a truly onerous and painful task, buying up the eighteen large tracts of mountain fastness from the big timber companies was, in some ways, even more excruciating.[44] The lumber companies had, of course, fought the park boosters tooth and nail, much preferring the land to be designated a national forest. As such, the U.S. government would be responsible for things like managing public access and fighting wildfires while the big corporations could still come in and clear-cut the mountainsides whenever the trees were ripe for harvest.[45]

The timber companies used a wide range of tactics in their fight, some of them not even remotely legal. Champion Fibre Company, which owned Mount Le Conte and some ninety thousand acres in the heart of the mountain range, paid an attorney $15,000 to tamper with a jury in one of the property assessment cases.[46] A propark rally in Waynesville, North Carolina, was hijacked by employees of the Champion and Suncrest companies and flipped to become an evangelical, antipark revival. Suncrest lawyer "Bat" Smathers even used the forum to rail against the whole concept of outdoor recreation, summer vacations, and the unproductive enjoyment of leisure time. "This mad age of pleasure and recreation [leads straight] to hell as fast as possible!"[47]

More conventionally, the companies hired public relations firms and took out full-page ads in local newspapers touting the benefits of national forest over national park. Champion Fibre's ads reminded readers that its company employed over two thousand workers in western North Carolina, and (ironically), because Champion and its fellow logging companies had already clear-cut 75 percent of the proposed park area, the landscape no longer met national park standards.[48]

During the late winter of 1924–1925, Anne Davis was bending the ears of her fellow legislators in hopes of persuading them to seize a rare opportunity offered by one of the larger timber barons—Colonel W. B. Townsend, president of Little River Lumber Company. Townsend had been talking with Tennessee governor Peay about possibly selling seventy-six thousand acres of Smoky Mountain land to the state. After all, the company had nearly finished "harvesting" the forest, why not pack up their equipment now, cash in their real estate chips, and head for greener pastures? From the state's perspective, the tract along the scenic Little River included vital sections like Elkmont and Tremont and would help convince Congress that Tennessee and the boosters were serious about their Smoky Mountain park. Unfortunately, the powerful state representatives from West Tennessee were feeling less than generous toward their country cousins four hundred miles to the east.[49]

To sway the naysayers, the Knoxville Chamber of Commerce chartered a special train to haul members of the legislature from Nashville to Knoxville and eastward for an inspection tour of the proposed purchase. The state officials, none of whom had ever seen the Great Smoky Mountains, were greeted in Knoxville by the famed University of Tennessee marching band and generally cajoled and entertained before their trip to the Smokies the following morning.[50] Sights were seen: the Little River gorge, Rich Mountain with its expansive views, and Cades Cove, the pastoral jewel of the western Smokies. At a luncheon in the Appalachian Club at Elkmont,

Chapman promised the elected officials that the creation of a Great Smoky Mountains National Park would bring not only fame and respect to the state, but also a relentless flow of tourist dollars destined to benefit state coffers. On Monday, the *Knoxville Journal* described the train trip through the Little River gorge as "one of unending pleasure" while the *Knoxville News* headlined "Legislators Are Much Impressed."[51] According to one report, the legislators were "amazed to find these wonderful mountains" in their state.[52]

Anne Davis introduced the bill to secure an option on the Little River tract in March 1925. It passed the state Senate but was two votes shy in the House of Representatives. Anne's ally, Governor Peay, responded by calling several of the holdouts into his office, one by one, and using his charm and personal guarantee that the city of Knoxville would pay a third of the tab for the Little River land as inducement.[53] The following day the bill passed the House 58 to 32. Anne Davis attended the governor's signing of the bill and was awarded the quill pen for her leading role in the act's fruition.[54]

*

The third challenge facing park advocates, which dwarfed the aforementioned obstacles, was capital. In order to create Great Smoky Mountains National Park, the boosters had to rake up the scratch for a minimum of 427,000 acres of land (up to a maximum of 700,000 acres). Estimates of the final tab rose as high as $10 million ($184 million in 2025 dollars) and threatened to creep ever higher as negotiations with the remaining lumber companies dragged on.[55] The federal government remained adamant in its "not one cent to scenery" stance, denying federal appropriations for national parks even while land purchases for the more utilitarian national forests were routine.[56]

Chapman, the Davises, and other proponents kicked off the fund-raising campaign by hiring a top-notch public relations firm and filling local newspapers with beautiful photographs by Thompson and Asheville hiker-photographer George Masa. Sunny prose by famed outdoor writer Horace Kephart touted the multitude of benefits a Great Smoky Mountains National Park would bring. North Carolina's director of the fund-raising campaign, L. W. Sprague, estimated "at least two million persons would annually visit the Great Smoky Mountains," producing a windfall of $200,000,000 for the two states each year.[57] Terms like "tourist Mecca" and "goose that laid the golden eggs" became common currency in the hyperbolic crusade. Following the P.R. company's advice, goals were established: $500,000 for Tennessee; ditto for North Carolina.[58]

It's doubtful that either East Tennessee or western North Carolina saw such an energetic fundraising campaign before or since. School kids in four East Tennessee counties sacrificed their lunch money and cracked open their piggy banks to contribute $1,391.[59] Church members were cajoled in their pews as special collection plates were passed. Bankers in Knoxville were patted down to the tune of $12,000. Bellhops assembled in front of their Knoxville hotels and ponied up their tips. Tiny Bryson City, North Carolina, which stood to lose property tax revenue because of the park, reached its $25,000 quota in the first twenty-four hours.[60] Ashevillians assembled two dozen vehicles and barnstormed the entire state from the mountains to the Atlantic handing out brochures and making propark stump speeches. Newspapers in Asheville and Knoxville stoked the effort with promises of "unprecedented growth, development, and prosperity." The local Chambers of Commerce put together interstate whistle-stop tours with as many as two hundred park boosters riding the rails to Florida, Louisiana, Texas, and Oklahoma.[61]

When the initial campaign fell short, the *Knoxville Journal* reported "Chapman Declares Park Campaign Is Nearing Failure."[62] The drumbeaters had come too far to admit defeat, however. They extended the deadline for a month and redoubled their efforts in a climactic do-or-die fundraising frenzy. Knoxville mayor Ben Morton declared March 16, 1926, as "Great Smoky Mountain National Park Day." A front-page editorial chastened the citizens that the day "is a municipal holiday, not for jollity, for sports of any kind, but for the . . . most momentous opportunity that has ever confronted the people of Knoxville."[63] A. A. Fielding, a prominent African American realtor and politician, wrote a check for $300 ($5,300 in 2025 dollars), "on behalf of my people," to match the contributions of a long list of other Black Knoxvillians who had pledged.[64] Students at Knoxville High School tendered nearly $2,500. Knoxville's banks doubled their collective pledge to $25,000. In twenty minutes, the Asheville Chamber of Commerce raised the final $35,000 needed for North Carolina to hit its goal.[65]

During the late spring and summer of 1926, citizens of Asheville and Knoxville took to the streets to celebrate the attainment of their fundraising goals. Against seemingly impossible odds, they had raised 10 percent of the funds necessary to purchase the land for the creation of Great Smoky Mountains National Park. The considerable balance would need to come from benefactors with much deeper pockets than the citizens of the proposed park's gateway communities.

Concurrent with the Smokies' 1926 fundraising milestone was the Rockefellers' second family summer trip to Mesa Verde, Yellowstone, the Grand Tetons, and other existing and potential western national parks. That summer was a turning point for Junior, as he was about to greatly multiply his commitments to public land philanthropy. This prodigious development was of keen interest to Cammerer, who considered the Smokies the most important project of his career.[66]

At this juncture in the Great Smoky Mountains origin story, it's also worth noting that, while the Smokies and Shenandoah were the victors in the 1925 national parks competition, the information gleaned from the Southern Appalachian National Park Committee's scouting trips into potential national park sites continues to have great value today. All of the more than two dozen surveyed sites exist within a day's drive of many of America's population centers and offer exceptional mountain scenery and rich and imperiled biodiversity. If conservationists were considering potential sites for additional national parks in the southern Appalachians, the SANPC report would be an excellent place to start. Stephen Mather's annual reports to the secretary of the interior would be another.

RICH PEOPLE ARE TEMPERAMENTAL

Great Smoky Mountains, 1926–1927

Having reached the first rung of their tall, $10 million fundraising ladder, the park boosters set their sights on the two state legislatures. In Tennessee, there was enthusiasm from representatives from the eastern third of the state, but only blank stares from the middle and western districts, where the bulk of the state's population resided. North Carolina was also an uphill battle with over two thousand logging company employees inhabiting the western fringe. In addition, many of the Carolina boosters were holding tight to grudges born when their Grandfather Mountain–Linville Falls, Asheville, and Brevard area potential parks finished as distant runners up in the SANPC pageant.[1]

Still, the states' elected officials could not help but be impressed by the monies raised by the citizenry and municipalities of their poorest burgs and villes. Supporters, including Cammerer, pressed North Carolina first, requesting a pledge of $2 million contingent upon Tennessee promising a like amount. Even more significant was the stipulation North Carolina would not have to write its check until *all* the money had been raised to buy the lands necessary for creation of the park.[2] Cammerer recalled, "I did more speech-making that one day than I have in the entire preceding month."[3] North Carolina governor Angus McLean, who always maintained an open door for timber company executives, remained staunchly neutral on the proposal.[4]

Yet, by loading the legislation with so many highly improbable requirements, the states' duly elected representatives gradually warmed to the proposal. They could still win favor from their supporters near the mountains while in all likelihood never confront the day when the $2 million bill would come due.[5] The passage by North Carolina thusly backed the Tennessee legislature into a corner from which it could only escape by passing its own provision-laden, unlikely-to-ever-be-paid pledge. Regardless, Chapman stated the Tennessee legislature engaged in "a terrific fight" before the vote.[6] In this manner, park boosters had reached the 50 percent mark on their goal, even though most of the money, from not only states but also towns, counties, businesses, and individuals, existed merely as promises awaiting a $5 million matching contribution.

Another sticky contingency froze much of the land-buying activity until the states had the cash sufficient to purchase practically all the lands. This created an increasingly uneasy and confusing situation between land buyers, farm families, and timber companies. Farmers were hesitant to make improvements or even plant crops until they had a clearer picture of their fates. An attorney for one of the lumber companies encouraged landowners to cut new roads through their property to improve their assessments.[7] Even more troubling was Colonel Townsend's threat to renege on the deal forged with Anne Davis and the state of Tennessee for the seventy-six thousand acres of Little River Lumber Company land. Townsend stated he had "been offered $40,000" for the virgin timber along the route of the main park transmountain road ("right at the front door of the Park").[8] The lumber company's offer was much higher than what the state had proffered for the timber.

As the year 1927 waned, Chapman and the other boosters grew increasingly anxious. Although they had come so very far, they were still only halfway up the fundraising mountain, and dusk was approaching. No doubt Chapman and the Davises pondered whether they had been chumps to pin all their hopes on finding some big-name donors to instantly match all the monies wrung from long, exhausting years of begging for donations. The timber companies, of course, rejoiced at every stumble and continued with their operations and vocal opposition. Owners of summer homes and some of the wealthier farm families were hiring lawyers, reaching out to elected officials, and refusing to budge. In the midst of all this, powerful park ally Tennessee governor Peay died unexpectedly. Pledgers breathed a sigh of relief, feeling increasingly confident that their bills—made during the heat of fundraising fervor—could conveniently go unpaid. Fundraisers had to mail eight thousand invoices to people and businesses just in Tennessee in an attempt to collect on unfulfilled pledges.[9]

Park proponents pinned all their hopes of dredging up the last $5 million ($92 million in 2025 dollars) on a Hail Mary they called the National Campaign. The premise of this maneuver was simple: find some extremely wealthy whales from out of town who had more money than they knew what to do with and convince them to give an astronomical amount of currency for lands that would immediately become the property of the federal government. During the 1920s in America (and even more often in the 1930s), it was not uncommon for ordinary folk from small towns to dream of salvation from their errors and misfortunes via some magical Daddy Warbucks. Yet, the fact that the Smokies campaign had involved so many prominent people and institutions, while relying on such an unlikely fairy-tale ending, is the most extraordinary piece of the entire crusade.

In 1926, the strategy for landing a munificent donor relied on one man: Major W. A. Welch. The New Yorker had worked indefatigably for the Smokies, most notably in securing the key pledges from the two state legislatures. His role in managing the Palisades Park in the Hudson River Valley pressed him into association with some of the world's wealthiest families, including the J. P. Morgans, the E. H. Harrimans, and the J. D. Rockefellers. As a youth, Junior enjoyed crossing the wide Hudson River by ferry and riding his horse along the banks beneath the Palisades. When quarrymen began dynamiting the diabase columns of the towering river bluffs for the purpose of building skyscrapers and roads in Manhattan, both Junior and Senior took notice.

Creation of the Palisades Interstate Park was instigated by the American Scenic and Historic Preservation Society and the New Jersey Federation of Women's Clubs. After a decade of wrangling, it was officially designated by both New Jersey and New York state governments. The Rockefeller Foundation and the Laura Spelman Rockefeller Memorial were generous donors during the early 1900s, footing the bills for land purchases, development of trails and picnic areas, and even two steamboats to carry pleasure seekers from the concrete canyons of New York City to the forested mountains and riverside beaches of the Palisades.[10] Years later, Junior also became involved with the design of the scenic parkway that now wends along the top of the bluffs northward for thirty-eight miles from the George Washington Bridge to Bear Mountain.

In the Great Smoky Mountains, just when the crucial national campaign needed to be hitting pay dirt, Welch went incommunicado. It's possible that SANPC's blunder of tethering Shenandoah to the Smokies for national fundraising purposes was coming home to roost and the whole dual-park effort was unraveling. The two proposed parks were in different states and had different potential benefactors. Those who favored one likely resented being forced to contribute to the other. It's also likely that the stress of trying to raise such a large sum of money for two big southern parks from dozens of distracted and self-important Yankee donors, caused the good Major to hit the wall. In any case, Welch's silence aggravated Chapman.

"I had concluded that it was hardly worthwhile for me to communicate with Major Welch," growled Chapman in a September 1927 letter to Cammerer. "He has ignored my letters and telegrams. Our people ask questions daily and the newspapers are clamoring for news about the National campaign. I have stalled until I have stalled out. Our people are wondering if anything is going to happen and as a natural result, have let up on paying their subscriptions [pledges]."[11]

Over in North Carolina, state senator Mark Squires was also growing anxious. "Our friend the Major does not seem to get busy. Think over the matter and advise me confidentially what can be done to accelerate the process."[12] Later, in a curt letter to Chapman, Squires implored, "Please give me a full account of what is going on in respect to our National Park, or rather what is not going on."[13]

The ever-cheerful Cammerer replied to Chapman immediately and tried to allay his fears. The assistant director reported that he was headed to Maine to work on some Acadia Park "problems of pressing importance" but also hoped to "sit down around the fireplace with several men of means" to chat about the Smokies. Cammerer explained that Welch was "running the biggest three-ring circus in parks the country has to offer" and was required to travel extensively.[14] Cammerer further relayed to Chapman and others that Welch had "broached the matter to some twenty whose names he gave me confidentially, who are interested." Happily, Welch estimated he had pledges good for $2 million.[15]

Any reassurance Chapman and Squires received from Cammerer's letters was short-lived. Three months later the assistant director informed the Smokies' top advocates: "You have been advised that Major W. A. Welch has severed his connection with the Great Smoky Mountains National Park movement."[16]

Welch's mysterious resignation from the national campaign not only removed him from the playing field, it subtracted $2 million and twenty prospective donors from the drive. "At present we are all a bit stunned and confused," Cammerer admitted to Chapman in January 1928. "Such a complete failure...is incomprehensible."[17] On the home front in North Carolina, Squires fretted, "Our collections [of pledges] are distressingly slow."[18] Yet, Cammerer urged patience and perseverance: "You fellows shouldn't admit failure at any point, —keep the campaign open and keep this situation for the present among ourselves."[19]

Cammerer did in fact know a bit more about the faltering national fundraising campaign than he could let on to Chapman, Fink, Morton, Squires, the Davises, and the other boosters. In late 1926 he had gotten busy cobbling together some basic facts to present to potential big donors, cheat sheets on how much land needed to be acquired and how much said land might cost. To park proponent Frank Maloney, he stated: "I have been talking with a number of very wealthy men recently." He asked Maloney and others to quickly compile maps of the proposed park area and estimates of the cost of land acquisitions.[20]

Since April of 1927 Cammerer had been courting Junior and prepping him for a big ask. He had lunched with him at the Whitehall Club in New York City and dangled the bait in a letter stating the Smokies contained "the only primitive forests still to be found in the east" and that these resources were threatened by "the lumberman, who is accelerating his activities." Aware of Junior's involvement in park museums, and his efforts in Jackson Hole, Cammerer also wisely asserted, "From a national education standpoint, this park here will be worth three or four together in the west."[21] Junior agreed to meet with Cammerer later that summer to talk about the Smokies and hopefully gain approval for additional carriage roads in Acadia.

Unfortunately, the Smoky Mountain and Shenandoah boosters had previously bungled the request to Junior by approaching him from different angles over a number of years in a scattergun, dissident manner. Cammerer, too, had only anticipated an ask of $1 million to Junior, to complement what Welch would bring in. All of the confusion risked making the boosters and Park Service look like higgledy-piggledy amateurs. Regarding the delicate situation, Cammerer confided to Chapman, "The trouble is that not a dollar has been paid in yet, nor a pledge entered. It takes weeks to get the psychological moment for an approach. Rich people are temperamental, and an unfortunate approach spoils the prospect as far as that individual is concerned."[22]

Years later, park proponent Carlos C. Campbell would refer to this chapter of Smokies history as "The Darkest Hour."[23] Campbell was correct that, in the falling darkness, it was impossible for anyone to foresee how formidable the synergy of businesspeople, conservationists, public servants, and philanthropists could be.

SIGHT UNSEEN

Great Smoky Mountains, 1927

Arno B. Cammerer, the son of a Lutheran minister who had been forced to drop out of high school in Nebraska and go to work to help support his parents and siblings, moved to Washington, DC, in 1904. Once in our nation's capital, he was able to land a job as a civil service clerk and continue his education. He completed high school and various business courses and went to night school at Georgetown University Law. National Park Service director Stephen Mather and his assistant Horace Albright discovered Cammerer in 1919 after he had worked his way up to assistant secretary for the U.S. Commission of Fine Arts. The odd-duck agency had been created primarily to design the parks and monuments of Washington, DC, but it also advised the Park Service on matters of esthetics and design for facilities in other national parks.

Albright and Mather decided they needed to hire Cammerer as assistant director of the NPS to replace Albright, who had stepped aside as Mather's number two man. The politically fraught position of assistant director had almost destroyed Albright's personal life and sent him on dips into depression.[1] However, Mather wisely convinced Albright to stay on with the Park Service, serving as superintendent of Yellowstone during the summer tourist season and field director for the western parks during the long Yellowstone winters.

To the senior Park Service leaders, Cammerer seemed a highly competent administrator and a man of integrity with experience working budgets and the halls of Congress. Albright described him as "a charming fellow" with "sensitive feelings," who was cultured and refined as well as "kind and considerate."[2] Cammerer also seemed like the type of earnest bureaucrat who would dutifully hold down the fort in Washington while Albright and Mather mixed business and pleasure in the national parks.

As usual, Albright and Mather were correct in their assessment of Cammerer: he was a worker and a doer. Unfortunately, his place in NPS history is sometimes overshadowed by Albright and Mather, more likely a result of the timing of his career than any malice on the part of the latter. Cammerer's credentials were also subdued by the constant friction between himself and the famously callous Secretary of Interior Harold L. Ickes, who reportedly "couldn't stand Cammerer."[3] Cammerer's humble, congenial nature not only irritated Ickes, it also restricted him from crowing his many accomplishments in park creation and management. Too often these major achievements (including Acadia, Shenandoah, Great Smoky Mountains, and other parks and NPS programs) have become conveniently lumped within Albright's and Mather's legacies.

Not surprisingly, like Mather and Albright, Cammerer spent too many hours at his desk, causing his personal life and mental health to suffer. At the time of Major Welch's abrupt resignation from the Southern Appalachian National Parks Committee in 1927, and the subsequent existential park establishment crisis, Cammerer had a wife sick in bed attended by a nurse.[4] He was also serving under (and often filling in for) Mather, who continued to suffer from debilitating bouts of depression.

During 1928, with many park issues boiling over on the stove, Cammerer reported to his good friend (and Rockefeller advisor) Kenneth Chorley, "I have had little sleep this past week, one time working straight through the night and coming to work right after breakfast. So now I go home to lay me down to sleep, and pray the Lord I will still find something left in the bottle in my private cache."[5] He had been frequently hospitalized and prescribed "minor operations and treatments with physicians" for unspecified ailments which were likely blood clots and heart problems.[6] Cammerer would die young, at age fifty-seven.

In August of 1927, as Cammerer prepared to make his crucial pitch to John D. Rockefeller Jr., he was painfully aware that asking the Rockefellers for money was an extremely popular national—even international—pastime.

One of the family's early routines, after breakfast prayers, was for Senior to go through a thick folder of requests for money from people around the world and assign them to his four young children for evaluation.[7]

As previously mentioned, even on the matter of the Great Smoky Mountains, there had already been several independent solicitations of Rockefeller monies. F. Roger Miller, Asheville Chamber of Commerce manager and treasurer of the Great Smoky Mountains National Park Purchase Fund, appealed to Junior in late 1926 for "a subscription which would express your interest in the enterprise."[8] Rockefeller associate Thomas Appleget replied with a polite "not inclined to contribute in accordance with your suggestion."[9] Appleget then ratted out Miller to Major Welch who lambasted Miller in a two-page letter for acting out of turn.[10]

Of the big three in the NPS, Albright had the closest relationship with Junior. Still, there was no question that for the Great Smoky Mountains proposal, Cammerer needed to do the courting. Albright's responsibilities as western field director encompassed 95 percent of the existing parks and monuments, but they did not include the Smokies or Acadia. Fortunately, Cammerer and Junior were not strangers. Cammerer had been highly supportive of Junior's work in Acadia National Park, both the millionaire's land purchases for inclusion in the park and his construction of artfully landscaped motor and carriage roads. Cammerer and Junior had also collaborated with future NPS director Newton Drury and others on various projects to save giant redwood trees. The Cammerers and Rockefellers were in fact becoming close friends, not only because of their common interests in conservation, but also their shared Christian faith which they often mentioned in personal correspondence.[11] There is no doubt the public-private collaboration between Junior and the NPS was as sincere and personal as it was professional and fruitful.

Cammerer and Chorley were especially chummy. "Cam" often invited Chorley to accompany him on trips to the parks and they frequently exchanged gifts such as Virginia country hams.[12] However, as far as funding for the Smokies went, early in 1927, Chorley admonished Cammerer "not to be too optimistic."[13]

The lead-up to the August 4, 1927, meeting in New York reveals that Cammerer and Junior had very different agendas for the conference. Junior had actually requested the meeting back in May to discuss "the matter of which I desired to speak" (undoubtedly roads in Acadia) which had nothing to do with the Smokies.[14] In fact, Junior had grown slightly miffed that a lead NPS landscape engineer had been unavailable that July to review a Rockefeller road project, thereby delaying progress.

After more than a decade of correspondence and in-person meetings with Junior over his plans for Acadia, Cammerer was likely suffering from carriage-road fatigue. Making matters worse, Cammerer's trip from Washington, DC, to New York City for the meeting had to be done with vacation leave from the NPS. His accommodations were reimbursed by the Great Smoky Mountains Conservation Association via Chapman. Although no other use of Cammerer's time would seem more important, fundraising from private individuals was apparently not in the assistant NPS director's federal job description.

The parley at Standard Oil headquarters did not go the way Cammerer had hoped. First, he was made to sit in a waiting room for an hour and a half because Junior was "unusually busy." Rockefeller finally squeezed him in over a hurried thirty-minute lunch. During the repast, Junior spoke "on matters *he* was interested in—then he had to run."[15]

Was Junior truly so busy as to justify leaving the number two person in the NPS cooling his heels in the waiting room for ninety minutes? Junior was fifty-three years old and had been largely retired from the dog-eat-dog world of oil refining and other business interests for some seventeen years. In place of personally toiling to make more money for the Rockefeller coffers, Junior had devoted himself to giving it away. By August 1927, when the future existence of a Great Smoky Mountains National Park teetered on Junior's whims, the overzealous benefactor was indeed swamped by a dizzying array of philanthropic and leadership endeavors.

In addition to the multitude of conservation and philanthropic projects listed earlier, Junior was busy buying swaths of expensive real estate in his Manhattan neighborhood to fend off the sprawl of speakeasys and potentially make room for a new opera house. Inadvertently, this effort led to his development of Rockefeller Center, a construction project compared in scope to the Pyramids.[16] He had also reluctantly made a $700,000 commitment to a project spearheaded by wife Abby to found the Museum of Modern Art in New York City and a $2 million ($36.5 million in 2025 dollars) project to create the Palestine Archaeology Museum. The latter museum rescued countless antiquities that had been left to decay in the wind and sun in the Middle East and established one of the finest archaeology institutions in the region.

In 1927 the cornerstone was laid for the interdenominational Riverside Church in Manhattan's Morningside Heights neighborhood. The landmark structure was part of Junior's quest to unite the diverse Protestant denominations into one, and to lure the progressive Reverend Harry E. Fosdick (brother of Raymond B. Fosdick) into becoming its pastor. Reverend Fosdick had refused Junior's offer to be pastor at the Rockefellers' Park Avenue Church, complaining the congregation was too wealthy. Consequently, Junior spent $4 million building the Riverside's Gothic cathedral in a less exclusive, more diverse neighborhood. Fosdick eventually accepted and preached before the six thousand people who turned out for the opening in 1930.[17] The liberal congregation has maintained its original goal of being "interdenominational, interracial, and international."[18] For many wealthy philanthropists, the creation of, and long-term support for, the Riverside Church in itself would be an outstanding legacy.

Also in late 1927, Junior had reached the point of no return in what would become one of the largest philanthropic undertakings of his career—transformation of a forlorn eastern Virginia hamlet into Colonial Williamsburg. Dr. William Goodwin, an Episcopalian rector bent on restoring Williamsburg's historic Bruton Parish Church, had been wooing Rockefeller's interest since 1924. Junior toured the decrepit town, met with Goodwin on multiple occasions, and slowly warmed to restoring the former capital of Colonial Virginia. He viewed the project as an enduring demonstration of a pivotal period in American history—one that involved George Washington, Thomas Jefferson, Patrick Henry, James Monroe, and James Madison and led to the nation's revolution and independence.[19] It also likely appealed to Junior's perfectionist side and his desire to create timeless, idealized landscapes.

Of all the gargantuan efforts that Junior and his father's money made possible, Williamsburg singularly demanded the lion's share of Junior's attention for the better part of thirty years. "It is my desire and purpose," he communicated to advisor Arthur Woods, "to carry out this enterprise completely and entirely."[20] By the time he called it quits, Junior had spent close to $80 million (topping $1.5 billion in 2025 dollars), tinkered with nearly every aspect of every restoration plan—from materials to gardens to furnishings—demolished 720 of the town's more contemporary structures, and built or revived 500 historically accurate colonial-era buildings.[21] He and Abby also claimed one of Williamsburg's residences, eighteenth-century Bassett Hall, which they quietly inhabited for two months every year.

There was also a very untidy matter that would become known as Teapot Dome, perhaps the greatest American political scandal of all time (at least until Watergate). Junior, as the largest single stockholder in Standard Oil of Indiana, unknowingly became a player in the ham-handed chisel. Even head NPS brass Mather and Albright were ultimately impacted by the affair.

Teapot Dome was a rich geological formation in Wyoming that was part of the U.S. Navy's petroleum reserves. The core of the scandal involved Secretary of the Interior Albert Fall who accepted sizable bribes in exchange for leasing oil rights to a subsidiary of Sinclair Oil and the Pan American Oil Company at extremely reasonable rates. Fall, who was Albright and Mather's boss, was fired and imprisoned. He was the first active cabinet member of a U.S. president (Harding) to be incarcerated.[22]

One subplot of Teapot Dome was an infamous bamboozle now referred to in the annals of Standard Oil history as "the Stewart Case." Perhaps because they needed a slush fund to bribe Fall and others in the Teapot Dome scam, Colonel Robert Stewart, chairman of the board of Standard Oil of Indiana, and some of his cronies colluded to purchase thirty-three million gallons of Texas oil at a fair price, pass it through a phony shell company, then resell it to one of the conspirator's own firms with a twenty-five-cent-per-barrel mark up. The $8 million "profit" ($150 million in 2025 dollars) from the effortless transaction was more than enough to cover the group's bribery bills with plenty of spillover into the pockets of the self-serving oil company mavens.[23]

Surprisingly, especially considering the extent to which John D. Rockefeller Sr. was vilified in the popular press, as well as the father and son's culpability in the Ludlow massacre, Junior gained the moral high ground in the case and was even considered something of a capitalist hero. As Stewart's guilt became increasingly evident, Junior very publicly called for his resignation from the Standard Oil of Indiana board. When Stewart did everything in his power to deny guilt and maintain his lucrative position, Junior rallied the company's other major stockholders and decisively voted him out. Junior's ouster of the crook was well received by political and business leaders and hailed in the press as a victory for the average person.

In conclusion, Junior declared "the highest ethical standards are as vital in business as they are in other relations of life."[24] Rumors of drafting Junior as a New York City mayoral candidate indicate some level of forgiveness, if not redemption, for the Ludlow disaster. The closest thing to redemption for the Rockefellers and their earlier antiunion sentiments—which contributed to the Ludlow massacre—came when Junior went deeply in dept to build Rockefeller Center with fifty thousand union workers during the Great Depression. The enormous construction project was also noteworthy for never being interrupted by a labor strike.[25]

All in all, it is safe to assume that on that August day when Cammerer was left twiddling his thumbs, Junior truly did have a lot of irons in the fire. Fortunately, though he preferred to piddle with the details, he had learned ways to get big things done. Chorley recalled that Junior "had a great capacity...for delegating work."[26] The senior Rockefeller also embraced the merits of delegation early on, which, coupled with the family's ability to attract the best legal, business, and administrative talent, was key to their success.

*

Cammerer, despite the brush-off, reported back to the park boosters that he did get in a word or two about the Smokies during that hurried lunch and he stuffed Junior's briefcase "with all the photographs of the Big Smokies I had collected." Rockefeller assured Cammerer that he was "tremendously interested" in the project and would like the assistant director to come visit him for "a week when we could talk things over."[27]

Cammerer, unfortunately, was also much too busy to spare an entire week for a visit, so the two men decided that Cammerer would write a letter detailing the need for the park and the resources required for its creation. After Junior and his advisers had digested the letter's contents, they would meet again in person.

Writing such a letter to an exacting philanthropist like Junior was no small undertaking, especially considering the entire fate of an eight-hundred-square-mile national park—currently under assault by a dozen termite-like lumber companies—relied on Cammerer's prose. The assistant NPS director first had to gin up some maps and crunch endless numbers, mostly estimates and conjectures from Chapman, Welch, Squires, and others regarding what tracts of land needed to be purchased and what they would cost. He also needed to decouple Shenandoah from the Smokies for the purpose of this specific request, without throwing the Virginia park entirely under the bus.

Regardless of how daunting the task was, only a week elapsed between the disappointing lunch in New York and the day Cammerer sat down with his secretary at her manual mill and began to bang out the letter. As surveyor of the park boundary, the assistant director knew better than anyone the wildlife, virgin forests, and crystal-clear mountain streams that would live or succumb to the axe depending on every slap of the typeheads. He knew from his own experiences, and likely from coaching by Albright, that the Rockefellers were fastidious people. When still in the oil business, Senior ascertained that his five-gallon cans of kerosene could be sealed with thirty-nine drops of solder rather than forty. He estimated the elimination of the superfluous drop of solder saved him hundreds of thousands of dollars over the coming years.[28]

And so it was to John D. Rockefeller Jr., the exceedingly busy, relentlessly solicited, and compulsively parsimonious philanthropist, that Cammerer addressed his letter and affixed the stamp requesting $5 million for a national park in the Great Smoky Mountains. As he was doing so, Cammerer was no doubt painfully aware that the letter's middle-aged New York recipient had never once laid eyes on the place he wished to protect.

ANGELS CAN'T DO MORE

Great Smoky Mountains, 1927–1928

In the first paragraph of his crucial letter to Junior, Cammerer described his own personal commitment to the Smokies campaign: "It appeals to me more than any national park in the West." Then, word by word, paragraph by paragraph, he laid the foundation for why it was essential that there be more national parks in the East. "It is obvious that only a very small portion of those living east of the Mississippi will ever have the time and money" to visit the parks in the West.[1]

Cammerer continued with some valiant prose. He predicted the new park would attract visitors "from the congested centers of population, the workers at the machines in the lofts and mills, the clerks at the desks, and the average fellow of the small towns." Time spent in the Smokies would help these ordinary folk unwind and recharge and conclude that "a country that is reserving such areas and maintaining them for the benefit of its citizens was a pretty good country to live in."[2]

The letter suggests that even in the mid-1920s, Cammerer was concerned that national parks were more available to affluent travelers and adventure seekers than to the bulk of working families. Historians have since fretted that protecting nature and wildlife often served the interests of an "urban leisure class" and penalized rural folk who might have previously used the preserves for hunting and other utilitarian purposes. Albright, Cammerer, and Mather were correct in believing that, especially with the growing popularity of the automobile, big eastern parks like the Smokies and Shenandoah would be more accessible to a broader socioeconomic swath of the public.[3]

Aware of the Rockefellers' attitude toward alcohol, Cammerer then sweetened his letter by promising the establishment of the park would help eliminate the "greatest moonshine distilling section of Tennessee." But the coup de grace was the paragraph in which Cammerer sounded the alarm on the timber cutting in the Smokies' virgin forests. "In my opinion there is an emergency situation," he declared. Cammerer warned that lumber and pulp companies were driving their operations "into the heart of the wonderful primitive forest stands that are the real justification for the establishment of the park." These companies, Cammerer stated, were keenly aware of the park movement and were accelerating their clearcutting in response.[4]

One can assume from the letter that Cammerer knew all about Junior's soft spot for big trees. Cammerer had worked with Junior and advisor Raymond Fosdick on the Bull Creek and Dyerville Flats Save-the-Redwoods League project since 1924. Stories were often told of Junior "taking pains" to save every tree possible as he designed and built his famous carriage roads in Maine and New York, sometimes rerouting an entire stretch of byway to spare a single maple or birch.

Junior's affection for trees went way back. The family's Forest Hill estate near Cleveland was a nurturing place for the shy, awkward boy who often shuddered in the shadow of his father's enormous responsibilities and less-than-sterling reputation. Forest Hill was where he retreated to recuperate from the two disabling nervous collapses that interrupted his youth. His home remedy of chopping wood and performing other types of rote physical labor in the outdoors was practiced well into adulthood.

Junior loved planting trees on the estate and one of his fondest memories was sitting in a large beech tree with his three sisters, each sibling in their own seat, each sharing passages from a favorite book. Forest Hill also featured deep ravines that sheltered old-growth groves of magnificent trees—maples, hickories, oaks, and sourwood. As a hobby, Senior, with Junior at his side, surveyed and constructed carefully crafted paths and carriage roads through the woods, crossing the ravines on arched bridges built of native stone.[5]

In the fall of 1926, when Junior was supposed to be inspecting the ragged remains of Williamsburg, Virginia, and assessing its potential for historic restoration, he was instead drawn into a grove of trees. As one of his guides reported, he "walked into the woods, past the gigantic oak tree which Mr. Rockefeller greatly admired. He enthused over the woods and the autumn paths and foliage and said, "If I come back some day, can we bring our lunch down and eat it under the oak tree?"[6]

*

After receiving acknowledgment from Junior that he had received his letter, Cammerer referred dramatically to his upcoming Manhattan trip paid for by Chapman. He confided in Chapman—who was becoming increasingly agitated by the lack of results from the national campaign—"I have never promised to do more than I can carry out, and my word especially with you fine fellows cannot be lightly given. I'll do my damndest, and angels can't do more."[7]

Junior, however, even before the coming tête-à-tête, had already been swayed. In his carefully crafted letter, Cammerer had checked all of Junior's philanthropic boxes. Junior realized that securing the Smokies was more urgent than creating Shenandoah; Shenandoah lands were not being cut over, but in the Smokies it was essential to buy the lumber company land and not prolong "for one unnecessary day the continuing destruction of primeval forest land."[8] The least impulsive person in the known world wanted the logging of the virgin forest halted, now. Still, his only documented affirmation to Cammerer's letter lacked a solid commitment: "I have read it with interest and am disposed to cooperate in the matter."[9] He also reiterated his desire to have more of his proposed Acadia carriage roads approved. Three weeks later Junior pledged $1 million to $1.5 million in a letter to Major Welch, emphasizing the need for others to pitch in. So even the high end of Junior's pledge left $3.5 million more to conjure.

Junior assigned two of his top aides to the project, Kenneth Chorley and Beardsley Ruml. The two confidants began their due diligence, checking Cammerer's numbers and likely ordering some sleuthing of the principals, including the timber companies involved in North Carolina and Tennessee. Fearing that word of Junior's interest in the project would cause land prices to jump, all parties kept things hush-hush. Even Chapman remained mostly out of the loop. Years later, the Great Smoky Mountains Conservation Association (GSMCA) would recall the winter of 1927–1928 as a "period of despondency, despair, and gloom."[10]

Over the winter, Junior concluded that, with a pledge of only $50,000 from friend Edsel Ford, and no other big donors in sight, it would be up to him to save the Smokies. He requested a second masterful letter from Cammerer, for presentation to the Laura Spelman Rockefeller Memorial (LSRM) trustees, explaining the whole scheme for creating a park from private lands in North Carolina and Tennessee. The letter included much of the same information as the August letter to Junior, but also made the crucial point: "Such appropriations [for purchasing land] have never been granted by Congress for the establishment of parks, and there is no probability of this being done."[11] Junior, who presided over the memorial fund, easily convinced the majority of trustees to commit the entire $5 million. The fact that the memorial was on course to be merged with the general Rockefeller Foundation was likely a factor, as Junior believed the preservation of the Great Smoky Mountains would be a grand and fitting final tribute to his mother.[12]

*

As the first inklings of spring arrived in the Great Smoky Mountains, as trailing arbutus and violet-hued hepatica bloomed on the sunniest slopes, the boosters' darkest hour was about to brighten. Chorley boarded a southbound train, not only to be on site when the $5 million donation was announced, but also to ensure the pledges made by the states of North Carolina and Tennessee were honored. Chorley, who was the most Doberman Pinscher–like of Junior's lieutenants, first checked in with the Tennessee governor, Henry Horton, and provided details on how the state's pledge had been matched by the LSRM. He and Cammerer then proceeded to Raleigh for a meeting with Governor McLean and the North Carolina park commissioners. En route, North Carolina state legislator Mark Squires tipped off the two men that the governor was in the pockets of the big lumber companies and advised them to be prepared for the governor's attempts to worm out of that state's official promise.[13]

When Chorley, Cammerer, and Governor McLean finally sat down, the governor did more hemming and hawing than a tired mule. Squires and Chorley both believed that the park project did not have McLean's "whole-hearted support."[14] The governor requested evidence that all other parties who'd made pledges were holding up their end of the bargain and that the required federal acceptance of the lands as a national park was a certainty. With so much happening so quickly, McLean protested, perhaps they should let the news settle in, then reconvene in Washington in a couple of weeks or so.

Chorley never flinched. He told the governor the LSRM trustees would not have passed the resolutions that rested in the governor's hands unless they were satisfied everything was in order and that all that remained was the issuance of the North Carolina state bonds.[15] Cammerer guaranteed that the National Park Service was anxious to welcome the Great Smoky Mountains into their fold. Governor McLean was not used to being outgunned, but he was experienced enough to recognize when he was. He turned to his attorney general and requested the release of $2 million ($37 million in 2025 dollars) from the state treasury.

Folks in Tennessee were jubilant. According to Chorley, "Knoxville went wild with excitement. It could be compared to nothing but Armistice Day." The declaration evoked "extra editions of the newspapers, the blowing of factory whistles, the ringing of church bells." Everyone from bankers to politicians asked Chorley "to convey to Mr. Rockefeller and the trustees of the Memorial their sincere appreciation."[16]

The *Knoxville News Sentinel* dedicated its entire front page to the Spelman-Rockefeller gift and park origin story. The double-decker banner headline blared "Smoky Mountain Park Assured as $5,000,000 Gift by Rockefeller Memorial Foundation Is Announced." Photos of David Chapman, Tennessee governor Henry Horton, and John D. Rockefeller Jr. ran with the subhead "The Three Men Who Helped Make It a Reality." Articles approached the event from every possible angle, announcing that church bells, school bells, factory and railroad whistles would all sound at four p.m. for five minutes. Another article (though not quite accurately) noted, "The appropriation to the Smoky Mountain Park is the first entrance of the [Laura Spelman Rockefeller] Memorial into the field of conservation of nature." In a burst of optimism, Knoxville realtors proclaimed that property values in their city had just jumped "by fifty percent."[17]

*

Naturally, not all reactions to the announcement were favorable. Letters purportedly written by forlorn mountain families, but actually concocted by lumber company stooges, began showing up at the Rockefeller offices in New York. They warned of corruption and nefarious behavior (including sampling moonshine in the Cosby area) by Cammerer, Chapman, and others within the GSMCA. Junior and his associates, always committed to due diligence, sent attorney Francis Christy down to Asheville and Knoxville to get to the bottom of the accusations.

Christy convened with two local newspaper editors to discuss the letters. A lumber company agent pounced on the New York attorney and baited him with offers of bootleg whiskey and dinner with "luscious maidens." When these temptations were declined, the lumber barons threatened to use their cohorts in the Tennessee legislature to introduce legislation formally discrediting the GSMCA with claims of "irregularities in expenditures."[18] All the underhanded ploys were obviously aimed at scaring Junior into yanking his support from the project. Christy and the propark editors fought back with threats to expose the crooked state legislators as frauds. When Christy returned to New York City, Junior was still committed to funding the creation of the national park and the most stubborn of the timber companies were still bent on cutting more trees.

Champion Fibre was one of the most deeply imbedded thorns in the side of Junior, Chapman, Cammerer, the Davises, and all the other park advocates. Years of negotiation, litigation, trials, appeals, jury tampering, and political shenanigans were not concluded until 1931 when the Rockefeller Foundation organized a conference in Washington, DC. The foundation convinced top National Park Service brass to serve as mediators between the states (who were purchasing the lands) and the timber company. After three contentious days, aided by the considerable muscle of the Rockefeller legal team, the two sides agreed upon a compromise price of $3 million for the spruce- and fir-cloaked heart of the Smokies. The deal gave a huge jolt of confidence to the battle-fatigued park boosters, something they sorely needed as the process of purchasing some eight hundred square miles of land slowly ground on.

Yet, into the early 1930s, John D. Rockefeller Jr. himself had yet to set foot in the Great Smokies. Albright and Cammerer certainly urged him to visit on more than one occasion, Albright proclaiming, "You know how enthusiastic Mr. Cammerer has been about the Great Smokies and I want to tell that I feel I am just as enthusiastic as he is. It is glorious country."[19] It was the autumn of 1934, four months after Great Smoky Mountains National Park was officially established on June 15 of that year, when Junior and Abby finally made it down to the Smokies. In a personal letter to the children, Junior recalled, "We drove down the mountain on the Tennessee side through the most gorgeous country we have almost ever seen. The wealth of rhododendron along the streams and covering the mountainsides is almost unimaginable." According to Junior, both he and Abby were "delighted beyond expression with the extraordinary beauty of the country."[20]

In exchange for the $5 million gift (later increased by $60,000) from the memorial, Junior asked for one thing: a plaque acknowledging his mother and the generous people of North Carolina and Tennessee in the establishment of the park. After a decade of wording and design changes, the monument was finally erected at Newfound Gap, barely in time for the formal dedication of Great Smoky Mountains National Park by President Franklin D. Roosevelt in 1940.

Of the ten thousand or more people who showed up for the September 2, 1940, event—headlined by FDR and First Lady Eleanor—Junior was not among them.[21] His absence was impossible to ignore, especially since the ceremony involved the unveiling of the Founders' Plaque at Newfound Gap, a ridgetop site straddling the North Carolina–Tennessee state line.

Junior had poured his characteristic attention to detail into the creation of the plaque. The wording was revised several times and the Olmsted landscape architecture firm was engaged in its design. Junior, who had by then lost over half his fortune to the 1929 stock market crash and ensuing Great Depression, made numerous suggestions as to the plaque's font and letter kerning.

> For the permanent enjoyment of the people, this park was given one half by the people and states of North Carolina and Tennessee and by the United States of America, and one half in memory of Laura Spelman Rockefeller by the Laura Spelman Rockefeller Memorial, founded by her husband, John D. Rockefeller.

Note, however, that "the people" and states received top billing and Junior's name is not included. This was one of the few instances in which Junior (or the memorial) requested a plaque in recognition of a donation. From the wording, it seems clear the plaque was intended to recognize all who had given for the cause and to serve as a final, tangible tribute to Laura Spelman Rockefeller (Junior's mother).

*

Junior declined personal invitations to the dedication ceremony from Secretary of the Interior Harold L. Ickes, National Park Service director Cammerer, and the governors of North Carolina and Tennessee. Junior told Cammerer he would not attend the event "for various reasons which I need not take your time to enumerate," perhaps having to do with the persistent lack of progress in the Grand Tetons and disappointment with Secretary Ickes on several park matters.[22] As suggested earlier, throughout his long and accomplished philanthropic career, Junior almost never attended a commemoration or similar ceremony. His declination became awkward when the dates for the Smokies ceremony and presidential visit kept changing, yet his unavailability remained fixed. In declining Ickes's invitation over a year before the actual ceremony, Junior did request that the following be read at the dedication:

> May the beautiful spirit of my mother, who was a loyal and devoted wife, a wise and loving parent and an earnest Christian woman, descend upon those who visit these mountain fastnesses, who find refreshment in the shaded valleys and new courage by the side of the sparkling streams![23]

Junior's primary reason for skipping the event is likely best summed up by his lifetime associate Chorley. "He did not want the spotlight," Chorley told writer Michael Frome. "He [Junior] was the most modest human being I've ever known."[24]

Happily, Cammerer did attend. Albright later said of his successor's role in the Smokies effort, "When it comes to discussing who convinced Rockefeller of the importance of the project, and who furnished him with the information necessary for him to make a decision, we must give credit to Arno B. Cammerer."[25] Unfortunately, by then "Cam" had resigned from the Park Service because of a heart attack the previous year. He died of heart failure the following spring. Acadia superintendent George Dorr once said about his own park's gratitude to Junior and Cammerer, "Had you [Cammerer] not then passed favorably upon it [Junior's road plans], there would doubtless be now no ... Great Smokies National Park—at least on its present scale."[26]

Anne Davis attended and sat behind President Roosevelt and the first lady in the V.I.P. section with other park founders. Unfortunately, her husband, Willis, had passed away by then. He did live long enough to celebrate the Rockefeller donation and therefore had confidence that the park he toiled so long and hard to create would in fact be born.

After her husband's death, Anne moved to Gatlinburg, Tennessee, and purchased a house as close to the park boundary as she could find. She spent her days hiking in the Smokies and cultivating a flower garden at her home. As a member of the Gatlinburg Garden Club she helped coordinate the famous week-long Spring Wildflower Pilgrimage, a much-revered event (which still occurs annually) attracting nature enthusiasts from around the world to take hikes in the park led by expert naturalists from area colleges.

Thanks to the Founder's Plaque, the Laura Spelman Rockefeller Memorial donation did not go entirely unnoticed by visitors to the park. For decades after the installation, thank-you notes written by park visitors were still coming into Junior's office in Manhattan.

In 1953 a construction engineer from Columbia, Mississippi, wrote, "A few weeks ago I returned from one of many visits through the Great Smokey [*sic*] Mountains. At this time, I want to take occasion to do what I should have done long ago, that is to thank you for the great gift, made in memory of your mother."[27]

In 1956, a buoyant visitor from La Mollie, Illinois, wrote, "We have just returned from a 10-day trip through 'The Smokies,' and as one golden day of happiness slipped into another, I found myself thinking of those who had made so much beauty available to so many of us."[28]

From Greenwich, Ohio: "This summer my wife and I took our first vacation away from members of our family in twenty-eight years of married life and traveled to the park where we spent several days enjoying the wonderful area. We hope to be able to return there next year."[29]

Each thank-you note received at Junior's office was acknowledged with a gracious reply, either from Junior himself or his assistant Janet Warfield. If the thank-you note writer mistakenly credited Junior with the contribution, he always corrected them by saying the money originated from the Laura Spelman Rockefeller Memorial in the beautiful spirit of his mother.

In 1953, surviving members of the GSMCA put together a celebration of the twenty-fifth anniversary of the Rockefeller gift. The event included this proclamation: "It [the gift] has brought inspiration, relaxation, education and enjoyment to many millions of our citizens. Our gratitude, appreciation and rejoicing for this great and timely gift is constant, sincere and undiminishing."[30]

Junior was invited to that event as well but politely declined. His munificent contribution for the creation of Great Smoky Mountains National Park affirms his full evolution from a large landowner on Mount Desert Island in Maine, who helped create and develop Acadia National Park (conveniently located in his own backyard), to a full-fledged conservationist who agreed to fulfill the dreams of NPS leaders Albright, Cammerer, and Mather to create America's most visited national park. Unlike the circumstances at Acadia, Junior had no particular attachment to the Smokies prior to his donation and had nothing personally to gain from the creation of the park.

While a cynic might assert that creating national parks boosted automobile travel and thereby supported the sale of Rockefeller gasoline, such motives would be incongruous with the history of Rockefeller philanthropy. Their contributions to education, health care, religious activities, conservation, historic preservation, social welfare, and the plight of the poor and stricken would have little direct benefit to their diversified investment portfolio of energy, railroads, real estate, banks, insurance, and government bonds, other than the fact that Rockefeller philanthropy benefitted humankind and all of their investments in some way related to the existence of human beings.[31] More explicitly, philanthropy clearly exasperated and exhausted Junior, just as it had his father. More wealth only required more philanthropy. "What do I want with more money?" Junior asked a reporter with the *New York Tribune* in 1919. "Nearly all my time . . . is devoted to how best and wisely to distribute the money accumulated."[32]

The best explanation for Junior's gigantic contribution (through the LSRM) to protect a place he'd never seen may be revealed in a memo between directors of the memorial. "A Park of this kind will be an enduring project which will be of increasing usefulness as the years pass." Of course, projects that are enduring, economically viable, and increase in value over time, also deserve the attention of philanthropists today.[33]

The park in the Great Smoky Mountains demonstrates that commerce and conservation are not inherent adversaries. Citizens and elected officials fight for NPS sites in their districts because the federal designations dependably create jobs and revenue. Thanks to consistent stewardship by the NPS and the incredible beauty of America's diverse landscapes, our national parks are a brand like no other. The Smokies generates 32,590 jobs and has a cumulative fiscal effect on the area of $3.3 billion annually.[34] The park is in fact much too popular, so that visitors grow frustrated because the parking areas at trailheads, visitor centers, and other destinations are full to overflowing and scenic roads become so congested they resemble parking lots themselves.

There are many other beautiful, wild, biodiverse potential national park areas in the Appalachian Mountains, and all across this great land. The Grandfather Mountain area of North Carolina was first proposed for national park protection over a century ago and Junior had offered to contribute half of the purchase price.[35] For some reason, North Carolina's Mount Mitchell, the highest peak in the East, is not within a national park. Nor is Roan Mountain in Tennessee, Canaan Valley in West Virginia, and numerous other spectacular areas in which park status would protect the mountains' world-renowned richness of flora and fauna, bestow a massive boost to the struggling local economies, and provide for the recreation, relaxation, inspiration, and elucidation of the people.

The Great Smoky Mountains and their characteristic summer haze. Great Smoky Mountains National Park photograph.

Proponents of Great Smoky Mountains National Park in Washington, DC. Front row, left to right: Ben Hooper, Willis Davis, E. E. Conner, David Chapman, Henry Horton, John Noland, James Fowler. Back row, left to right: Kenneth Chorley, Arno Cammerer, Wiley Brownlee, J. M. Clark, Marguerite Preston, Ben Morton, Frank Maloney, Cary Spence, Russell Hanlon, Cowan Rogers. Great Smoky Mountains National Park photograph.

Arno Cammerer holding down the fort in Washington, DC. National Archives.

The Walker sisters lived in the Little Greenbrier area of the Great Smoky Mountains and were famous for their self-reliance. They grew their own food and made their clothes from the cotton, flax, and wool they grew. E. E. Exline photograph. Great Smoky Mountains National Park.

The trail to Mount Le Conte led up and beyond Rainbow Falls. Dick Burns photograph. Great Smoky Mountains National Park.

John Nicholson peeling apples in Shenandoah, 1935. A. Rothstein photograph. Library of Congress.

Scenic overlook in Shenandoah National Park. Virginia Conservation Commission photograph. Library of Congress.

Senator Charles Kramer and Gordon McDonough, a Los Angeles county supervisor, in the Yosemite sugar pines, 1937. Ralph H. Anderson photograph. Yosemite Historic Photo Collections.

Yosemite Lumber Company logging sugar pines, 1926. Yosemite Historic Photo Collections.

Old-growth redwood forest in California. Schuyler U. Bunnel photograph. Library of Congress.

Members of the Bohemian Club in their grove of pristine redwood trees. Arnold Genthe photograph. Library of Congress.

Loggers and fallen redwood tree in Humboldt County, California, 1905. Jesse A. Meiser photograph. Library of Congress.

Scene from the Blue Ridge Parkway near Bluffs (Doughton) Park in North Carolina, 1940. National Archives.

View of Linville Gorge in 1951. National Archives.

View of Linville Falls in North Carolina. National Archives.

The Smokies' first superintendent, J. Ross Eakin (left), and associates in the virgin forest John D. Rockefeller Jr. helped save, 1931. Great Smoky Mountains National Park photograph.

Mount Moran with Jackson Lake in the foreground. Ansel Adams photograph. National Archives.

The restoration of Menor's Ferry was key to resolving the twenty-four-year-long battle for Jackson Hole. John N. DeHaas Jr. photograph. Library of Congress.

THE LINGERING SORROW OF THE DISPOSSESSED

Shenandoah, 1924–1935

Some fifty years after the state of Virginia condemned his family farm in the Blue Ridge Mountains, Reverend John Bradley was still smarting from the loss. "Our home . . . had been in the family for years and years. I don't know of a neighbor up there that was in favor of it [sacrificing home for a national park]."[1]

The Bradley farm, near Jewel Hollow, was larger than most—around six hundred acres—enough to support John, his parents, and John's seven siblings. They grew corn, wheat, rye, and barley and tended a big garden. They raised chickens, pigs, sheep, and cattle. Their log home with frame addition had the unusual feature of running water—a spring that ran through the cellar under the kitchen. It was a marvel to the tourists who began showing up at their place in the 1920s in need of water for their overheated cars and parched throats.

It wasn't that Bradley felt the national park was a bad idea. "The park has benefitted Luray [Virginia], it has benefitted all of the United States." Yet, Shenandoah's creation had also "inconvenienced a lot of people." In particular, Bradley believed the Virginia Commission on Conservation and Development underpaid his father for the family farm. "Had they a-paid what—the worth," Bradley stated, the people who had to sell and relocate "would a been satisfied."[2]

Owners of farms the size of the Bradleys were compensated between $4,000 and $21,170 ($6.67–$34.31 per acre). Records indicate the state commission was not cavalier in its land appraisals.[3] Evaluating the value of a mountain farm was in fact grueling work, not only due to the rugged terrain that needed to be surveyed, but because fewer than half of the 465 families that were affected actually owned the land they lived on. More than half the families worked their absentee landlord's property by an odd arrangement in which they dwelled upon and worked the land in exchange for perhaps minding cattle, fighting forest fires, guarding against poaching by timber companies, and simply occupying the place.[4]

Properties were appraised according to the quality of the homes and outbuildings and the value of the timber, orchards, and cropland. Verdant, relatively flat lands in the vicinity of Big Meadow might fetch fifty dollars per acre; rocky outcrops and near-vertical slopes as little as one dollar.[5] According to former Shenandoah ranger, historian, and author Darwin Lambert, 133 residents asked for arbitration from the courts on the offers made for their lands; some of these were able to negotiate higher prices.[6] There is no evidence of a deliberate effort on the part of the state commission to underpay landowners, although they did have limited funds and appraising mountain property has never been an exact science. Some had to initially rely on records of land acquisitions made in the 1700s that had been surveyed with handheld compasses. It is easy to understand how a landowner whose beautiful, third-generation family farm had been condemned by the state would feel aggravated and inadequately compensated.

Many, including members of the press and even the president of the United States, sympathized with the plight of the mountain families. Reporters were eager to cover the forceful eviction of Melanchon Cliser and family by a reluctant evictor, Page County sheriff E. L. Lucas. Cliser believed the state of Virginia's use of condemnation by eminent domain was unconstitutional (despite a 1935 Supreme Court ruling to the contrary). He refused the state's offer of $4,865 for his forty-six acres that contained both his home and retail business. Lucas had to handcuff Cliser while deputies emptied the home and piled all their furniture and other worldly possessions beside the road. The deputies then boarded the home's windows and doors to prevent re-entry. At least in the case of the Clisers, the evictors did not torch the home as the former owners trudged down the road and looked back over their shoulders.[7]

Photos and the story of the expulsion were widely publicized. "One grows attached to the land which has been owned for generations by one's ancestors, and it cannot be denied that most of us would feel just as CLISER did, if we were dispossessed," the *Richmond Times-Dispatch* noted. Yet, the editorial writer concluded, "the whole state of Virginia, the whole eastern half of the United States will benefit from the Shenandoah National Park."[8]

*

One of the primary movers and shakers behind the creation of Shenandoah National Park was a man whom P. T. Barnum would have greatly admired: George F. Pollock. Frequently described as "flamboyant," he was the owner of the quasirespectable Skyland Resort in the Shenandoah high country. Pollock's neighbors and guests often recall a man of short stature and wildly fluctuating wealth mounted on horseback and blaring a bugle, either to awaken slumbering guests, organize an outing, or simply to enunciate his presence. Locals often climbed the trails to Skyland on weekends to listen to the live music, watch the square dancers, and perhaps partake in the abundance of moonshine likely provided by George Corbin or another highland entrepreneur.

Skyland, situated near the summit of Stony Man Mountain, attracted more-or-less upper-crust guests from Washington, DC, Baltimore, Philadelphia, and New York. Some owned cabins on the grounds as seasonal retreats while others rented from Pollock. A few of the guests lingered for months at a time, escaping summer's heat, humidity, mosquitos, and allergens, as well as the sooty air and invisible pathogens (polio in particular) of the cities. On work trips, Arno Cammerer often stayed at Skyland and seemed to enjoy Pollock's eccentric company, hospitality, and knowledge of the area. Pollock employed a small army of locals, as well as those from farther afield, to staff the resort. For the dining room, where Pollock (sans wife) shared a table with his dog (seated in a chair), the owner employed professional waiters of African American descent. Most of the locals worked as landscapers, handymen, firefighters, guides, launderers, or housekeepers. They also made baskets and other crafts and sold them to Skyland guests for much-needed cash. In addition, Pollock purchased a good amount of food from local farms and hired skilled tradesmen for stonework and other construction projects. Although some of Pollock's employees complained he was stingy and overly strict, he was at least tolerated—if not liked—by others, both for his paychecks as well as his antics.[9]

Pollock's support for the national park was characteristically impulsive. He was ambivalent toward the prospect until the creation of a national park in his mountains became a competition against the Great Smokies, Grandfather Mountain, Tallulah Gorge, and over a dozen other scenic areas in the southern Appalachians. Suddenly, his limitless energy for promoting Skyland was transferred to the creation of Shenandoah National Park. Such was his devotion that he felt no visible remorse for exaggerating nearly all of the area's attributes when completing the government's park eligibility forms. According to Pollock's prose, Shenandoah's mountains were higher, its forests older, and its landscape safer from wildfires than any other sources had previously documented.[10]

Along with Pollock, the boosters behind the movement to create Shenandoah were almost carbon copies of those pushing for a Great Smoky Mountains park in Tennessee and North Carolina. Both efforts were born in the mid 1920s and were led by energetic businesspeople more concerned with commerce than conservation. Both jumbles of promoters led members of the Southern Appalachian National Park Committee (the group created by Stephen Mather to evaluate potential parks in the region) on selective tours of their most scenic landscapes. Both the Smokies and Shenandoah had a difficult time raising money to purchase the private lands necessary for park creation and both had to scale back their ambitions to fit their bank accounts, especially Shenandoah.

Pollock's crusade was joined by eminently powerful Governor Harry F. Byrd of Virginia and William E. Carson, chairman of the Virginia State Commission on Conservation and Development, which, despite the seemingly paradoxical title, was created to do exactly what its name implied. Through the dollars generated by the burgeoning tourism industry, Carson eventually proved that creating Shenandoah National Park was a good example of boosting tax revenue through the preservation of Virginia's rich and diverse natural areas.[11]

Pollock, Carson, and Byrd, along with National Park Service brass Albright, Cammerer, and Mather, soon realized that Shenandoah and the Smokies were not only competing for the first-place nod from the Southern Appalachian National Park Committee (SANPC), they were also vying for donations from John D. Rockefeller Jr. and other potential benefactors.

Much to the chagrin of the Smokies advocates, Shenandoah won the first round with the SANPC, not for its superior scenic value, but because it was close to Washington and its more modest topography made road building and other development more feasible. The SANPC blue ribbon, however, only counted as a recommendation to the Department of the Interior, and included no prize money.

Originally, the national fundraising campaign commanded by Major Welch was intended to benefit both Shenandoah and the Smokies. However, the Shenandoah campaign bogged down at several points, hampered by legal challenges (all the way to the Supreme Court) and debates on how to handle the purchasing of land and relocation of residents.[12] Meanwhile, the situation in the Smokies became increasingly urgent. Recognizing their leverage, the big lumber and paper companies threatened to expedite timber cutting if they weren't paid exorbitant prices. The only way to stop them from clearcutting the last of the virgin timber in the Smokies and then selling off the denuded land was to aggressively negotiate and purchase. At the very last moment in Major Welch's national fundraising campaign, the NPS and Smokies boosters decided to decouple Shenandoah from the Great Smoky Mountains and divert the Rockefeller donations to the Smokies. While the residents of Knoxville were celebrating with the blowing of train and factory whistles and the tolling of church bells, Pollock, Carson, Byrd, and the other Shenandoah advocates realized they had just been hung out to dry. The Shenandoah boosters had hoped for $2 million from Welch's national campaign. In March of 1928, Governor Byrd, obviously perturbed, sent a terse congratulations to Cammerer: "Delighted to see that the Great Smoky Park has been provided for. I hope the Shenandoah Park is progressing satisfactorily."[13]

Cammerer did much of the heavy lifting in the national campaign to raise funds to purchase lands for Shenandoah. He asked well-positioned friends to "jot down a few names of those who in your broad experience and contacts you might consider favorably inclined toward a contribution."[14] His office sent out over a hundred four-page personal letters to well-to-do persons beyond Virginia's borders. In the appeals, Cammerer recycled some of his best lines from his effective letter to Junior two years earlier regarding a contribution to the Smokies. "It [Shenandoah National Park] will add to the happiness of our people, enabling the individual and the average

American family to get a breath of outdoors...and a realization that our country is a pretty good place to live in after all."[15]

Junior helped with a donation of $163,631 ($3.7 million in 2025 dollars) and convinced his buddy Edsel Ford to part with another $50,000.[16] The amounts satisfied Junior's original pledge to Shenandoah, but he did not offer to go above and beyond, as he did in the Smokies, apparently because there was no imminent threat to Shenandoah's forests. As philanthropists, the Rockefellers had always favored partnerships, but Junior was willing to switch to "whatever it takes" mode when a potential park's essential natural or cultural resources were at immediate risk of irrevocable harm.

Even prior to the October 1929 stock market crash, very few from America's other wealthiest families rose to the cause. The Ball Brothers of Indiana contributed $5,000 and Thomas A. Edison parted with $250.[17] Cammerer bemoaned the shortfall, warning, "Virginia is facing failure in this park project." The underperformance forced the advocates to scale back their ambitions in terms of Shenandoah's acreage and to consider settling for "the minimum area that would in any way be acceptable."[18] The deficit in the national campaign, along with higher-than-anticipated costs for purchasing cut-over lumber company lands and other properties, caused further trimming of the park boundaries. From an initial goal of 521,000 acres (roughly the size of Great Smoky Mountains), the Shenandoah boosters were forced to settle for 181,000.

*

Shenandoah's ultimate salvation (like its SANPC blue ribbon) came from its proximity to Washington, DC. President Herbert Hoover, a native Iowan and farm-raised Quaker, was lured to the Blue Ridge Mountains by Carson. His bait for Hoover was trout fishing, the president's passion, along with the prospect of a peaceful riverside retreat, close enough to Washington, DC, to be convenient, yet far enough away to feel like the antidote to high humidity, politicking, and city living. Carson and Madison County even paid to punch a crude road into the Rapidan River site so Hoover could see it with his own eyes.[19] The ploy was successful; the Hoovers fell in love with the place (now called Rapidan Camp or Camp Hoover) located west of Culpeper, Virginia, in the middle of today's Shenandoah National Park. The Hoovers were enchanted by the river, the landscape, and the mountain people. So much so, Herbert tapped his own bank account to purchase 164 acres and outfit the rustic "Country Whitehouse." First Lady Lou Henry Hoover remarked that her husband "likes to be near enough to hear the water murmuring."[20]

The president and first lady visited their camp often. Lou rode horseback far and wide on the mountain trails and visited neighbors while the president menaced the "speckled mountain trout." As a publicity stunt, a boy named Ray Buracker gifted the president with a live opossum. The Hoovers reciprocated with a first-rate elementary school which they built and staffed on their own dime.[21] And even though there was still no Shenandoah National Park, Hoover ordered work to begin on the federally funded Skyline Drive, a beautifully engineered scenic ridgetop road that absorbed some of the expense of buying land for the park, employed local tradespeople, and bolstered the number of park proponents.

Also because of the Hoovers' presence in the mountains and the widespread awareness (in Washington, DC, and beyond) of the displacement of the Shenandoah families, several important concessions were made in the process of dispossession. Over forty of the older "and especially meritorious" residents were granted lifetime leases by the secretary of the interior to remain in their old homeplace. The privilege had many strings attached, however; most importantly, their farming activities were sharply curtailed so their fields and pastures could be rewilded. Descendants of the leaser were also deprived of inheriting the land, a serious disruption to traditions of family farm succession.[22]

Another 170 residents, some of whom held no deed to their Shenandoah lands, were moved to federally funded resettlement communities outside the park. In some instances, Civilian Conservation Corps enrollees used their trucks to help with the move. Despite objections from then senator Byrd, who called the communities "a permanent monument to . . . [government] waste and extravagance," funding was provided by various New Deal institutions, including the Farm Security Administration and Federal Emergency Relief Administration.[23] President Franklin D. Roosevelt allotted $900,000 for the resettlement in 1935.[24] An unusual feature of the resettlements was many offered beautiful, nostalgia-invoking views of the mountains from which the resettled had just been exiled.

Still, all-in-all, the communities, consisting mostly of small, brand new homes, were pretty decent. The settlements offered hobby-size farms for families who wished to keep a few chickens or dairy cows and raise a big garden while holding down wage-paying jobs in town. There were also larger tracts for families who chose to continue pursuing agriculture as their primary occupation. The land for the resettlements was purchased mostly from farmers who sold their lands to the government through options (during the Great Depression, when prices for agricultural products had plummeted).[25]

There were seven resettlement communities in all constructed for Shenandoah's dispossessed. Those who moved in had a fair chance of being familiar with their new neighbors, either as extended family members or old friends or acquaintances from the mountains. The most unusual of the resettlements was in Greene County and was modeled after a socialistic European cooperative. Families lived in homes clustered together from which farmers departed each morning to work in their jointly managed fields. Likely because mountain folk typically prided themselves on their individualism, the Greene County commune was the least successful of the seven communities.[26]

The resettled were charged a reasonable rent, starting at around $5 ($114 in 2025 dollars) per month and had the option to purchase their new abodes with attractively financed mortgages. According to the *Staunton (VA) News-Leader*, the homesteads would cost at least $2,000 for a house on fifteen acres. "Ten of the fifteen acres were to be good crop land," with the remainder composed of pasture and forest.[27]

Dorothy Housh, whose husband, Chester, was employed by the Farm Security Administration as a resettlement community manager, spent a lot of time with the women and children in the new communities. "I just talked with the people, exchanged ideals," Housh recalled. She described the residents as "real generous good-hearted people," who were happy "their children were getting an education." It's also clear, however, that they retained deep attachments to their mountain homes. Many transplanted boxwood shrubs from their old home to the new. "They cut sprigs of boxwood off and stuck them in potatoes, and brought them down." Some of the resettled stayed in their new homes for the rest of their lives and some moved. Vacated homes were sold on the open market, so the communities gradually became less artificial and more mainstream.[28]

The Great Depression and the election of FDR as president (who chose Catoctin over Camp Rapidan as his weekend getaway because of his limited mobility), ironically, opened up a slew of opportunities for displaced mountain families. Because of their self-sufficient rural-living skills and knowledge of the mountain landscape, many were hired by the Civilian Conservation Corps as foremen or skilled tradespeople.[29] Construction of the Skyline Drive, the scenic ridgetop road that has become Shenandoah's hallmark, also offered good-paying jobs with private contractors.[30] Some former residents found jobs with the NPS as park rangers or maintenance workers or were absorbed by the rapidly growing tourism economy.[31]

Despite government efforts to soften the blow of dispossession and New Deal economic opportunities, many mountain families remained less than enthusiastic about losing their farms. James G. Burner, whose Shenandoah roots went back to the eighteenth century and who found work with the CCC and the NPS post-removal, recalled that for older residents, losing their land "was tremendously hard on them. There was no way they could make a new life for themselves." However, Burner also believed "the younger ones—yeah it was good. It was the best thing that ever happened to future generations."[32]

Ralph Cave, who was forced out of his home in Dark Hollow, expressed similar sentiments. Dispossession "really hurt" many of his neighbors. "They didn't like it at all." Yet he admitted the prospects for the small farmer working marginal mountain land, even without formation of the national park, were dimming. "The people made it all right then, but I don't think they could make it in there now."[33]

Better schools would prove to be a sound reason for younger families to leave Shenandoah. Miriam Sizer, a Virginia woman who taught vacation school for two months at Old Rag, reported her classes were held in a church with "no patent desks ... no maps, charts, or library books except those which I provided."[34]

*

In retrospect, it's easy to note many of the mistakes that were made in the creation of Shenandoah, as well as other parks like Great Smokies and the Blue Ridge Parkway where eminent domain and mass dispossession were used. "There was a lack of communication between the State and the Federal government and the mountain people," Burner recalled.[35] Poor, opaque communications meant hearsay and rumors of malicious behaviors were rampant. Generally speaking, the state did not offer the kind of long conversations and public meetings that would have allowed residents to air their concerns and for compromises to be forged. Residents never clearly understood their options and the benefits of national parks for current and future generations were never fully explained.

The methods for assessing land values could have also been openly discussed and debated. Land purchasers should have erred on the high side for compensating owners since they were not just purchasing their homes, they were taking their livelihoods, communities, and ways of life. A little more cash up front might have saved money in the long run and reduced the legal expenses and decades of resentment of parks and the federal government by descendants of the dispossessed.

It also would have behooved the land buyers to be more generous with lifetime (or long-term) leases that would have eased the transition for older and less adaptable residents. The purchasers of city land for Colonial Williamsburg successfully used a "life tenure" system in which the property was purchased (with Rockefeller monies), then leased back to the owner for one dollar per year. The agreement endured for the lifetime of the owner.[36] Future park developers should consider arrangements employed by New York's Adirondack State Park and parks in Canada and Europe in which a limited number of residents and businesses are adopted as long-term components of the park. Although this sets up enduring frictions between park in-holders and park managers, it is preferable to the bitterness created by mass dispossession in some cases. Park planners should also accept that dispossession and resettlements should not be hurried.

Things that were done right include lifetime leases for older residents and the resettlement effort, especially the assignment of quasi social workers like Chester and Dorothy Housh, who offered assistance to the resettled. Little if any such support was offered the people displaced by Great Smoky Mountains National Park. Doing so in an organized, timely fashion might have expedited creation of the park by reducing the number of lawsuits and extended negotiations.

For perspective, it is worth noting that the painful displacement of families to make way for the new national parks was much smaller in scale than the exodus forced by the federally incorporated Tennessee Valley Authority (TVA), which began in 1933. TVA's purpose was to build dams and create massive reservoirs for flood control, improved river navigation, and the production of hydroelectricity, all intended to uplift the economy of Appalachia. The organization, which represented just one of many mammoth dam-building efforts during the era, often used eminent domain in its removal of over seventy-two thousand residents from their farms and riverside towns. TVA also used several New Deal programs, like the one employed near Shenandoah, to ease the transition for displaced families.[37]

One should also acknowledge that, in most cases, mountain farms were not large enough to be divided among all the children. Consequently, by primogeniture and the limitations of the small-farm economy, most of the children would eventually need to leave. Also, it may come as some consolation to the descendants of the dispossessed, that, unlike most other Americans, they may have the opportunity to return to the lands once owned by their ancestors and see them rewilded, perhaps to a state resembling the beautiful landscape first witnessed by their settler forebearers.

*

As with other national parks described in this volume, even the most optimistic grassroots advocates, philanthropists, civic boosters, and state and federal governments underestimated the immediate and long-term popularity of their new parks. As early as 1935, before Shenandoah had even been officially established, the *Richmond Times-Dispatch* crowed, "Shenandoah National Park, in its tremendous number of visitors during the past summer and autumn, has shown a popularity which even its sponsors did not anticipate. Already it has the greatest drawing power of any such park in the United States."[38]

It's important to remember that, like the Great Smoky Mountains, Shenandoah was conceived by businesspeople and elected officials seeking a tourist attraction to stimulate jobs and commerce. Most Virginians today would find it hard to imagine their state without Shenandoah National Park and Skyline Drive. The park is a permanent enrichment of their lives, which bolsters pride in their mountain landscapes and cultural heritage. People in Washington, DC, and elsewhere in the mid-Atlantic states apparently feel much the same. Each year, visitors to the park spend $104 million in local communities which supports 1,240 jobs with a cumulative economic benefit of $145 million.[39]

And even though Albright and Cammerer may have been disappointed that Junior didn't pony up more than the $163,631.05 he had pledged for

Shenandoah (and there was some heated debate as to the actual sum of his original pledge), they sent a heartfelt thank you upon its completion: "Entirely aside from the substantial financial support you have given in the past years . . . there has been nothing that has given us such courage to carry on our work as the sympathetic understanding and cooperation, so generously given by you."[40]

THE SUGAR PINES' LAST STAND

Yosemite, 1928

While Kenneth Chorley was in Knoxville, Tennessee, presenting the $5 million check needed for the creation of Great Smoky Mountains National Park, he also carried a letter from Junior to personally deliver to Arno Cammerer.

> February 28, 1928
>
> I have been reading in the papers lately about the possible further destruction of some of the trees in the Yosemite Valley. Is there any prospect of saving these trees? What are the facts in the matter? Is this a situation to which I might well give consideration?[1]

Even before the Smokies was a done deal, the ever-curious Junior had stumbled across another national park crisis in need of a conservation superhero. Between February 21 and February 26, 1928, the *New York Times* published at least four articles or editorials regarding the cutting of sugar pine trees in faraway Yosemite National Park. The first article was both an exposé implicating National Park Service director Stephen Mather for failing to fend off the California lumber companies and a bombshell revealing the felling of magnificent trees *inside* a national park—adjacent to some famous groves of giant sequoias! The smartly reported story, with the subhead "Park Officials Powerless," declared, "Logging within the park area has proceeded rapidly, lumber interests evidently fearing the public would discover the destruction of the park forests."[2] The article, published without a byline, made that fear of discovery become real.

On February 22, the *Times* published a second article on the topic, quoting John Muir, the granddaddy of California conservationists, at length: "No traveler, whether a tree lover or not, will ever forget his first walk in a sugar-pine forest.... [They are] the largest, noblest and most beautiful of all the seventy or eighty species of pine trees in the world, and of all the conifers, second only to 'King Sequoia.'"[3]

Two days later, Mather, an accomplished corporate public relations guy before becoming Park Service director, spun a statement apparently designed to shift blame away from himself and his agency and weigh it upon the shoulders of the apathetic public. Mather declared that past efforts to save the Yosemite trees "have failed to arouse the nation." He maintained the Park Service could not do the saving alone and the public's assistance was necessary to help the NPS acquire the privately owned lands (inholdings) within Yosemite.[4]

The *Times* journalists, as well as letter writer Willard G. Van Name, a renowned scientist and conservationist affiliated with the American Museum of Natural History, were not exaggerating; Mather and the Park Service had gotten themselves in a serious pickle. Van Name accused the Park Service of hiding the inholdings from Yosemite tourists and letting the "lumbermen have all the sugar pines." He criticized Mather specifically for concealing the actions by burying documentation in fine print within the supplements of his annual reports.[5]

The sugar pines (*Pinus lambertiana*) were a rather uncommon species, growing mostly in the mountains of the Sierras and Cascades. They attained heights of over two hundred feet and lived up to five hundred years. Bears and birds favored their sweet cones for sustenance. According to Albright, the Yosemite sugar pines were "the best remaining in the world and should never be cut."[6] While clearcutting of the pines had been going on for years in Yosemite's inholdings, it was only now that tourists and journalists were peeking behind the thin roadside veneers of intact forest and expressing dismay at what lay behind the curtain. No doubt the recent failure by environmentalists to save Hetch Hetchy Valley, also in Yosemite, had something to do with the volume of the outrage.

Sugar pines were not only noble giants, they were also valuable commodities. Per board foot they fetched a substantial premium to yellow pine, the second priciest of the commonly logged trees in the Sierras. Their phenomenal heights and massive volume—second only to "King Sequoia"—meant there was a lot of standing lumber per acre. Another catalyst for cutting was the convenient proximity of the big sugar pines to essential logging infrastructure, including roads, railroads, and mills.[7]

The roots of the conflict can be traced to Yosemite's erratic origin story that involved a tug-of-war between the state of California and the federal government resulting in boxy, arbitrary, and highly fluid boundary lines. Because the borders had been originally drawn with a pencil and straight edge on a crude map without much consideration for private land holdings, topography, or tourist access, the modern boundaries frequently needed to be adjusted. Excisions made to boundaries by opportunistic mining interests also muddied the picture. These revisions inevitably created inholdings: islands of private lands that frequently vexed the superintendents tasked with managing their parks for the benefit of current and future generations. In Yosemite, in 1928, most of these malignant inholdings were the property of big lumber conglomerates such as the Yosemite Lumber Company.[8]

In a confidential memo to Junior, Albright summarized the situation thusly: "As far back as 1912 the private land situation had become an acute problem." The NPS and Yosemite Lumber Company had already completed several timber exchanges which saved trees along Wawona Road but sacrificed other parklands in their stead.[9] Most of the timber company lands inside Yosemite were on the park's west side, south of Hetch Hetchy. Others were beyond current park borders but on lands proposed for inclusion under proposed boundary adjustments; still others were both inside and outside the park along the routes of new roads in the planning or construction phase, particularly the Big Oak Flat, Mather, White Wolf, and Harden Lake roads. This checkerboard of properties made the possibilities for negotiations nearly infinite and the need for expensive surveys, timber cruises, and legal services equally unbounded.

Cries of outrage were heard in the press and the tourist lodges of Yosemite as early as 1925. Soon they were echoing in the government offices of Sacramento and Washington, DC. Yosemite tourist Sidney H. Buckham took snapshots of one of the inholdings in its post-logging state and railed in a widely circulated letter, "The cutting already has ruined the once beautiful setting of the Merced Grove of Big Trees [sequoias], and the company [Yosemite Lumber Company] figures on thirty more years of logging that vicinity." Famously colorful Carl "C.B." Bachem, an executive with the Yosemite Lumber Company, forwarded the photos and letter to Yosemite superintendent W. B. Lewis, labeling Buckham and his ilk "cranks and agitators."[10]

Park aficionado and nurseryman Roy Elliot from Los Gatos, California, wrote his state congressman to complain that the Yosemite Lumber Company was "cutting an average of 450,000 feet of logs per day in the park" and "leaving behind a barren desert." He correctly assumed the park's sugar pines would soon be decimated save for a remnant screen along the roadsides. "Is it necessary for the richest Nation in the World to destroy such an asset for the few dollars the lumbermen pay for stumpage?"[11]

In late 1926, May W. Mount reported in the *New York Herald Tribune* that Yosemite tourists traveling the Wawona Road need only wander "a few rods on either side, to see vast wastes of denuded land crisscrossed by abandoned logging railroad beds and enormous stumps that bore witness to a thousand years of growth."[12]

As the public outrage grew, it eventually stirred the Park Service from its resigned complacency. While Mather considered Yosemite his favorite national park by far, and Albright was almost equally devoted, both had surrendered themselves to the fact that the inholdings were owned by the logging companies and would eventually be logged (but then turned over to the Park Service). Yosemite's acting superintendent Ernest P. Leavitt and NPS director Mather and field director Albright had to do quite a bit of archival digging and consultation with federal solicitors before they could wrap their arms around the complexities of the situation. Their conclusion was that when Yosemite's stewards had fated the conflict decades earlier, their hope had been that the NPS, U.S. Forest Service, and timber companies would negotiate a land swap that would spare the Yosemite sugar pines in exchange for timber harvests on less revered National Forest lands far, far away. In the spring of 1928, Albright advised Junior: "It seems to me that the Department of Agriculture ought to cooperate in this matter but should its officers refuse to do so within a year the lumber companies will doubtless begin their operations."[13]

One would assume that Albright was well suited for negotiating just such a land and timber exchange. He was a smart, affable University of California alum with a law degree and life-long ambition to work with mineral rights and mining claims. He initiated the campaign by paying a visit to the head of the Forest Service, Major Robert Y. Stuart, and did his best to be friendly and grease the skids for future negotiations. This was a prickly task since the Park Service and Forest Service fought bitterly over turf in the Grand Tetons and elsewhere. As Albright recalled late in life, "From the moment an independent Park Service was organized, the Forest Service was jealous of it and never failed to fight it whenever their land was involved."[14]

In the matter of the Yosemite sugar pines, Albright supposed "the Forest Service would prefer not to give up their holdings, although willing to do so if pressed."[15] Yet the Forest Service had numerous options for dodging any deal that would be conceived by the Park Service. The Forest Service had even planted a type of "poison pill" into an old agreement that allocated 25 percent of proceeds from Yosemite timber harvests to the counties from which the trees were chopped.[16] Not harvesting the sugar pines would therefore deprive the citizenry of significant revenue and consequently the generous provision conveniently turned the locals' sympathies toward the Forest Service. A similar caveat held up the expansion of Grand Teton National Park for at least a decade.

The timber companies, however, presented a much more formidable barrier to an amicable land swap. A slump in timber prices and the high cost of lumbering in arduous terrain had left two of the companies that owned Yosemite inholdings insolvent and the third strapped for cash. Since the federal government required companies to post significant bonds and make other expenditures that they couldn't afford as part of any exchange, the timber cutters offered every excuse (other than the truth) to undermine negotiations. And then there was the ultimate deal breaker: the Forest Service claimed not to have enough harvest-ready timber equivalent to the Yosemite sugar pines with which to barter.

By the spring of 1928, it had become obvious to Mather, Albright, and the leadership staff at Yosemite that all negotiations for a land exchange were "deadlocked."[17] Yet the clock was ticking. The Yosemite Sugar Pine Company hoped to start falling the remaining trees within weeks, purportedly because of the season's increasing fire danger.[18]

Albright appealed to the timber company executives, stating that "the National Park Service had its back up against the wall," mainly because of "the many editorials that have appeared in papers all over the country, demanding that the *forests* of Yosemite be saved."[19] To Mather, Albright remarked that if not for the public outcry, the deal offered by the lumbermen to selectively cut the best sugar pines (under Forest Service supervision) and leave a screen of remnant giants along the roads would have been "tremendously attractive" to the NPS.[20]

Fortunately for the sugar pines, Junior had given the articles in the *Times* more than a casual glance. From their years of work together, Albright and Cammerer were well aware of Junior's love for trees. They may have even known that Junior's tree-lover roots came from his father, the great industrialist and monopolist. Though Senior was a busy man, he, like his son, was also good at delegating and managed to take off "two or three afternoons each week" to putter in the woods.[21] His self-proclaimed avocation was as an "amateur landscape architect" defined as one who enjoys "the art of laying out [carriage] roads and paths." In his autobiography, Senior stated, "I have spent many delightful hours, studying the beautiful views, the trees, and fine landscape effects." He also recalled that at his Pocantico Hills estate, "All the old big trees were personal friends of mine."[22] A longtime archivist at the Rockefeller Archive Center in Pocantico Hills, New York, related that in Rockefeller lore, Senior would "disappear for hours" to wander about in the forests of the sprawling estate.

Senior spent a good deal of his valuable time overseeing the transplanting of trees around his estates and bragged that the survival rate of his transplanted trees was 90 percent greater than that of other tree relocators. He recoiled against losing an elm tree at Forest Hill in Ohio because of a public sidewalk. "Please let me know if we can save the tree," he urged.[23]

Consequently, after reading the stories in the *Times*, Junior dashed off the aforementioned letter to NPS associate director Cammerer proclaiming his availability to embark on another punishing quixotic crusade. It's noteworthy that Junior initiated the conservation effort in this instance, rather than being called upon by Albright, Cammerer, Dorr, or Mather. Junior had also initiated the roadside improvements in Yellowstone in 1924 and set the high standards for roads in Acadia. One could assert that Junior was setting a higher bar than even the NPS brass in the effort to save the Yosemite sugar pines and landscape Yellowstone and Acadia roads. This remarkable willingness by Junior and the NPS leadership to either follow or lead in their collaboration is one of the reasons their conservation partnership was so productive.

Cammerer quickly responded to Junior's offer and proposed that he and Albright come to New York in a couple of weeks to explain the dilemma and "lay the [Yosemite] exhibits before you."[24]

FRIENDS IN HIGH PLACES

Yosemite, 1928–1930

As the battle to protect Yosemite's beleaguered sugar pine trees wore on, conservation journalists around the country continued to turn up the heat on the park's appointed stewards. Secretary of Interior Hubert Work wrote a letter to Secretary of Agriculture William Jardine (overseer of the Forest Service) and stated: "There is a situation in Yosemite National Park which I believe demands our personal attention at this time."[1]

Secretary Work envisioned four possible solutions to the sugar pines threat, most of which had already been attempted to no avail by Work and Jardine's underlings. Two included land swaps, one was to persuade Congress to buy the land, and the last was to encourage a rescue by a wealthy patron of the parks.[2]

Of the choices, Work favored trading the Yosemite sugar pines for other trees on Forest Service lands outside the park. His communication was buttressed by a letter from U.S. congressman Louis C. Cramton of Michigan, a friend to Albright, Mather, and the national parks. Known for his "patience, persistence, and persuasion" Cramton had made a long speech to the House of Representatives in 1924 on protecting the Grand Canyon from private interests.[3] "The Grand Canyon belongs to the world," Cramton boldly declared.[4] As a member of the House Appropriations Committee and chairman of the Subcommittee on Interior Department Deficiencies, his letter to Chief Forester Colonel W. B. Greeley, and copied to the president and the cabinet-level secretaries of interior and agriculture, had some heft. Succinctly, the letter pressed the Forest Service to conjure up timber somewhere for a swap.

The Forest Service again responded that they didn't have comparable timber to trade, and that even if they did, what about the 25 percent of gross timber proceeds owed to Yosemite's neighboring communities?

In May, Junior was once again stirred by an editorial in the *Times*. Although two of the lumber companies had agreed to a "ceasefire" for the 1928 harvest season in Yosemite, a third, the Sugar Pine Lumber Company, announced plans to begin cutting immediately. The writer conceded there was no point in waiting for the Forest Service to make a deal since they were in the business of harvesting trees, not preserving them for esthetic purposes. Consequently, the only hope was to acquire the land for the government.[5] An impassioned Arthur N. Pack, son of a wealthy timber baron and founder of the American Nature Association, declared: "Every outdoor-loving man and woman must cry aloud with pen, typewriter and voice, 'Save the Yosemite!' and perhaps it can be saved."[6] In another *Times* article on the same day, the editorial writer, quoting a story by Robert Sterling Yard (editor of the *National Parks Bulletin*) as well as Pack, suggested the first step needed to be an independent assessment of the value of the lands and timber.[7]

Also on that very same day, Junior ordered an associate to send a copy of the *Times* editorial to Cammerer and remind him and Albright that he was ready and willing to do what he could to stop further sugar pine cutting. Cammerer responded with a personal and confidential five-page memo (likely written by Albright and Cammerer) outlining the whole "deplorable" history of the Yosemite private inholding problem. The memo revealed the startling information that the Forest Service and Park Service had previously worked out a plan to exchange timber company inholdings in Yosemite National Park for other inside-the-park lands that would be given to the Forest Service. Those former parklands would then be harvested by the lumber companies (though under more conservation-minded Forest Service timber management policies).[8]

According to the memo, that plan only faltered because "newspapers all over the country took up the plea to save Yosemite Park forests." Considering this and the failure of secretaries of agriculture and interior to push through a deal, Cammerer listed three possible ways forward. Unfortunately, he labeled the likelihoods of the first two passing as "remote" and "most highly improbable." The third had a fighting chance: "Congress may be persuaded to appropriate 50% of the funds needed provided that the other 50% can be obtained through private donation."[9]

Cammerer, who at the time was once again filling in for Mather as acting NPS director, then laid the bulk of the responsibility for saving Yosemite's sugar pines upon Junior. Cammerer requested a survey be made "as quickly as possible" with the aim of determining "an honest, fair valuation of the timbered acreage involved." Like a Gotham police chief pointing his emergency spotlight toward the clouds, Cammerer requested "the financial assistance of a public-spirited citizen greatly interested in the Yosemite forests. The Service is now awaiting word as to whether or not this cooperation will be extended and the survey made."[10]

Using the old Standard Oil network developed by Senior, Junior and his top associates—Chorley, Raymond Fosdick, Thomas M. Debevoise, Vanderbilt Webb, and Arthur Woods, among others—responded to the call. They selected the San Francisco practice of McCutchen, Olney, Mannon, and Greene, the largest law firm west of Chicago, to manage the survey. The firm's task was formidable: analyze the vague boundaries of the private inholdings and the value of the timber, investigate the health of the timber companies, and scrutinize the sometimes-conflicting legislation that had been passed decades prior regarding timber rights. The work called for supreme accuracy and expedience, and of course it would cost a fortune.[11] McCutchen, Olney, Mannon, and Greene, working in conjunction with Wall Street attorney Webb, hired the best surveyors and timber cruisers in the region and immediately dug into the task.

During the same summer, Secretary of Interior Hubert Work resigned to help with Herbert Hoover's presidential campaign. Work was replaced by Roy O. West, who had little interest in national parks and was more-or-less a stranger to Albright and Mather. With their careers thusly dangling in the political winds, Albright and Mather, along with Chorley and other interested parties, embarked on a tour of the Yosemite inholdings. Along the way, Albright formulated what he believed to be a crafty plan to resolve the sugar pines conflict. By far the most valuable stands of sugar pines were owned by White & Friant, one of the two overextended timber companies teetering on the brink of bankruptcy. Albright schemed that if the government could buy the White & Friant timber rights first (for an estimated cost of $1 million), they would save the largest and oldest trees and encumber access to most of the other timber by the Yosemite Lumber Company and Sugar Pine Lumber Company.[12] The Yosemite and Sugar Pine companies would then be forced to surrender their holdings to the government at bargain basement prices.[13] And although Chorley concurred with the strategy initially, he later warned Arthur Woods and Junior that "this is the kind of undertaking that Albright would perhaps not handle as well as he would other matters of less size. You may remember how we ran into some of his shortcomings in the Jackson Hole matter when that grew in size beyond what he was used to."[14] It should be noted, however, that Chorley was also critical of the way almost everyone handled almost everything.

When Junior was informed of the White & Friant timber purchase plan, he concluded that immediate action was necessary. To Chorley and Woods, Junior pledged "not only the million dollars, but whatever sum may be necessary, even if a second million is involved for the purchase of these lands," as long as his contribution remained anonymous.[15] When asked if Junior would also pony up the 25 percent of timber revenues that would theoretically be owed the neighboring communities (even though their timber was not cut), Chorley replied that he "didn't think there would be any objections."[16]

Negotiations for the White & Friant lands went on through the summer and fall with company president John R. Ball and his representatives squaring off against Vanderbilt Webb, Chorley, Cammerer, Mather, and others. An apparent sticking point was the timber cruise carried out by subcontractors Thomas & Meservey of Portland, Oregon, which concluded the value of the timber was significantly less than the 1927 assessment cited by White & Friant. Communications related to the potential deal were so sensitive that Yosemite staff used coded messages to communicate with NPS brass and Junior's people. Chorley, for example, was referred to as "TYELROP" and the Forest Service was "SELOHD HOLJATO."[17] While the various parties were working to iron out the discrepancies in timber value, John R. Ball abruptly announced he had sold the key White & Friant lands to Yosemite Lumber Company (YLC). The deal gave the YLC the option to cut the mother lode of sugar pines themselves or leverage the government to pay an exorbitant amount to prevent the carnage.[18]

Attorney Webb was livid. His scathing fourteen-page letter to Ball went over every detail of the negotiations, including personal promises by Ball to Mather, and concluded that Ball had been deceitful toward the Park Service and Junior the whole time and never planned to sell them the land. In November, Chorley wrote Debevoise that they had been promised by Ball that "he would not negotiate with anyone else for the sale of this property before the end of the year." Chorley estimated losses of $15,000 ($277,000 in 2025 dollars) due to the unnecessary bargaining and left open the possibility of a lawsuit if negotiations with the timber companies collapsed.[19]

It's likely the YLC had seen Albright's maneuver coming a mile away. Since the YLC, Sugar Pine Company, and White & Friant had been competing in the same market for many years, each knew the others' strengths, weaknesses, and strategic advantages. With the plan's implosion, the NPS and Junior lost the leverage they had hoped to gain and, worse still, were forced to negotiate with YLC president Arthur H. Fleming, a wealthy timber baron.

Herbert Fleishhacker, a prominent California businessman, told Albright that Fleming was "one of the hardest men he had ever had to deal with," that he was "first, last and all the time a trader." According to Fleishhaker, Fleming once "spent a week trading for $1.50."[20] As negotiations for preserving the thousands of acres of Yosemite sugar pines proceeded, Fleishhaker's assessment proved prescient. By the summer of 1929, reports were coming in of Fleming cutting timber in the park not far from the Merced Grove of giant redwood trees.[21]

Around the same time, in Wyoming, another of Albright's land acquisition strategies had backfired. Albright, who had first suggested that Junior purchase land in Jackson Hole for an addition to Grand Teton National Park, also recommended he do so surreptitiously with the use of a shell company ("a recreation and hunting club, or some other hypothetical organization"), eventually named the Snake River Land Company.[22] The purpose of the subterfuge was to prevent a run up in land prices which the Rockefeller name might provoke. It was also designed to dampen the risk of an upswell of antipark sentiment among locals and the Forest Service. During the fall of 1928, there was discussion among Junior's advisors of using the Snake River Land Company to purchase the Yosemite timberlands. That idea was squelched by Debevoise, who disclosed, "We have recently received word . . . that our [Rockefeller] connection with the Snake River Land Company is suspected. If the suspicion grows and publicity follows, the California negotiations might be seriously affected. . . . I have never liked the idea of using the Snake River Land Company and learning of the suspicion mentioned above only confirms my feeling that it would be dangerous to do so."[23]

Fortunately for Yosemite's sugar pines, Congressman Cramton and the NPS were already working on Cammerer's option number three. In fact, Cramton and Work's actions, which led the Forest Service to formally reject the land swap option, cleared the way for pushing option three through Congress. If Cramton did not have documentation that the Forest Service had already rejected the logical timber exchange choice, Congress, who thus far had always refused to buy lands for national parks, would likely have handily rejected the fifty-fifty public-private tender.

Cramton, who represented a state void of national parks, was, nonetheless, one of Albright and Mather's strongest allies. He frequently shared train cars, automobiles, and pack animals with Albright, Cammerer, and Mather on inspection tours of the expanding park system. He had long fulminated against private inholdings in national parks and made their surgical removal one of his committee's priorities. Unfortunately, Cramton's tenure in Congress was near its end; he would soon be unseated by a candidate who favored the repeal of prohibition, something Cramton stubbornly opposed.

While Congress refused to purchase lands for new national park lands (as evidenced in the Grand Tetons and Great Smokies), the well-positioned and passionate Cramton did scrape together enough votes to permit purchase of some of the private lands festering inside existing parks with the contingency that federal funds be matched dollar for dollar with private contributions. Declaring "there is nothing of greater importance to our national park system today" and "the most acute situation in the national parks is in Yosemite," Cramton persuaded his fellow representatives to appropriate up to $3 million to eliminate inholdings.[24] Albright called the bill "one of the most difficult pieces of work" ever achieved.[25]

Unfortunately, Cramton's bill was sidetracked in the Senate by Senators Thomas J. Walsh and Burton K. Wheeler, both of Montana, and—low and behold—both owners of family inholdings in Glacier National Park. An epic debate ensued with the two senators squaring off against Cammerer and fellow NPS administrator Arthur E. Demaray on whether the government should be able to condemn private lands in national parks. The

Montanans' other bone to pick in the appropriations bill was securing funding for the completion of the Going-to-the-Sun Road, the main route through Glacier National Park, which was also awkwardly tied to resolving private inholdings very close to the senators' properties on picturesque Lake McDonald. The *New York Times* astutely called out the senators' ethics, stating, "The question narrows itself down to whether we should have a privileged class who may own land in our national parks or whether such lands shall be bought by the government and placed under the same regulations as the rest of the parks, with special privileges for none." The *Times* could have also mentioned that Walsh, in other circumstances (including when he investigated the Teapot Dome scam), was noted for defending public lands and properties from corporate abuses.[26]

One of Albright's motives for getting a bargain on the timber rights was to fend off likely criticism directed toward himself and Mather by the California press and members of Congress.[27] In 1928, the vast majority of Americans had never visited a national park and the muckraking press was always on the lookout for stories of elected officials squandering taxpayer dollars. When negotiations between the timber company and San Francisco law firm commenced, Albright labeled Yosemite Lumber Company's opening gambit of $3.6 million as "preposterously high." And though he vowed not to "give an inch," to appease the timber interests, his stance softened considerably when his attempt to work directly with White & Friant was undermined.[28]

The NPS-Rockefeller team learned that Fleming had paid approximately $2.8 million for all the lands the conservationists hoped to protect. Therefore, they offered Fleming $2.8 million. Fleming asked for $6.2 million, citing the increased cost of harvesting his timberlands outside the park caused by the protection of the parklands. He then lowered his price to $5.4 million, and his son-in-law hinted that he might go for $3.8 million. Junior and his associates offered $3.5 million. Fleming countered with an offer of $10 million and the rights to cut all the sugar pines. Junior replied that $3.5 million was his best and last offer.[29]

During the summer of 1929, Albright immediately sounded the alarm after touring the Yosemite timber with Bachem, American Forestry Association president George D. Pratt, and others. "Section 20 is only vital area to be saved now [stop] Next week section 20 will be ruined. He [Fleming] also violated his agreement by cutting 64 acres in Section 24." In regard to several hundred acres of timber that Fleming claimed to own, but which the party failed to locate, Albright decried, "We suggest Fleming is running a monstrous bluff."[30]

In the meantime, public outrage over the cutting of the increasingly rare sugar pines within Yosemite National Park grew louder still. The American Nature Association, through its *Nature Magazine*, "sent out the call to its 120,000 members." The *New York Times* and Sierra Club continued to cry foul. *The Live Oak*, a California "nature monthly," declared: "If these areas in Yosemite are not preserved, future generations will not know the Sugar Pine."[31]

The brutal negotiations, primarily between A. Crawford Greene of McCutchen, Olney, Mannon, and Greene of San Francisco and Fleming and his attorneys dragged on through the end of 1929 and into 1930. The case was complicated by the lumber company disputes of timber values and Cramton's stalled legislation. The countdown had definitely started since Cramton's bill needed to pass through Congress by June or the whole deal could end up in the woodchipper. As unfruitful discussions persisted, and the legal fees mounted, the NPS and the new secretary of interior, Ray L. Wilbur, leaned toward condemnation.

In June of 1929, acting NPS director Arthur E. Demaray met with the Department of Justice to tentatively pursue condemnation of Yosemite timber lands. The justice department informed Demaray that an intimidating amount of research and documentation would be required before court proceedings could begin. Demaray responded by passing the document requests to Chorley. Condemnation looked less and less appealing as local experts advised the Park Service that the California courts would likely sympathize with the timber company and award a hefty premium on top of the true value of the land and lumber.[32]

Shortly thereafter, Junior learned that the $30,000 ($554,000 in 2025 dollars) he had approved for the survey and legal counsel had been nearly expended and another $20,000 would be required to do the technical work. Junior approved "further sum as may be necessary."[33]

*

On January 22, 1930, Stephen Mather died suddenly from a heart attack. Although the first director of the NPS had been out of commission for nearly a year following a severe stroke, most people believed he would recover, at least enough to enjoy a well-deserved retirement. Mather's career had been deeply scarred by the mental illness that was no fault of his own, yet he was still the colossus of the national parks and his unselfish contributions and accomplishments would never be equaled. Everyone who knew Mather knew that Yosemite was his favorite national park. He had given hundreds of thousands of dollars from his own pockets to see that it was protected, appreciated, and enjoyed.

Six months prior to Mather's death (after the director's stroke and retirement), Junior wrote a touching and insightful tribute to Mather that could serve as a eulogy:

> Before I came to have the pleasure of knowing you, I had met a number of the superintendents of national parks—among the first, Mr. Nusbaum, later Mr. Albright and others. Thus some years ago I had come to realize what extraordinarily fine, able, unselfish men had been brought into the park service. Knowing how meagre the salaries which these positions carry, I wondered that men of such high caliber had been attracted to them. This was made clear to me when I came to know you and to learn something of what you have been doing in the department of the parks these many years. These young men have been drawn into the service because of their admiration and affection for you and because of the fine example of unselfish public service which you have set them. The result of all this has been that there is perhaps no other department in the national government run on so high a plane and so wholly in the interest of the public which it serves as the Park Service.[34]

In his will, Mather remembered Albright and Cammerer generously.[35] His long-time colleagues and friends would succeed him as the second and third directors of the NPS.

America had Mather to mourn, as well as the collapse of its economy. In the fall of 1929, the stock market crashed and the Roaring '20s dissolved into the deepest economic depression anyone could imagine. Unemployment was on its way up to 25 percent and Junior's fortune was reduced by at least half.

*

The first rays of hope appeared in Yosemite's evergreen future when Fleming let his option to buy expire on 2,080 acres of virgin timber known as the Steinmetz inholdings. Albright immediately plunked down a deposit and Junior agreed to foot his 50 percent share of the $250,000 required to buy the lands.[36] Then, most providentially, Albright discovered that two well-known and upstanding Californians, Herbert Fleishhacker and Paul Shoup, were major stockholders in the Sugar Pine Lumber Company which had been acquired by Fleming. Fleishhaker alone owned $200,000 in Sugar Pine Lumber stock. He was a wealthy banker and San Francisco philanthropist who served on the local park commission and was a trustee for Stanford University. Shoup was president of the Southern Pacific Railroad and a founding board member for Stanford's school of business. As Albright reported, both were "intimate friends" of Secretary of Interior Wilbur and neither wanted or could afford "to be mixed up in adverse publicity." Upon learning of the timber cutting by the Yosemite Lumber Company near the Merced Grove, Shoup picked up the phone and gave Fleming a jingle.[37]

Following Albright's contacts with Fleishhaker and Shoup, Yosemite superintendent Charles G. Thomson reported a sudden change in Fleming's deportment.[38] Records show Fleishhaker and Fleming met several times in San Francisco while negotiations were underway.[39] Attorney Greene, in a decisive maneuver, suggested all parties convene for a final showdown in Washington, DC, in early March 1930. This allowed all the attorneys, as well as representatives from the Park Service and Forest Service to negotiate face to face and pore over their reams of maps, deeds, and timber cruise ledgers. To Fleming's delight, the Park Service finally dropped all pretense of a timber swap, sparing the lumberman the large cash deposits such a government deal required. Soon thereafter, Fleming accepted an offer proffered by the San Francisco attorneys, not of $3.5 million, but $3.15 million, a sum that had been significantly reduced thanks to a more thorough timber cruise sponsored by Junior.[40] In an interview conducted several decades after the fact, Greene claimed it was his bluff of condemnation, conceived of during a private meeting with Secretary Wilbur, that ultimately caused Fleming to cave. However, knowing the difficulties associated with condemnation, it is more likely that the concerns of Fleishhaker and Shoup tipped the scales.[41]

Assisted by pressure from the *New York Times*, the American Nature Association, and a plea to "all conservationists" from Arthur N. Pack,[42]

Cramton was finally able to secure $5 million in the 1930 Department of Interior appropriations bill for matching funds to purchase private lands (inholdings) within Yosemite, Glacier, Rocky Mountain, Lassen, Mount Rainier, Crater Lake, and other parks. Legislation also confirmed the federal government's right to condemn private lands inside national parks. Not surprisingly, the powerful Montana senators Walsh and Wheeler were able to hang onto their private lodges in Glacier and see construction proceed on the Going-to-the-Sun Road.[43]

On April 14, 1930, President Herbert Hoover signed the bill expanding the boundaries of Yosemite National Park and saving the crucial sugar pines. All told, 11,818 acres (18.5 square miles) of old-growth forest were saved. Junior's final share of the expense was $1,709,238 ($31.7 million in 2025 dollars).[44] As it turned out, his due diligence and attention to detail paid off. He and the U.S. government later split a refund from Fleming for $119,790 when their 1928 check cruise of White & Friant lands proved they had been overcharged.[45]

Most importantly, many of the majestic sugar pines remained standing beside the Tioga Pass Road and in other areas that are highly visible to today's park visitors. After the agonizing defeat that John Muir and other Yosemite lovers suffered with the flooding of Hetch-Hetchy Valley in 1913, the successful sugar pines rescue helped reinvigorate America's nascent conservation movement. It also proved that when individual citizens, conservation organizations, government, and the press flexed their collective muscles, they could help make the difference in conservation crusades.

Yosemite's new superintendent C. G. Thomson could not have been more pleased. "I am at a loss in this attempt to express my appreciation," he told Junior, "of your wonderful gift to Yosemite and to present and future generations of Americans." As to the peoples' response to the news of conifer salvation, Thomson related, "I have never known such warmth and spontaneity in public reaction."[46]

The *New York Times* gave credit for the rescue first to Junior. "Thanks to Mr. John D. Rockefeller, Jr.... the splendid sugar pines and other stands of virgin timber within the boundaries of the park will be saved." The article also stated: "To Representative Louis C. Cramton of Michigan and to Mr. Horace M. Albright, Director of the National Park Service, are due thanks for having engineered the passage of the necessary laws and having conducted the delicate negotiations with the owners of the property which the park has now obtained."[47]

Albright was also effusive with his praise: "We feel that he [Junior] has done a lion's share in working out what is perhaps our worst private land problems."[48] And in his annual report to the secretary of the interior, Albright declared, "It is impossible to overestimate the importance of this Yosemite forest acquisition."[49]

As always, Junior volleyed the compliment straight back to Albright. "I am constantly impressed by your own patience, foresight, and breadth of vision in matters of this kind," Junior wrote. "My congratulations on what you have accomplished."[50] Harry S. Truman's adage, "It is amazing what you can accomplish if you do not care who gets the credit," was certainly proven by Junior and Albright's consistent deferral of acclaim.

The fight to save the sugar pines reinforces some important aspects of the collaboration between Junior and the NPS. One is the significance of the legal and administrative support provided by Junior and his associates. Just as Junior was called upon to help manage and bankroll the tedious deed research necessary for the establishment of Sieur de Monts National Monument in Maine, the Yosemite sugar pines would never have been rescued without the expensive expertise of lawyers, accountants, timber cruisers, consultants, and other professionals recruited and paid by Junior. As ANA president Arthur Pack stated, "Government red tape forbids the expending of funds" for a decent timber survey.[51] The government's weakness was Junior and his associates' strength, and this conservation symbiosis is just as viable today.

Another component is the importance of persistence. There were a lot of heavy hitters opposed or indifferent to the expansion of Yosemite National Park, including the secretary of agriculture, much of the U.S. Congress, well-connected congressmen who owned inholdings in other national parks, and several powerful timber barons. Albright and Junior suffered defeat after defeat in the early going yet never let up on their effort. The resolve of Junior, who never relented or compromised once he made up his mind to accomplish something, matched by highly determined NPS leaders, is what made the partnership such an extraordinary example for future conservationists to consider.

The fight to save the sugar pines was a relatively short one, but a lot had changed over the course of the two-year rescue. The dream of a park to protect the Great Smoky Mountains had become a reality at great expense to the Laura Spelman Rockefeller Memorial, and the stock market crash of 1929 had substantially reduced the wealth of Junior and his fellow Gilded Age tycoons. Consequently, in the years and decades ahead, the depth of Junior's commitment to conservation philanthropy would be tested. Likewise, the partnership between Albright and Junior would soon be substantially altered.

A HELL MADE FROM HEAVEN

Grand Tetons, 1927–1933

Following Junior's pledge to Horace Albright to purchase most of Jackson Hole and donate it to the federal government for a national park, the Rockefeller juggernaut quickly prepared to take action. Before approaching the first landowner, however, Albright advised Junior and his associates Kenneth Chorley and Arthur Woods to form a bogus front organization that resembled a hunting club or cattle company.[1] Albright reasoned the ruse would conceal Junior as the purchaser and avoid the inevitable runup in prices and potential obstructive behaviors. It also obfuscated the federal government as the eventual owner and manager of the lands, a prospect that many independent-minded ranchers and other locals found repugnant.

Albright also suggested the Salt Lake City law firm of Fabian and Clendenin as the primary land acquisition attorneys for the Snake River Land Company (SRLC). Albright had known Beverly Clendenin since they were kids and the two had been college classmates. Albright considered Harold Fabian and Beverly Clendenin as "men of unimpeachable integrity and judgement." Albright also maintained the two would "act in this matter on a very narrow margin of cost."[2] As it turned out, Fabian assumed a lead role in the project and became one its most prodigious and charismatic participants. Although the creation of the SRLC was suggested by Albright, it was similar to land acquisition practices used by Junior for the creation of Colonial Williamsburg and Acadia National Park. In those circumstances, Junior hired local realtors and attorneys to buy up properties that were for sale or, when the property was not on the market, contact the property owners, assess the values, make offers, and when amenable, close deals. In Williamsburg, Virginia, he sometimes purchased historic homes, then rented them back to their elderly residents for $1 per year for the rest of their lifetimes.[3] He delegated all day-to-day administrative tasks to his associates Chorley and Woods who supervised the transactions and kept Junior apprised of progress and roadblocks.

The subterfuge in Jackson Hole, which began in 1927, not only misdirected private landowners as to the goal of the SRLC, it also kept the U.S. Forest Service in the dark. Albright correctly forecast the Forest Service would do everything in its power to block the national park once they caught on.[4] He also knew the sympathies of many of the local ranchers leaned toward the Forest Service (a Department of Agriculture agency), which allowed livestock grazing, firewood cutting, and hunting on its lands.

Although the Snake River Land Company's purported purpose was surreptitious, its manner of mass land acquisition was softer than methods used by the states of North Carolina, Tennessee, and Virginia for the creation of Great Smoky Mountains and Shenandoah National Parks and the Blue Ridge Parkway. In those instances, the states had the option to use eminent domain—or the threat of eminent domain—to force reluctant landowners to sell. Such strong-arm tactics used by the states in the Appalachians inevitably led to some resentment of the NPS and federal government that festered for generations.[5]

The manner in Jackson Hole, while sneaky, was more akin to an ambitious rancher extending his rangeland or a developer snapping up private properties for a new commercial enterprise. It also clearly resembled the old Rockefeller pattern of expanding their country estates by purchasing neighboring tracts of land as they became available. And the SRLC probably paid landowners more generously than these other potential acquirers would have. According to Chorley, dude ranchers were offered lifetime leases allowing them to stay on their land and run their property while the SRLC paid their taxes.[6] No doubt the dude ranchers, livestock growers, owners of tourist businesses, and summer homeowners were deceived into thinking they were selling out to a large sportsmen's club or cattle operation rather than an eastern millionaire bent on expanding a national park; yet their transactions were voluntary and, most often, beneficial to the seller.

According to the *New York Times*, "Mr. Rockefeller instructed his agents when they went into the Teton country to deal 'most liberally' with the owners of the property. . . . The land in question has little value for homesteading purposes, and much of it had been abandoned by the original settlers."[7] It is also important to note that during the 1920s the local newspaper ran a long list of properties advertised for sale because of nonpayment of taxes.[8] Between 1923 and 1932, the *Jackson's Hole Courier* listed thirty-four foreclosures on agricultural properties in Teton County. Besides the short growing season and arctic-like winters, occasional withering droughts also made farming and ranching in Jackson Hole tenuous.[9]

As the land purchasing commenced, Fabian reported that many landowners "took the money and went out and got themselves ranches that were in a little more favorable ranching climate." Other Jackson Hole residents, however, dug in their heels.[10] Geraldine L. Lucas, a divorced mother from Iowa, homesteaded one of the most scenic (and isolated) spots in the Rockies: a flat, open, wildflower-filled meadow on the banks of a meandering stream directly at the foot of the Grand Teton. Albright described the property as "the most beautiful place in the country" and advised Fabian its 428 acres would be a bargain at $50,000 ($924,000 in 2025 dollars).[11] To the Snake River Land Company's increasingly generous offers, Lucas responded, "You stack up those silver dollars as high as the Grand Teton and I might talk to you."[12]

Lucas's steadfast refusal to sell her sanctuary to any entity resembling the federal government continued until 1938 when she died from heart failure at age seventy-two. Her only child inherited the mountain paradise and almost immediately sold it for $60,000 to J. D. Kimmel, owner of the Kimmel Kabins resort and a would-be developer of a sprawling community of homes in the vicinity of picture-perfect Jenny Lake. Fortunately, Kimmel and Fabian were becoming friends and one day the two decided to take a drive into the Gros Ventre Mountains on the far side of Jackson Hole. Kimmel was well aware that Fabian and the SRLC desperately needed his properties for the establishment of the national park, that his lands were in fact the crème de la crème of the Grand Tetons, and that without his cooperation his subdivision would become one of the most egregious inholdings in America's entire national park system. He also knew he was sitting on a gold mine.

Yet westerners are a different lot. Kimmel, who had made his fortune trading oil royalties, and who always wore two loaded pistols on his belt, had also recently befriended park superintendent Charles "White Mountain" Smith and his family. No doubt hanging out with the Smiths influenced his attitude toward national parks. As Fabian and Kimmel motored along in Kimmel's open Packard, winding through the gently contoured Gros Ventre Mountains and catching sublime views of the Grand Tetons across Jackson Hole, Kimmel asserted: "Fabian, I can ruin your whole damn project."

"I know you can," Fabian replied.[13]

The men drove on, smelling pine, dust, and cattle in the distance. When the laconic Kimmel felt that Fabian had ruminated sufficiently over his threat, he spoke again. "But I ain't a goin' to. I believe this is a great thing you're doing." He then offered to sell the SRLC his land for exactly what he had paid for it, as long as he could retain his store and cabins on twenty acres at Jenny Lake.

Fabian replied, "I don't need to ask anybody, it's a deal right now."[14]

The next summer, when Junior was in Jackson Hole, he insisted on thanking the Kimmels personally. He, Fabian, and Chorley stopped by their cabin on Jenny Lake. "Mr. Kimmel, I want to thank you particularly, because you are the only man in Jackson Hole that has offered me tangible help for what I am doing. I can't tell you how much I appreciate it."[15]

Some journalists, politicians, and disgruntled citizens recall instances in which the Snake River Land Company used coercion to force reluctant landowners to sell. One case was Homer C. Richards, a barber who wandered into Jackson Hole with his family in 1924. Richards was fortunate enough to land a job guiding tourists on boat and horseback trips in the vicinity of Jenny Lake and eventually homesteaded 320 acres not far from the Lucas and Kimmel properties on free land as equally gorgeous as his neighbors'.[16] "People laughed at me for wanting that homestead," Richards recalled later. "They said the soil was too rocky to raise anything. But I knew what I was doing."[17] He built tourist cabins and a barbershop on the shores of Jenny Lake. When the Snake River Land Company offered him $25,000 for the land, Richards countered with a price of $35,000.[18] As far as scenery and outdoor recreational opportunities go, one would be unlikely to find a more valuable place in the world, and today, even coins stacked as high as the Grand Teton could not buy 320 acres on Jenny Lake.

In order to persuade Richards to sell, Fabian sent a friend out to Jenny Lake to spread false rumors about a reroute of the Jenny Lake access road that would "throw the present collection of cabins and stores off the highway."[19] The ploy proved to be unnecessary, however. During the negotiations, Richards became a close friend of the Albrights.[20] According to later interviews with Richards, it was this friendship, rather than the threat of being marooned by the rerouted road, that convinced the entrepreneur to sell for $25,000 ($460,000 in 2025 dollars).[21] Even at that price, the *Jackson Hole Guide* noted in the 1970s that the long-lived Richards had made a "healthy profit" off the sale.[22] "That was quite a bit of money in those days," Richards affirmed.[23]

The historian Robert W. Righter notes, "There is little evidence to suggest that this type of cooperative subterfuge was widely used, yet it does help to explain why the Snake River Land Company and the National Park Service made enemies in Jackson Hole. And for every documented incident there were ten unsubstantiated rumors."[24]

With cash in hand, Richards moved his family to the town of Jackson and started the Ideal Motel at the base of Snow King Mountain. The first drive-up tourist motel in town, it included twenty-five cabins and was therefore a much larger operation than the one he ran on Jenny Lake. Tourism continued to flourish in Jackson over the coming decades and Richards sold the Ideal and bought the Flame Motel as well as partial interests in several Jackson landmarks, including the Wort Hotel and Cowboy Bar. He lived into his nineties, was always photographed in a clean white Stetson hat, and became the town's most beloved booster. Richards served on the town council for a dozen years and enjoyed membership in every civic organization from the chamber of commerce to the Masons, Elks, Rotarians, and Legionnaires. He was also involved with founding the Jackson Hole Museum and the town's popular rodeo grounds.[25]

Richards's friendship with Horace and Grace Albright lasted for at least forty-eight years.[26] As an early visionary in the area's tourism industry, he also became a major supporter of the national park. Looking back, Richards claimed he would have given his Jenny Lake properties to the Park Service "rather than see it developed."[27]

"I could see the Park coming," Richards said, "and I felt it was a good thing. If the Park wasn't made, the road between Yellowstone Park and Jackson would be lined with junk and people wouldn't be able to enjoy the mountains at all."[28]

Other instances of coercion or perceived coercion may have been facilitated by Robert E. Miller, president of the Jackson State Bank, and former U.S. Forest Service supervisor. Miller was perhaps unwisely hired by the Snake River Land Company to head up the land purchases, partially because he had allegiances to the Forest Service and because, according to Fabian, he "really hated Horace Albright" and the Park Service.[29] Also, Miller's bank owned the mortgages of several of the properties targeted for acquisition, giving Miller unfair leverage in negotiations. Miller's compensation also incentivized him not to overpay for properties. However, Miller, a longtime resident and rancher himself, was sympathetic with most of the landowners he was dealing with. The ranchers, especially the most downtrodden, knew Miller because he and his bank had carried them through hard times. "They were very faithful to him," Fabian recalled.[30]

For the most part, landowners were offered a fair or more-than-fair price for their properties.[31] This was facilitated by the fact that prices for cattle and other agricultural products had cratered after the end of World War I and that slump continued into the 1920s. Junior and Albright felt fair payment was necessary both to move the process along as swiftly as possible and to prevent sowing seeds of ill will. According to Fabian, Junior had always said he would rather "pay more than less" for lands he was purchasing. Junior did not want any of his philanthropies "to be predicated on the misfortune of other people."[32] It also stands to reason that landowners generally don't voluntarily sell their properties and move unless they perceive they are receiving a good offer.

Albright, though certainly biased, remembered a discussion with longtime rancher Bill Crawford in 1920. "He said nobody would make a living on these properties, that the climate was too cold, the soil too barren, and that people were destroying the lives of themselves and their families by trying to ranch in this country." Struthers Burt, in his *Diary of a Dude-Wrangler*, noted that "the first farmers came into my valley about ten years ago and today are broken and ruined men."[33] In 1925, ninety-seven Jackson Hole landowners signed a petition favoring a buyout of their properties by state or federal governments for the creation of a recreation area.[34]

Also in 1925, President Calvin Coolidge formed the Coordinating Commission on National Parks and Forests to settle a number of boundary disputes between the Forest Service and Park Service. The commission looked at multiple ways to extend Yellowstone National Park down into the Tetons, but the Forest Service bitterly opposed most options.[35] In 1929 a compromise was struck which gave the Park Service ninety-six thousand acres in the Grand Teton Mountains, but none of the lakes at the base of the mountains. The minipark only served to protect the vertical portion of the area, an alpine realm that had no practical economic use (other than tourism) and therefore really required no protection anyway. Historian John Ise called the 1929 park "stingy" and "skimpy" because the boundaries ended near the eastern shores of Jenny, String, and Leigh Lakes, leaving these gemlike remnants of the last ice age ripe for commercial development.[36] Jackson Lake and the Snake River were almost entirely outside the park.

*

Any sense of temporary, semiresolution to the Jackson Hole conflict was shattered in the spring of 1930 when the rumor mill finally churned up the Rockefeller connection to all the recent land purchases. In an attempt at damage control, Fabian decided to preemptively reveal the wizards behind the SRLC curtain to the press. The April 10 edition of the *Jackson's Hole Courier* proclaimed, "Snake River Land Company Officials Break Secrecy."[37] Other than a handful of conservation-minded locals, the majority of Jackson residents were outraged by the disclosure. It was as reflexive as swatting a horsefly for hardworking ranchers in the American West to be infuriated when they learned they had been hoodwinked by the federal government *and* a starchy New York millionaire. Suddenly the premium prices they had been paid for insolvent ranch land seemed grossly inadequate. More importantly, the men and women in the pointed-toe boots and Stetson hats saw their picturesque way of life suddenly wavering and dimming like a desert mirage.

The South Park Community Women's Club of Jackson passed a resolution opposing expansion of the new national park, as did the Jackson Lions Club. The women's club resolution called on other organizations to likewise "assist in protecting Wyoming homes and communities from further aggression of private interests and eastern capitalists who, under the guise of philanthropic motives are securing all the concessions within the national park."[38]

The fury at the deception of the SRLC was as strong or stronger beyond the valley of Jackson Hole as within. "Dubois [Wyoming] is definitely against any further enlargement of the park," declared the editor of the nearby town's newspaper. "Dubois sees in it the ultimate end of the big game industry."[39]

"Once the land passes into a national park system, the state loses all control of the game," complained employees of the Wyoming game commission.[40]

Writers for the *Sheridan (WY) Journal* made some hyperbolic statements, including: "Should this land be taken over by the government it will mean Teton must cease to function as a county."[41]

In Jackson, the controversy led to the creation of a second newspaper in the small Wyoming town. The *Grand Teton*'s almost singular purpose was the incessant and bombastic criticism of Junior, Albright, Fabian, Chorley, and anyone else involved with the SRLC, expansion of Grand Teton National Park, or conservation in Jackson Hole. "With the wealth of a Croesus and the power of a super-government the plan is plain. A citizenship and a community must die," the paper reported.[42]

These attitudes were also loudly articulated in a column on the front page of the *Casper (WY) Tribune-Herald* titled "Brass Tacks" and headlined "Another Syphon to Drain Wyoming." The editorial accuses the SRLC of removing "the most attractive territory bordering on the new national park for commercial purposes" and monopolizing the concessions within the park, including lodging and transportation. "Conservation and commercial monopoly seem to be synonymous," the editorial concluded.[43] Had the writer of the piece known that Junior was also involved with concessions in Acadia National Park, he likely would have had a field day with that as well.

On the other hand, the Grand Teton National Park expansion plan did have its enthusiasts. Struthers Burt waxed with his characteristic folksy prose in the generally pro-park *Jackson's Hole Courier*.

> Jackson Hole has experienced through the coming of the Snake River Company an almost unbelievable stroke of luck.... The Snake River Company has bought about 35,000 acres of land at least three fourths of which would not have been worth one cent to its owners, the county, or the state.... On the fingers of two hands you can count the ranches of any value.... Anyone who cares to do so can find out what a National Park means financially to a country.[44]

Burt went on to emphasize that he knew the people behind the land company well and that their motives were "altruistic" and not in any way self-serving. He also cheered on the effort in personal letters to Albright.

> I doubt if even you, not actually living in Jackson's Hole, can completely grasp the sickness of heart with which so many people who love the country have watched the wrong-headed, ruthless destruction going on. Under current conditions, any attempt at control, any attempt at decency gets nowhere. In short, you are supposed to sit by and see idiots make a hell out of heaven.[45]

Of course, Burt and the other dude ranchers differed from the real cowboys of Jackson Hole. Dude ranches were places where rich easterners came to play cowboy (and maybe do some hunting) and paid a hefty sum for the privilege (up to seventy-five dollars per week in 1930).[46] Dude ranchers, most of whom were wealthy easterners themselves, usually closed up shop in fall and headed to Florida or Philadelphia or somewhere else cushy for the winter. Most importantly, dude ranchers had already made the transition from agriculture to tourism and therefore tended to be economically allied with the propark people.[47]

*

By 1933, the fervor had reached such a pitch that U.S. Senator Robert D. Carey, a Republican from Wyoming, seized the political opportunity to call a congressional investigation of the SRLC and the NPS "to sift to the bottom of reports and charges that the National Park Service and companies owned by John D. Rockefeller Jr. are pushing homesteaders off their land in Wyoming."[48] Carey jumped on the antipark bandwagon when he was running for the Senate in 1930 and was casting about for ways to pick up groups of voters. He also bore a grudge against Albright, likely related to a fight over damming Yellowstone lakes for irrigation purposes, dating back to Carey's days as governor of Wyoming and Albright's time in Yellowstone.

Carey was a typical western politician in his distrust of the federal government and a champion of state's rights. Three weeks before the hearings began, he pitched gunpowder on the investigation's fire by telling a reporter from the *Denver Post* that the SRLC burned fences and ranch houses to force property owners into submission. Newspapers reported "houses and equipment were burned when settlers refused to sell" and "the work of despoiling the State ... by Eastern millionaires has continued for years."[49] Senator Peter Norbeck of South Dakota told Albright that Carey's accusation was "just dirty" and an attempt to pander to Wyoming Senator

John B. Kendrick "without whom Carey can accomplish little."[50] Local opponents to the SRLC were able to attract reporters from the *Christian Science Monitor*, *New York Times*, and *Salt Lake City Tribune*, promising a scandal on the scale of Teapot Dome which had brought down a former secretary of the interior. The possibilities of humiliating and even indicting a Rockefeller were no doubt the main reasons for such broad media interest.

Conveniently, the Senate hearings were scheduled for Jackson in August, a place and season where the weather was much more amenable than Washington's summer sauna. The "prosecutors" called witnesses who had sold their land to the SRLC and those who had refused. Mrs. S. A. R. Dale, the county school superintendent, told the committee that the Rockefeller plan had created a controversy in Jackson Hole that would "make a Kentucky feud look like a tame thing."[51] Wyoming state senator William C. Deloney testified that he and the rest of the town had always felt Grand Teton National Park "was shoved down our throat."[52]

Yet, when the long-simmering sense of betrayal and exaggerated accusations were exposed to the light of day, they appeared to have little substance. Senator Carey backed off from his claims of arson, clarifying that he had really told the reporter that "fences, barns, etc. had been burned after ranch property was purchased." One rancher complained that the rural schools were closing due to the land purchases. However, upon questioning by Senator Norbeck, it became clear that the land company was still paying taxes on the properties they had purchased, so the problem with the schools was not lack of funds but a dearth of children. Norbeck, a long-time ally of the NPS, also cleverly revealed a little-known windfall to stockmen who had sold to the SRLC but were allowed to continue using the land. "You mean to say that the cattle men now have free use of lands on which Rockefeller pays the taxes?"[53]

"Most of them testified [that] prices paid by John D. Rockefeller Jr. through the Snake River Land Company, were fair," reported the *Billings Gazette*, a conservative, pro-stockman Montana paper.[54] Norbeck characterized the whole kerfuffle as "Rockefeller is having a hard time trying to do nothing more than give away a couple million dollars."[55]

By the third day, reporters were packing their bags and heading for the train stations. The committee chairman, Gerald P. Nye, a Republican from North Dakota, determined that "the Snake River Land Company has concluded its purchases in a most unselfish manner." He added that the hearing had uncovered no evidence of unfair dealings by agents of Junior.[56] A wag from the *Denver Post* supposedly quipped that there was no tempest in a teapot, not even a "squall in a thimble." Norbeck thundered that the hearings were "an outrageous expenditure of funds. It cost $5,000 to run down three unfounded rumors."[57]

The final day of the Senate hearings was also Albright's last official day with the National Park Service. The cocreator of the agency in 1916 and its director from 1929–1933, Albright had announced his resignation earlier that summer. Nearly everyone was surprised if not shocked. The antipark *Grand Teton* newspaper, which referred to park proponents as "the Snakes," attempted to make hay from the announcement by declaring, "No honest public official ever resigns under fire."[58] Many years later, Albright told an interviewer he had resigned primarily for financial reasons. The transfer of power from Republican president Herbert Hoover to the Democrat Franklin D. Roosevelt had highlighted the vulnerability of his position as director of the NPS (even though FDR had decided to keep Albright on). After twenty years of government service, Horace and Grace had little if any savings and their "children were growing up and must be educated." For the sake of the family, Albright "concluded the move to the business world would be more than justified."[59]

Regardless of his sterling reputation as one of the greatest NPS directors to ever wear the gray and green, Albright was always a somewhat reluctant public servant. In law school his uncomplicated dream was to marry Grace and secure a good-paying job as an attorney. When he was recruited to Washington by Secretary of the Interior Franklin K. Lane (a fellow UC–Berkeley grad) in 1913, he was only willing to commit one year of his young life to helping the national parks. Likewise, premier NPS director Stephen Mather only promised to give a year himself. As each work anniversary neared, however, Mather was strung along by the secretary and managed to talk Albright into "just one more," as well, usually by promising a pay raise and larger salary supplement from Director Mather's own pocket.

Albright threatened to quit frequently. The legal clerk's job he had originally accepted in 1913 at $1,200 per year ($38,000 in 2025 dollars) became more and more onerous as Mather's health deteriorated. During Mather's long recuperations from a breakdown and stroke, Albright had to delicately cover for Mather while still fulfilling the very difficult duties of his own position. When Mather's health problems derailed Albright's intention to resign from the NPS and return to California in 1918, Albright reported feeling "depressed, unhappy, and in a dreadful quandary."[60] In 1919, at age twenty-nine, Albright sent his letter of resignation to Mather. He reasoned that he had put off his law career long enough and it was time for him, Grace, and son Robert Mather Albright to start a more normal and comfortable life. Mather immediately protested and invited him to Washington to fill in while he embarked on a six-week junket to the Hawaii parks. Mather sweetened the deal by agreeing to grant Albright the job of his dreams: superintendent of Yellowstone in summer and NPS field director over the winters. The Albrights accepted.

While on a trip to New York to attend the wedding of John D. Rockefeller III and Blanchette Hooker in 1932, Albright finally received an offer from the United States Potash Company he couldn't refuse. The new position of executive vice president and general manager would more than double his salary to $20,000 per year ($458,000 in 2025 dollars) and provide a pathway to his eventually becoming company president. Albright stalled for a couple of months, then accepted the position on the condition that he not leave the NPS at an inopportune time. He also received permission from U.S. Potash to continue to do conservation work while in the company's employ, a sideline that certainly took up an extraordinary amount of his time and energy for many, many years. In his new job, Albright would work with the Department of the Interior in obtaining permits for mining on public lands, so his experience with the NPS was valuable to the company. Though always conflicted about his resignation, Albright felt at the time that he had accomplished most of the items on his NPS to-do list and that Arno Cammerer, his preferred successor, would capably continue where he had left off.

Albright's parting words to his staff in 1933 are relevant to park managers, conservationists, and philanthropists today.

> We are not here to simply protect what we have been given so far; we are here to try to be the future guardians of those areas as well as to sweep our protective arms around the vast lands which may well need us as man and his industrial world expand and encroach on the last bastions of wilderness.
>
> Some precious park areas can easily be destroyed by the concentration of too many visitors.... The parks, while theoretically are for

> everyone to use and enjoy, should be so managed that only those numbers of visitors that can enjoy them while at the same time not overuse and harm them would be admitted at a given time. We must keep elements of our crowded civilization to a minimum in our parks.[61]

Albright wrote Junior on his first official day with the mining company:

> The privilege of enjoying your friendship and your confidence, and the inspiration of your character and ideals have meant more to me personally than I can ever express.
>
> It has been impossible for me to repay, even in small measure, my indebtedness to you for your very great contributions to the success of my administration, and I deeply regret that our Jackson Hole project which you adopted on my representations has brought you difficult problems to solve.[62]

Although Junior responded warmly and appreciatively, there is a faint undertone of disappointment in his correspondence. With Mather dead and Albright leaving the Park Service—and the fight for Jackson Hole raging—Junior must have felt abandoned. Characteristically, though, he nearly smothered himself with humility. "These happy years of cooperative endeavor in the public interest which it has been my pleasure to share in a very modest way with you, gave me the greatest pleasure." Junior then delivered the slightest of snubs: "I am glad that you are to be in and about New York [Albright occupied a corporate office in Rockefeller Center for many years], and shall look forward to seeing you from time to time. But whether we meet often or not, please know that I shall always retain for you the highest esteem and warmest friendship."[63]

By 1933, Albright and Junior had been performing conservation magic for some eighteen years. It was a complementary and remarkably productive partnership. Junior had nearly unlimited resources (both fiscal and human), the desire to see "ideal" projects completed perfectly, and nearly superhuman persistence. Albright had a clear vision of what the national park system needed to become and how the public would benefit. He was also willing to fight incessantly against rigid opposition and toxic criticism to protect existing parks and create new ones. And, fortunately for the parks, the Albright–Rockefeller conservation collaboration was far from finished.

THROWING IN THE TOWEL

Grand Tetons, 1943–1947

If December 7, 1941, was a day that would "live in infamy" for millions of outraged Americans lurching toward another world war, then May 2, 1943, was a day that could also be called infamous in the seemingly endless struggle to create an expanded Grand Teton National Park in northwest Wyoming. On that cool May morning, a small group of Wyoming stockmen toting Winchester rifles and holstered pistols climbed onto their horses and drove 549 yearling cattle across the federal boundary and directly through the newly declared, hotly contested Grand Teton National Monument. Leading the drive was Academy Award–winning film star Wallace F. Beery. The actor owned a cabin on Jackson Lake and, although he only spent a couple weeks there every year, he was adored by the locals. He was also one of Hollywood's highest paid leading men of the era and had recently completed a series of Westerns, including *Wyoming* (filmed in Jackson Hole), *20 Mule Team*, *The Bad Man*, and *Jackass Mail*.

Beery, despite having been a suspect in a 1937 alleged murder in California, was considered the ideal figurehead for the well-orchestrated publicity stunt.[1] The cowboy actor and the local stockmen perfectly symbolized the conflict between freedom loving, don't-fence-me-in westerners and what was perceived as an overreach by the federal government. National Park Service director Newton B. Drury, obviously smarting from the stunt, claimed that Beery, "described in some publications as a 'prominent stockman,'" simply had a summer house on half an acre in Jackson Hole and owned but one cow. "During the controversy this one cow died."[2]

According to Conrad L. Wirth, who had joined the NPS as assistant director for land planning in 1931, the fifty-eight-year-old Beery needed a stepladder to mount his steed. "He was getting pretty well along [in years] and pretty heavy in those days," Wirth maintained.[3]

Beery and his posse's act of civil disobedience, which was wisely ignored by park rangers, could hardly have been more effective. Media large and small, from *Time Magazine* to the *Jackson's Hole Courier*, picked up on the story and delivered its iconic cowboy symbolism to living rooms from coast to coast. Wyoming governor Lester C. Hunt expressed his allegiance with the protestors by belligerently threatening to evict any federal official who attempted to assume authority in the newly declared national monument.[4] Other Wyoming representatives piled on, including U.S. Congressman Frank A. Barrett, who professed his allegiance to Jackson Hole residents by stating, "Their boys were away on the battlefields of the world fighting to preserve and protect their homes. And now their government...had suddenly taken action that ultimately would deprive them of their homes and their means of livelihood."[5] Writers, pundits, and politicians across the political spectrum, nearly all seemed to ally themselves with the cowboys, those romantic symbols of a simpler, more genuine way of life. Journalist Westbrook Pegler reported: "President Roosevelt and Harold Ickes have recently perpetrated in the State of Wyoming an Act of Annexation which follows the general lines of Adolf Hitler's seizure of Austria."[6]

The well-orchestrated protest ride certainly turned up the heat on the long-simmering conservation controversy, but perhaps not as much as media hyperbole exclaimed. Wirth recalled being in Jackson Hole around that time with several other NPS bigwigs and park proponents and receiving hearsay reports that if the "Eastern dudes" showed their faces in Jackson, "something terrible was going to happen." Chorley, a towering man who seemed to thrive on controversy, was among them. "They know we're here. I think we ought to go downtown and make our appearance," Chorley challenged the Park Service folks.[7]

And so they did. "We made our appearance. We walked around the square [in Jackson], went into one saloon and looked around, bought a drink and then walked out and around, and went in another." Unlike the climactic scene in *Gunfight at the O.K. Corral*, Wirth reported: "Nothing ever happened. People hardly even noticed you."[8]

*

The catalyst for President Franklin D. Roosevelt's decision to declare Jackson Hole a national monument was a letter of exasperation written by Junior. The long letter was conveyed to the president, via Secretary of Interior Harold L. Ickes, with Junior stating he'd had it up to here with the political wrangling and foot dragging over the last decade and a half that had delayed the expansion of Grand Teton National Park. Junior warned that he planned to sell off the more than thirty thousand acres he had acquired in Jackson Hole unless the government accepted it as a gift, ASAP. He reminded the president that he had purchased the land in good faith based on the recommendation of Yellowstone superintendent Horace M. Albright[9] to

> provide winter feeding for the great quantities of game which were gradually being exterminated by starvation, and to preserve the superlative scenery of the Grand Teton approaches, confidently expecting the Federal Government would gladly accept the land as a gift to be added to its National Park System. Fifteen years have passed. The government has not accepted the property. It cost me a million dollars [$18.5 million in 2025 dollars]. Taxes, maintenance and other costs have increased the figure roughly by half a million dollars. I have now determined to dispose of the property, selling it, if necessary, in the market to any satisfactory buyer.[10]

As to Junior's motives for the ultimatum, Chorley recalled, "I think he was fed up." Chorley also believed Junior was aware of the possibility of FDR declaring a national monument, and he was willing to put it to the test.[11]

Wirth believed Junior simply decided "he'd rather spend money on buying more land than on paying taxes on the land he has," and that he was exasperated with "sitting around and thinking what to do."[12] Because he was from a family often heralded for their patience, it is odd he chose to press the matter to the president during the darkest days of World War II, at the tail end of the Great Depression. There were no personal financial crises affecting the Rockefellers or other events in late 1942 or 1943 that one could point to as a spur. In fact, other than the universal trials of World War II (all Rockefeller sons were serving their country in either combat or noncombat roles and Abby and Junior were heavily involved in the USO and massive relief efforts), things were going well for the ultrawealthy couple. More likely, though, after fifteen years of frustration—most of it unnecessary and attributable to Wyoming politicians pandering to a very small number of constituents—Junior, Albright, and Secretary of Interior Harold Ickes simply decided their only remaining resort was what we might call today "the nuclear option." It's also possible that the three opportunistically viewed the war as a distraction that might work to the conservationists' advantage.

The question of whether or not Junior was bluffing in his threat to sell off Jackson Hole to the highest bidder has been raised in other histories of the national parks.[13] While Chorley and others claim Junior was not a bluffer, it is extremely difficult to imagine him selling the properties to the types of commercial tourism enterprises he had worked so hard to exclude.[14] If one recalls his actions in Acadia National Park, particularly the sudden (and temporary) withdrawal of his offer to fund construction of Ocean Drive because of opposition by a vocal minority of wealthy summer residents, one would have to recognize Junior as a person willing to bluff, or at least call another's bluff. In a letter to Cammerer in 1932 referring to the road in Maine, Junior stated, "People are interested in blocking something which they think I have my heart set on doing. If they know I am indifferent, then the proposal begins to be considered on its merits."[15]

It's possible to infer, then, that Junior hoped to force a similar awakening upon the antipark Jackson Hole residents and the state and federal politicians who represented them. The logic would lead them to imagine a future, despoiled Jackson Hole with views of the majestic Grand Teton Mountains obscured by billboards, utility lines, traffic lights, hotdog stands, tall hotels, souvenir shops, diners, dance halls, shooting galleries, haunted houses, gas stations, and bars. Lakeshores would sprout shoulder-to-shoulder summer homes and lodges three rows deep.

After Junior's challenge was sent and received, Secretary Ickes "took the bull by the horns" and managed to wrangle two lunch meetings with President Roosevelt.[16] The president had recently returned from a crucial mid-January 1943 wartime meeting with Prime Minister Winston Churchill in Casablanca and was no doubt under unimaginable stress. Regardless, Ickes was able to detour the president's mind to matters of conservation and urge him to sidestep Congress and write an executive order creating Grand Teton National Monument. Ickes warned Roosevelt that he would make enemies near and far by the action.[17] Regardless, on the Ides of March 1943, FDR used the Antiquities Act of 1906 to unilaterally create the monument.[18]

Why the president was willing to risk the political fallout and even possible revocation of the Antiquities Act is an interesting question. In 1937, while touring the Yellowstone country, FDR reportedly asked his traveling secretary, Marvin McIntyre, "if he could extend the Teton Park by executive order." The secretary replied to the negative and the president made plans to examine his options when he returned to Washington.[19]

Ickes (nicknamed "Honest Harold," as well as "The Old Curmudgeon") must also have been a factor. He had worked with FDR on Theodore Roosevelt's presidential campaign and both men adored stamp collecting and protecting the environment. Ickes had served on Roosevelt's cabinet since 1933 and was known as "Roosevelt's environmental conscience." Ickes also successfully headed the president's hallmark Depression-era Public Works Administration. Since he had survived as secretary of interior longer than anyone before (or since), it seems probable that FDR respected Ickes's opinions. Apparently, the families had a personal friendship as well; the Roosevelts often spent weekends at the Ickes's country house in Maryland.[20]

Albright, who had been working for over twenty-five years to add the Tetons and Jackson Hole to the national park system, may have also figured in the decision. The former director of the NPS was admired by both Ickes and FDR. In 1930, when Franklin (then governor of New York) and Eleanor Roosevelt had to withdraw from a planned trip to Bryce, Zion, and the North Rim of the Grand Canyon, they asked Albright to escort their son, FDR Jr., on the post-conference junket.[21] Shortly after FDR was first elected president, Albright had the good fortune to accompany Ickes, FDR, and others on a drive to Camp Hoover and a tour of Skyline Drive in Virginia. With hours to spend chatting along the way, Albright and FDR enjoyed sharing their impressive knowledge of many geeky aspects of American history and discussed battlefields in need of protection. Remarkably, Albright was successful in convincing the president to transfer historical military parks from the U.S. Army to the NPS, an important goal that had eluded Albright and Mather for years.[22] The creation of the Blue Ridge Parkway was also endorsed by Roosevelt and Senator Harry F. Byrd during the same motor excursion.[23]

FDR's decision on Jackson Hole was certainly influenced by his respect for Ickes and Albright, but also by his recognition of Junior's extraordinary efforts and perseverance. By his own words, the president, in a letter to U.S. House Speaker Sam Rayburn, stated his reason for creating Grand Teton National Monument was "to avoid losing Mr. Rockefeller's gift."[24]

In studying the prospect of wielding the powerful and controversial Antiquities Act, he was aware that the previous President Roosevelt, Theodore, had signed the act into law in 1906 and that Devils Tower, America's first national monument, resided in the same state as Grand Teton. Theodore Roosevelt went on to use the act to create seventeen additional national monuments, including Chaco Canyon, Grand Canyon, and Montezuma's Castle. Yet because many interpret the intent of the Antiquities Act as protection of "areas of importance in history, prehistory, or science," using the act for a big scenic area like Jackson Hole could be perceived as a stretch.[25]

Since the Antiquities Act circumvents Congress, it was then, and still is, one of the most controversial and powerful implements in a president's toolbox. When Congressman John Lacey from Iowa first drafted the weighty act, he intended it for the purpose of saving archaeological sites like Mesa Verde and Chaco Canyon from annihilation by pot hunters. It was designed for quick action to address imminent emergencies on federal lands; therefore, the president was allowed to bypass the often endless bickering, posturing, and lengthy recesses of Congress.

Though overshadowed by the public lands' legacy of the previous President Roosevelt, Franklin was no conservation slouch. Both FDR and Junior grew up with the beauty of the Hudson River Valley, had ridden horses and mucked around in the woods, and had been exposed to gardens and landscapes created by Andrew Jackson Downing. As governor, FDR worked with Gifford Pinchot, the first chief of the U.S. Forest Service, to protect watersheds and forestland in New York. While president, he played a role in adding over one hundred sites (many of them smaller historic areas, but also major parks like Great Smoky Mountains, Shenandoah, the Blue Ridge Parkway, and Grand Tetons) to the national park system. Most notably, however, FDR created the Civilian Conservation Corps, a New Deal initiative that not only provided jobs for millions of unemployed young men, it accomplished the planting of over two billion trees, the building of thirteen thousand miles of hiking trails, and the creation of campgrounds, picnic areas, and roads in hundreds of state and national parks.[26]

According to biographer Douglas Brinkley, after the second lunch meeting with Ickes, FDR "cheerfully agreed to use the Antiquities Act" to create Grand Teton National Monument.[27] The next day, Ickes called Junior to deliver the good news and Junior responded formally with a letter of thanks which included a rare bit of exuberance: "I'm sure we are both singing in our hearts 'Praise God from Whom All Blessings Flow.'" The letter then went on, more characteristically, to heap praise upon Ickes and attribute to him full credit for the accomplishment.[28] Junior also wrote the president and congratulated him for making "possible the preservation for all time of the most uniquely beautiful and dramatic of all areas set aside for national park purposes."[29]

According to Brinkley, FDR believed "this tapping of wealthy patrons to invest in helping to protect America's landscapes held great promise for enlarging the National Park Service in the future."[30]

The national monument was a huge expansion of the 1929 "mini" Grand Teton National Park that had been created simply by transferring U.S. National Forest Service lands to the National Park Service. Even that straightforward bureaucratic action took four long, quarrelsome years to accomplish with the Forest Service and timber interests fighting tooth and nail to keep the mountains open to hunting, lumbering, livestock grazing, mining, and other purposes allowed under the agency's more lenient multiple-use management guidelines.

To no one's particular surprise, FDR's national monument declaration was not at all popular with Wyoming's elected officials or some residents of Jackson Hole. According to NPS associate director Conrad Wirth, the action "practically killed the Antiquities Act" and definitely led to restrictions on its future use.[31] There was also considerable gnashing of teeth by county officials over the loss of taxable properties and subsequent revenue for schools and other services. Wyoming governor Lester Hunt was especially outraged and did his best to rally other western governors to fight the federal action and to repeal the Antiquities Act.

Yet it was the civil-disobedient cattle drive that elevated the issue to the national stage. Former Forest Service employee and stockman ally Felix Buchenroth was quoted in *Time Magazine*, "It may be a monument to Ickes, but it's a tombstone to me."[32] By the end of the month, Wyoming representative Frank Barrett had introduced H.R. 2241 to the U.S. House of Representatives, a bill to abolish Jackson Hole National Monument.

Wyoming senator Joseph O'Mahoney led the fight to nullify the monument in the U.S. Senate. Both houses of Congress handily passed the bill, which was then vetoed by FDR.[33] First Lady Eleanor Roosevelt later said of the close call, "It is disturbing to find how little real enthusiasm there seems to be among our people for the preservation of our national parks."[34]

O'Mahoney also recruited a senator from Nevada to cosponsor a bill to repeal the Antiquities Act. Although it failed, the state of Wyoming pressed forward with a court case stating that Jackson Hole was without significant historic or scientific resources and therefore could not be protected under the Antiquities Act. As Conrad Wirth intimated, FDR's controversial use of the Antiquities Act and the subsequent threats to its existence caused future presidents to use the act much more cautiously, and usually after consultation with the House Interior Committee.[35]

Historian Robert W. Righter perhaps said it best when he wrote of the seemingly endless battle for a full-fledged Grand Teton National Park: "Man has always had a possessory inclination toward a particular piece of land or even a whole mountain valley, and perhaps the greater the beauty the more determined the attachment becomes."[36]

*

Since the new monument continued to exist during the relentless state and national political wrangling, the U.S. Forest Service was required to transfer its ranger stations and other holdings to the NPS. To say the Forest Service did so reluctantly is an understatement on a tectonic scale. Forest Service employees gutted the Jackson Lake Ranger Station, indelicately removing plumbing, doors, bathroom fixtures, and an underground water tank, which "required cutting a four-foot-square hole in the floor." There were also allegations "someone placed a live skunk in the Jackson Lake Ranger Station, which died in the building."[37] The NPS eventually declared the structure uninhabitable.

Senator O'Mahoney continued his antipark crusade by attaching an amendment to the Department of Interior appropriations bill that severely limited funding to Jackson Hole National Monument for at least three years. The death of FDR in 1945 and the seating of President Truman re-energized the elected officials from western states to push for abolishment of the monument and repeal of the Antiquities Act.[38] Junior and Chorley were clearly frustrated, not only with the Jackson Hole situation, but with other long-term national park efforts.

In early 1947, the conservationists requested a meeting between Chorley, Albright (although he had resigned from the NPS over a decade earlier), and the new secretary of the interior, Julius A. Krug. Chorley, who did not share the patience gene with Junior, vented a long list of grievances. "After holding the property in Jackson for over twenty years," Chorley declared, "paying taxes and more or less carrying the load without much help from anyone, we were beginning to wonder whether the game was really worth the candle."[39] He also complained to the secretary about the largely symbolic and futile exploration for oil in Jackson Hole by the Sinclair Oil Company, the general "lack of interest on the part of the Park Service," and the fact that two of the national monument's orneriest opponents—Wyoming congressman Barrett and Wyoming senator Robinson—had been elected chairs of the subcommittees on public lands in their two respective halls of the U.S. Congress.[40]

Secretary Krug replied that he could well understand the conservationists' discouragement and that "if we did pull out he really could not blame us." The meeting ended, however, with Krug's significant pledge to contact the Wyoming delegation and push for a resolution.[41] In describing the meeting later to Harold Fabian, Chorley said, "I am inclined to think our strategy of putting him [Krug] on the defensive and making him show whether or not he was willing to go fight for us was sound."[42]

A few months later, propark Wyoming governor Lester C. Hunt warned Krug that unless the issue was settled before the next election, "the action of the Interior Department [creating the national monument] could be, by legislative action, completely reversed, and the Park Service might see the entire monument restored to its previous state [private ownership] together with the repeal of the Antiquities Act."[43]

Yet it is important to note—among so much short-sighted political glowering and gloom—that preserving Jackson Hole also had many proponents. In response to the antimonument, procowboy cattle trailing, Olaus J. Murie, famed wildlife biologist, environmentalist, and (along with his equally famed naturalist-conservationist wife, Margaret Murie) long-time Jackson resident, wrote an article for *National Parks Magazine* that was widely reprinted and distributed. Olaus concluded: "It is our government. It is our National Park Service. We have asked them to care for something precious that belongs to all of us.... It should be our ambition to assist all agencies to keep intact this one segment of America that we boast of as 'the last of the Old West.'"[44]

For the *Atlantic Monthly*, writer Katharine Boyd parlayed with a thoughtful, long-time Jackson Hole horseman and wilderness guide nicknamed "the Virginian." "Here they are talking this stuff... about parks spoiling things and them wanting to keep this country like it is," the Virginian growled. "And the rock bottom reason for parks is to keep the country like it is."[45]

And the May 1945 issue of the ever-popular *Sunset* magazine featured a cover photo of the Grand Teton Mountains with Jackson Lake in the foreground. The lead story was a poignant piece titled "Destination: Our Land," written by the editor, Walter L. Doty. The tragic context of the (unfortunately true) story was the editor visiting his younger son in the hospital after the soldier received severe burns while serving in the South Pacific. The son is (presently) dying and suffering horrible pain. The two reminisce about their trips to Olympic, Mount Rainier, Crater Lake, and Yosemite National Parks and plan a journey to Yellowstone and the Grand Tetons. "In the days and nights at the hospital, almost to the day of his death, the sure escape from pain was through memories and plans for the high country.[46]

"I've given your boots another saddle soaping in anticipation of a trip to Wyoming," their conversation continued. "This great landscape is part of the new Jackson Hole National Monument created in 1943, but still exciting controversy in Congress. Some Congressmen think that this shouldn't be 'our land.'"[47]

NPS director Drury wrote Doty, praising the article, and Doty's reply explained how the story evolved. Drury shared the article and excerpts of Doty's letter with Junior, whose son, Winthrop, suffered similar—though less severe—injuries while serving on a ship in the Pacific. Doty's letter to Drury concluded, "It was probably the hard way to learn of the spiritual values [of national parks], but it made them extremely important, and it fastened on me a responsibility I can never completely fulfill."[48]

NEVER SAY NEVER

Grand Tetons, 1947–1950

Time and time again journalists, area businesspeople, and Wyoming's elected officials raised suspicions of Junior monopolizing Jackson Hole's tourist concessions. And not without reason. The accusations were well-aimed shots at Junior's Achilles' heel, not only due to his father's demonization in popular culture for monopolizing the oil business, but because Junior, through his acquisition of Jackson Hole lands, had indeed become the reluctant landlord for nearly all the tourist facilities near the mountains. The muckraking journalists tied to the *Grand Teton* newspaper frequently attacked his "foreign corporation" for its monopolistic lodging operations in the village of Moran and elsewhere.[1] The Casper, Wyoming, "Brass Tacks" editorial stated the Teton Investment Company was positioned "to reap a golden harvest from development of the park and its environs. The monopoly seems to be complete."[2]

In truth, Junior and the subsidiaries of the Snake River Land Company had little choice but to keep operating the businesses they had acquired during their land-purchasing blitz.[3] The growing number of tourists visiting the area had no other accommodations outside of the town of Jackson where they could lay their tired heads. This policy of maintaining services cost the Teton Investment Company and Teton Lodge Company dearly, as most of the accommodations inherited in the land buys were in poor condition and required maintenance and upgrades costing tens of thousands of dollars. In order to steer the SRLC away from directly operating concessions, Junior and his associates leased most of the dude ranches they had acquired back to the original owners so they could continue hosting guests.[4]

In his lengthy essay in the *Jackson's Hole Courier*, Struthers Burt addressed the tourist business accusation by pointing out that "enormously rich men do not engage in picayune business. It is not likely that Mr. Rockefeller would enter the store or hotel business at Moran unless forced temporarily to do so. What does the *Casper Tribune-Herald* expect? That the Snake River Land Company abandon these properties?"[5] And though Burt was correct in his public statements, Junior and his son Laurance did in fact have more than an accidental interest in national park concessions, though their motives were not what the *Tribune-Herald* editors insinuated.

In 1928, Junior purchased one of Acadia National Park's few private concessions, the Jordan Pond Tea House. Junior and Abby were two of the restaurant's most loyal customers, relishing their carriage rides out to the pond for tea, toast, and jam while gazing across the rippling waters toward a pair of iconic mountains known as the Bubbles. After the purchase, Junior promptly leased the tea house back to its long-time owners who continued to operate it in the same manner for another decade. Junior's motives were fairly easy to discern, he feared the restaurant would be sold to new owners who might be oblivious to the tea house's traditions. In fact, Junior's attitude toward many things in the world was that he hoped they would never change, or, better yet, revert to how they had been.

Accommodations in national parks operated by concessionaires under contract with the NPS have been contentious for as long as national parks have existed. The best examples, including the Old Faithful Inn and Lake Hotel (Yellowstone), El Tovar (Grand Canyon), Ahwahnee (Yosemite), Many Glacier (Glacier), and Zion Lodge are ambitious, noble landmarks which complement the scenery, much as a tasteful setting enhances a gemstone. Others, often due to the prevailing architectural styles during the Conrad Wirth–inspired Mission 66 national park building boom (circa 1956–1966), slouch toward mediocrity.

Several conditions make operating a national park concession a challenging and often unprofitable endeavor. The season for providing guest accommodations in most parks is short—perhaps June to early-September. Retaining quality staff and housing them in a remote locale can be a human resource nightmare. Natural disasters—from forest fires to floods, avalanches, landslides, earthquakes, and volcanic eruptions—are almost routine in many parks. These tantrums of nature often result in the closing of park facilities or the entire park for extended periods. It would be difficult to defend the argument that Junior got into the park concessions business primarily to make money. We are reminded once more of a remark Junior made to a newspaper reporter in 1919: "What do I want with more money? Nearly all my time and nearly all the time that my father gives to financial affairs is devoted to studying how best and wisely to distribute the money accumulated."[6]

On the plus side, Stephen Mather and the Park Service decided early on that the federal government should steer clear of any direct involvement in the hospitality industry.[7] This opened the door, at least initially, to entrepreneurs with a love for the place to work in partnership with the NPS and provide quality services appropriate to their spectacular settings and middle-class clientele.

Despite the downsides to operating a business in a national park, Junior—and later, son Laurance—did willingly get involved in park concessions. The reasons for such involvement were complex and easily misunderstood. To the cowboy branding calves in Jackson Hole, or the congressman from a western state working in Washington, Rockefeller ownership of lodges and restaurants in the Tetons was understandably perceived as proof that the millionaire from back East was up to no good. A member of the Wyoming Izaak Walton League surmised: "The recurring difficulty in Wyoming…is the implication that Mr. Rockefeller has some ulterior motive. To many it is inconceivable that anyone would want to make a gift without expectations of some personal gain."[8]

In Acadia, as well as the Grand Tetons, Junior's willingness to purchase land and donate it to the federal government was directly related to his distaste for tacky tourist developments (e.g., Niagara Falls and Coney Island) that were often described by Junior, Abby, Albright, Dorr, and others as unsightly "amusements" or "hot dog stands."[9] Abby, in fact, launched her own crusade against dilapidated "wayside refreshment stands" along America's byways and worked with the Art Center of New York to encourage more pleasing esthetics.[10]

Although Junior seemed to have no aversion to tourist enterprises in park gateway communities such as Bar Harbor or Jackson (perhaps because at the time many leaned more toward historic and rustic than cheesy), nor with artfully designed grand lodges within the park, he was undoubtedly repulsed by the tacky stuff that distracted from the tourist's enjoyment of scenery and diminished the hallowed stature of national park landscapes. According to national parks historian Alfred Runte, similar sentiments about hucksters charging tourists for a view of Niagara Falls and spoiling the scenery with bric-a-brac inspired the creation of Yellowstone National Park.[11] Junior reported that one of a series of hotels built on the very top of Cadillac Mountain in Acadia was "more or less of a disorderly and disreputable resort, the character of which gave great concern to the summer people."[12] As a diehard perfectionist, Junior, along with Albright and Mather, felt strongly that the quality of lodging, dining, and retail facilities in any national park should be on par with the grandeur of the landscape.

"Why is this ramshackle old building allowed to stand over there where it blocks the view?" Junior inquired of Albright on his first trip into Jackson Hole in 1926. Albright recalled, "The Rockefellers expressed great concern that this spectacular country was rapidly going the way of development and destruction."[13]

Looking back in 1956, Junior wrote a letter to remind his children of his work on behalf of the park. "I had purchased the Jackson Hole area to rescue that incomparable region from being over-run by hot-dog stands and cheap commercialism."[14]

Little did Junior and the SRLC know when they were buying dude ranches, cattle ranches, and tourist cabins that they would remain overseers of these properties for more than twenty years. Since tourists in the 1930s and 1940s had few choices in the wilds of Wyoming for accommodations (for all the reasons stated above), Junior, Albright, and others felt it was essential to keep these basic services in operation during the unexpectedly long transition to parkhood. Revenues went to helping the SRLC pay property taxes on lands they were trying desperately to give to the federal government and American people.

One of the more unusual eggs in the SRLC's basket of properties was the "town" of Moran, an aging collection of dated-to-decrepit tourist cabins, a post office, gas station, restaurant, and other amenities situated just downstream from the Jackson Lake dam. The crossroads got its start in the 1890s as a hunting and fishing camp and expanded to house a few of the U.S. Cavalry troops guarding Yellowstone, as well as construction workers for the dam project. Other booms occurred when Hollywood film companies were shooting a western and needed long-term lodging for their crews. The town's Teton Lodge featured ninety-four cabins—six with showers and a bath. SRLC executive Harold Fabian, who was invested in the Teton Lodge and Teton Transportation Company, provided his perspective on the SRLC's Moran improvements to Secretary of the Interior Wilbur. "In short, the entire operation in this section is being transformed from a dilapidated, unsightly, and even unsanitary condition to a class of operation in every way consistent with National Park Service standards."[15] The Park Service encouraged the land company to maintain its facilities because the last thing they wanted was to frustrate tourists interested in visiting the Grand Tetons.

Tourism in the Teton country screeched to a halt during World War II as gas rationing and other austerity measures made travel next to impossible. After the war finally ended, Chorley complained to Laurance that those who had leased the lodges "had let the properties run down" and some had fallen into "deplorable condition." Fabian was equally dismayed: "Moran, as you know, is an impossible place to do anything.... I wish to heaven we could tear it down this fall but of course we cannot until we have something to take its place."[16]

Even though Junior remained the owner of Moran for many years, it represented the antithesis of what he wanted accommodations in a national park to be. To his great embarrassment and consternation, it resembled the "dilapidated cabins" that had stirred him to purchase Jackson Hole in 1926. After consultation with leading national park facility architect Gilbert Stanley Underwood, Junior, Laurance, Chorley, and others concluded their best solution was to salvage the old Grand Teton Lodge at Moran for parts and construct two major new facilities—Jackson Lake Lodge and Colter Bay—both near Jackson Lake and both affordable (in the day) to middle-class tourists and even those of more modest means. No doubt the part of the plan that pleased Junior the most was demolishing the sagging Grand Teton Lodge and the eventual return of the area known as Moran to its absolutely sublime natural state.

Albright, who knew Junior better than most anyone and understood his motivations, encouraged Junior and Laurance to get involved with park concessions. "A concession frankly run on a break-even basis," Albright stated, "might get excellent opportunities in national parks to conduct business in line with the objectives of the Park." A privately operated concession, he believed could do "a better job than the Gov't itself could do under the most favorable circumstances."[17] A federal investigation in 1936, possibly spurred by disgruntled former Teton concessionaires, scrutinized the relationship between the Rockefeller holdings, their concessions, and the NPS. The director of investigations reported to the secretary of the interior: "There is no evidence of unfair treatment of land owners and homesteaders." Likewise, "there do not appear to be any questionable irregularities between any of the companies mentioned [Teton Lodge Company, Teton Transportation Company, Jackson Lake Lodge Company, and SRLC] and the officials of the Government."[18] Previous federal investigations similarly had found "no evidence that the concessionaires are exercising an undue influence on park affairs." Special Agent Ramsey emphasized in his report the need for more lodging in the park.[19]

Before Junior dedicated himself thoroughly to the new lodges, however, his successful restoration work at Colonial Williamsburg led him to dabble in historic preservation in Jackson Hole. In 1942, Junior, Laurance, Fabian, and Williamsburg architect Andrew E. Kendrew initiated a two-month inventory of Jackson Hole properties acquired by the SRLC with the objective of preserving sites of cultural value. The survey identified the Cunningham Cabin (circa 1888), Manges Cabin (1911), Chapel of the Transfiguration (1925), and Menor's Ferry (including the Maud Noble Cabin, 1896–1916).[20] Although the youngest of the sites had been constructed only seventeen years prior to the inventory, all the structures did have important stories to tell about the history of ranching, religion, tourism, and America's westward expansion.

The darling of the bunch was the ferry complex, both because it was crucial to transportation around the valley prior to construction of a bridge over the meandering Snake River, and because its engineering was interesting enough to warrant an article in *Popular Mechanics Magazine*. The raft, which ferried people, livestock, and agricultural products across the broad, swift waterway, used the river's current for propulsion for crossings in both directions. Fabian pitched the idea of ferry restoration to Laurance, to which Laurance responded, "Yes. You go ahead and I'll give you the money to do it."[21]

Fabian and wife Josephine capably took charge of the ferry restoration. Both had the wisdom to involve as many old-timers and other residents as possible in the process: "Different families dug up old pictures, and people told their stories of how they'd ridden on the ferry," Harold reported, which "just changed the whole complexion of things." Locals told Josephine and him, "We didn't know you were interested in us. We didn't think you had any interest in us and our history."[22]

By the time of the grand reopening on August 20, 1949, the restoration had received national attention. According to the *Jackson's Hole Courier*, "several hundred people" attended the event, including the former governor of Wyoming, president of the Church of Jesus Christ of Latter-Day Saints, and the mayor of Jackson.[23] Laurance, author Struthers Burt, and Charles E. Menor, the grandnephew of the ferry's originator, were also involved in the ceremony. Local radio stations broadcast the event, and newspapers from Idaho to New York City heralded the restored ferry as "one of the most colorful mementos of the old west."[24] Struthers Burt described his ride across the Snake in the reborn ferry poetically, "I feel I've been riding with ghosts out of western America's pioneer past."[25] Harold Fabian recalled the emotions of the first fifty old-timers who crossed the river on the restored ferry: "The tears were streaming down their eyes."[26]

This was a significant turn from the dismal days of 1947 and early 1948 when Chorley had to report that, with the introduction of Wyoming senator Edward V. Robertson's legislation, "there are now two bills in the Congress to abolish the [Grand Teton National] Monument."[27] Albright fought the bills with all his might, even though he had officially resigned from the NPS long before. Commiserating with Laurance, Albright stated: "Dealing with Wyoming is like dealing with the Russians. You never get anywhere by trying to cooperate.... The state officials will take anything and everything they can get" and then "oppose anything that their benefactors and friends want."[28]

When Albright was finally able to assure Junior that the latest of Congressman Barrett's monument-annulling bills would fail, Junior gave a reply that should qualify him for the conservation hall of fame. "What you say about the Barrett Bill is most reassuring. That wonderful vision of yours which we discussed in Jackson Hole so many years ago and have been both working on ever since, must be realized some day. We will neither of us stop hoping or working for that end."[29]

Tangible evidence that the political tides had turned materialized on March 21, 1949, in Washington. Short-timer Wyoming Governor Arthur

G. Crane and the Izaak Walton League (supported by Olaus and Margaret Murie and Albright) pressed hard for negotiations that could finally end the standoff. All the staunch antimonument officials were there: U.S. senators from Wyoming Joseph C. O'Mahoney and Lester C. Hunt and Wyoming congressman Frank A. Barrett. The propark representatives were Chorley and Undersecretary of the Interior Oscar L. Chapman.[30] Senator Hunt stated that he believed "sentiment in Wyoming was swinging steadily in favor of the monument." Chorley reported the luncheon meeting was "extremely friendly" with "no tension or unpleasantness." He also related that "Congressman Barrett, I feel, realizes that he is pretty much licked."[31] By the end of the afternoon, the only issue left unresolved was how to deal with culling the overpopulated elk herd, a conundrum that continues to occupy the thoughts of area land managers to this day.

Most importantly, the economy of Jackson Hole was rapidly changing from one based on livestock and agriculture to tourism. Combined with the end of World War II and the subsequent stampede of automobile-enabled tourists, the twenty-three-year-old controversy quickly became little more than a fading historic footnote. During the summer of 1949, a "tidal wave of tourists" had swept across Jackson and the controversial national monument. "Eating and sleeping accommodations were taxed to capacity. Campgrounds and trailer parks were also full."[32] For the remaining cattle ranchers struggling against volatile commodity prices, seemingly endless winters, and high transportation costs, the writing on the bunkhouse walls couldn't be plainer.

It is also important to note that during the decades since Junior had acquired most of Jackson Hole, the SRLC had, in many ways, managed the land as the NPS would. The utility lines, "hot dog stands," moribund dance halls, billboards, and other shoddy disfigurations of one of the world's most dramatic landscapes were removed. Albright, President Hoover, and many others couldn't help but exclaim how good the valley looked, even without fully funded NPS governance. The Department of Interior even published a book of photos by the overachieving Harold Fabian called *Before and After Pictures of Jackson Hole National Monument*. The book persuasively demonstrated the positive changes to the scenic qualities of the valley since acquisition by the Snake River Land Company. NPS director Drury told Junior the book was "eloquent testimony as to the lasting significance of the contribution you have made down there to the cause of park conservation."[33] Ironically, Junior had fretted to Drury that newcomers to Jackson Hole assumed the valley had always looked this good and "therefore, the value of park control is lost on them."[34]

While some continued to complain that the national monument lands deprived Teton County of $13,000 in annual property tax revenue, their moans were drowned out by the squeals of delight from the tourist hordes and those who provided their services and accommodations. Local retail sales tax collections doubled between 1943 and 1946. Deposits in the local bank doubled as well. The price to purchase lots in Jackson tripled in value in as few as four years.[35] An NPS promotional brochure emphasized that tourists were already spending $4.5 million per year in Yellowstone and [the mini] Grand Teton National Park.[36] A University of Wyoming study concluded, "The additional tourist facilities [mainly Jackson Lake Lodge] have more than made up for the loss in land values (used for taxation). To date [1959] it appears that the county has actually come out ahead on the matter of tax base."[37]

The game finally ended like a college football match with bruised and previously bitter opponents shaking hands and slapping backs. Compromise suddenly came easily and led to the final transition of hotly contested Grand Teton National Monument to a bigger, better, more widely accepted Grand Teton National Park. The federal government promised to compensate Teton County for lost property taxes for twenty-five years.[38] Remaining ranchers were given lifetime leases (transferable to heirs) to graze cattle in the national park as well as the right to trail livestock through "stock driveways" and "open right-of-ways."[39] Several of the dude ranches were allowed to continue being dude ranches. Wyoming's congressmen and congresswomen were rewarded with a prohibition on any further establishment of national monuments in their state without congressional approval (ignoring the fact that America's first national monument, Devils Tower, was a much-celebrated presence in their state).[40]

Seeing that the stars were finally aligning, Junior (then seventy-five years old) deeded the 33,562 acres he had acquired in Jackson Hole to the federal government on December 15, 1949. Laurance officially turned over the deeds for Jackson Hole to the new secretary of interior, Oscar L. Chapman, in a ceremony attended by Senator Hunt of Wyoming.[41] The fresh face at the Interior Department also seemed to have helped grease the skids with Wyoming's elected officials. The eloquent Chapman called the transfer of deeds "a high water mark in national park history." He also quoted Thomas Jefferson when he complimented Junior: "It takes time to persuade men to do even what is for their own good."[42]

On September 14, 1950, President Harry S. Truman signed the bill dissolving the old national monument and uniting the SRLC lands with other public lands to create a new, improved Grand Teton National Park that included the entire mountain range, the lakes, the Snake River, and the valley where the bison, elk, deer, and antelope played. Junior, in his usual self-effacing way, congratulated Albright and others for *their* accomplishments. Setting aside twenty-four years of frustration, vicious criticism, investigations, and major expense, and overlooking the fact that Albright had bailed on his leadership position with the NPS seventeen years before the Jackson Hole battle was won, Junior told Albright, "It was a happy day for me when our paths crossed. How much of interest and enrichment you have brought into my life."[43] Albright in turn tipped his hat to Junior with his own gracious prose: "I hope now that you will go out to Jackson Hole ... and get all possible pleasure out of traveling about that wonderful region, unspoiled as it is due to your intercession just at the right time."[44] Grand Teton superintendent John S. McLaughlin, on behalf of his rangers, added to the piling on of appreciation. "We know that adequate protection and preservation of the scene depended upon the lands you have given."[45]

In a response to Wyoming Governor Miller's letter of congratulations, Junior said, "My father had more patience than any man I ever knew. While I have, I fear, far less than he, the training and experience of my life made it clear to me what a valuable quality it is."[46]

Junior wrote to thank the new secretary of the interior Oscar Chapman for his help. "Had I not believed that ultimately those who have opposed what has now been done would come to appreciate its significance and value not only locally, but nationally, I would not have lent myself to the undertaking."[47]

Decades later, Chorley was asked if he thought Junior's commitment to protecting Jackson Hole was related in any way to fulfilling the wishes of Junior's father. "No, I don't think so," Chorley replied. "It was inspired by the beauty of the place, the majesty of these mountains."[48]

Junior remained in possession of over three thousand prime acres on Phelps Lake, the former JY dude ranch property, and several subsequent annexations, throughout his long life. This choice Rocky Mountain haven would remain a beloved family retreat into the twenty-first century. "Oh he [Junior] just loved it!" Chorley recalled. He "rode every morning for a couple of hours, even when he was in his seventies. Then he took a nap after lunch," and at Phelps Lake he would "catch the fish for us both."[49] The old JY Ranch would be the site chosen by Laurance and Mary French for their honeymoon. It nurtured an ongoing relationship between the Rockefellers and Jackson Hole residents, both financial and personal, which led to significant further support for the area (including a new hospital). Piece by piece, over subsequent decades and generations, Laurance whittled down the Phelps Lake inholding and moved the family retreat to a site outside the national park. He donated the last of the Phelps Lake property to the NPS in 2001.[50]

During the summer of 1950, Junior was sitting with the Fabians on the sunny front porch of a log cabin with a view of the snowcapped Grand Teton. "Now Mr. Rockefeller, we've made this beautiful park, we must have some accommodations that'll take care of the people," Harold Fabian remarked. And even though any sane seventy-six-year-old philanthropist would have run, not walked (perhaps with his hair on fire), as fast and far as he possibly could, Junior calmly replied "Why not, Mr. Fabian?"[51]

Twenty-four years had passed since Albright sat the Rockefellers down on the prostrate logs to watch the sun descend behind the Grand Tetons and listen to his vision of adding Jackson Hole and the magnificent mountains to the national park system. Approximately half of Junior's adult life had been spent acquiring land in Jackson Hole and managing it until the federal government finally warmed to accepting his gift. Eight Wyoming governors and eight U.S. senators from Wyoming had come and gone during the period. There had been six secretaries of the interior and four NPS directors. For Junior's legacy as a conservationist, perseverance may have been his most important attribute. In fact, despite his insecurities and lack of business instincts, as a conservationist, John D. Rockefeller Jr. was all but invincible.

Contemporary public lands conservationists should take note of the importance of the restoration of historic Menor's Ferry in the winning of the hard-fought battle for hearts, minds, and political power in Jackson Hole. Conservationists will always face opposition; therefore, empathy and understanding of the opponents' deepest concerns is crucial to favorable outcomes. For most of the ranchers and residents who fought the expanded national park, the preservation of their way of life and heritage was one of their chief concerns. Although not a calculated move, Junior's preservation and restoration of the valley's most important historic sites transformed the attitudes of many toward the eastern millionaire and the national park. The fact that the Fabians involved locals in the restoration process was crucial to the project's success.

Conversely, the lingering complaints of the descendants of those displaced by the creation of Great Smoky Mountains and Shenandoah National Parks (other than inadequate compensation) center on the obliteration of their homes, churches, schools, and other traces of their heritage. Families will never commend the erasure of their history by government action, especially when condemnation is involved. Involvement of the displaced in historic preservation programs is another way to help ease the pain of dispossession. It also makes for the best outcomes in museums, outdoor exhibits, and other historic preservation projects.

OF TREES AND MEN

The Redwood Forest, 1917–1968

According to Horace Albright, the movement to save California's redwood trees really got rolling in 1917 during his participation in the Bohemian Club's annual midsummer bacchanal at the organization's private grove of redwoods seventy miles north of San Francisco. The males-only club was founded in 1872 as a retreat for journalists, artists, and musicians but later admitted businesspeople, politicians, and others who considered themselves sophisticated and longed for reprieve from the hurly-burly boorishness of the Wild West. Albright coyly described the club as a place "where national businessmen and political leaders gather under the redwoods,"[1] neglecting to mention the morning-to-midnight drinking, public urination, and live musical and dramatic performances featuring prominent American men dressed in drag.

While club property sprawled for 2,000 acres along the Russian River, the Bohemian Grove proper was 160 acres of towering, pristine redwood trees. The grove was divided into more than a hundred camps with names like Pleasant Isle of Aves, Woof, Pow Wow, and Valhalla, reminiscent of youth summer camps enjoyed by the children of the well-heeled. And though the Bohemians' cardinal rule was "weaving spiders come not here" (interpreted from Shakespeare to mean "no networking or conducting of business allowed"), Albright, serving as acting director of the National Park Service, did engage in some important conservation business.[2]

As a guest of the club, Albright soon connected with John C. Merriam, University of California–Berkeley professor of paleontology; Henry F. Osborn, president of the American Museum of Natural History in New York; and Madison Grant, head of the New York Zoological Society. The three men of science and letters had gathered at the Pleasant Isle of Aves camp to discuss the World War I–related timber boom that was leading to the mowing down of an alarming number of coastal redwoods (*Sequoia sempervirens*), the tallest trees on earth. At least partly inspired by Stephen Mather's long-time interest in protecting redwoods, the trio resolved to take a tour of some of the redwood forests at greatest risk from loggers. They invited Albright, but he was committed to visiting southern Utah on a journey that would eventually lead to the creation of Zion National Park.[3]

No doubt the glad-handing Mather would have been a better fit with the Bohemian Club crowd than the more reserved Albright, but in 1917 the NPS director was still recovering from his catastrophic breakdown and subsequent depression during the park superintendents' conference. Regardless, the meeting at the Pleasant Isle of Aves camp is viewed by many as the beginning of the Save-the-Redwoods League, an organization championed by Mather and other prominent conservationists including Newton B. Drury, future director of the NPS.[4] It was also the beginning of one of the toughest conservation battles ever fought, and one in which victory remains incomplete. In retrospect, Albright called it "one of the most difficult projects ever attempted."[5]

On their driving tour, the trio beheld magnificent stands of redwoods and philosophized that "living trees like these connect us as by hand-touch with all the centuries they have known."[6] They also observed that the new roads and improved accessibility were a boon to the timber interests like the Pacific Lumber Company. While the mountain redwoods in places like Yosemite and Sequoia National Parks (*Sequoiadendron giganteum*) were too brittle for most commercial uses, coastal redwoods were quite valuable, offering myriad uses from railroad ties to home building to pencils. One whopper redwood, it was said, could provide the lumber for about twenty homes.[7]

By the end of their tour, as Albright later recalled, the trio "had fire in their eyes."[8] They were determined to join with other likeminded persons to raise money to purchase vulnerable lands and promptly wrote a letter to California governor William B. Stevens urging action to save the last best redwood groves. Soon thereafter, Mather was sufficiently recovered (and filled with manic energy) to lend his hand to the cause. The Save-the-Redwoods League was formed in 1918 and became a bona fide California nonprofit corporation in 1920 with the progressive Franklin K. Lane, secretary of the interior, as president, and Drury as secretary.

Grant, Merriam, and Osborn, the founders of the league, had more in common than simply being Bohemian Club enthusiasts, lovers of redwoods, and influential scientists; they were also passionate leaders in the field of eugenics. Viewed today through the lens of Nazi Germany's eugenics-influenced slaughter of some eleven million Jews, homosexuals, eastern Europeans, and other innocent people, eugenics is synonymous with the worst imaginable evil. However, during the early twentieth century, eugenics was, sadly and naively, widely accepted "science" in America and Europe.

In 1924 Madison Grant wrote Raymond Fosdick (one of Junior's chief advisors and a trustee for the Rockefeller Foundation) about touring the redwood country and helping the league save a pristine redwood forest in Bull Creek Flat. In closing, Grant made an odd reference to the Galton Society, a group of scientists headed by Grant devoted to eugenics research and forwarding eugenics policies. Grant stated the Bull Creek Flat redwoods were "worth far more than any cathedral or gallery of paintings or sculpture. Any of the last can conceivably be replaced by greater artists—that is, if the Galton Society is enabled to carry out its plans for a super-race."[9]

The Galton Society was named for Sir Francis Galton, statistician and cousin to Charles Darwin. Galton took Darwin's knowledge of genetics and theory of natural selection and somewhat logically combined it with the human race's long-time propensity for breeding livestock and hybridizing plants to create the most productive commodities. Galton was most excited about pairing up human geniuses (not unlike Ivy League colleges had been doing for years) to theoretically create super-smart people who could use their superior intellect to solve all the world's problems. This was known as positive eugenics. On the much darker side, negative eugenics discouraged those deemed less brilliant (usually by junk science intelligence tests and quacky psychological exams) from reproducing.

Of the many eugenicists of the pre–World War II era, Grant was one of the most overtly racist. His popular book, published by Simon and Schuster in 1916, *The Passing of the Great Race* (with an introduction by Osborn), divides the peoples of the world into numerous regional groups, with select northern Europeans (Grant was of English and French descent) ranked as the planet's fittest. Grant fretted that America's pure Nordic (derived from a French word meaning "north") stock was being diluted by immigrants from southern and eastern Europe (especially Jews). Adolf Hitler wrote Madison a note from the German prison in which he resided, thanking him for authoring the book and stating he considered the German translation his "Bible."[10]

The eugenicists were bolstered by Americans' growing fear of the wave of immigrants swelling America's cities in the 1920s. Consequently, Grant served as vice-president of the Immigration Restriction League and helped push through America's tragic Immigration Act of 1924, restricting the number of eastern European immigrants. Eugenicists also supported the involuntary sterilization of some sixty thousand men and women in mental institutions and other facilities, a policy approved of by a majority of Americans and affirmed in the Buck v. Bell Supreme Court decision in which Oliver Wendell Holmes Jr. callously wrote, "Three generations of imbeciles are enough."[11]

Junior had a number of friends and colleagues who were avid eugenicists, including Richard Corwin of the Colorado Fuel and Iron Works, Osborn, and Charles W. Eliot, president emeritus of Harvard and one of Acadia's "triumvirate" of founding conservationists. Junior was also a member of the American Eugenics Society, and the Rockefeller Foundation funded eugenics research in America and Europe, including Germany.

Yet, to conclude that Junior was a racist or that he condoned Nazi genocide would be a difficult assertion to support. Without the perspective of Nazi atrocities, in the context of the era, eugenics was a widely accepted branch of science. It was taught in the science departments of Harvard, Yale, and Princeton as well as in public schools. It was backed by the U.S. Department of Agriculture, U.S. Forest Service director Gifford Pinchot, and President Theodore Roosevelt. County fairs hosted eugenics-based "fitter family" and "better baby" competitions. Participating families were evaluated for physical appearance, intelligence, health, and behavior to predict which could produce superior offspring.[12]

Gifford Pinchot and Theodore Roosevelt collaborated to insert eugenics into the Progressive agenda. In the same 1909 progressive government vitality report that called for restaurant health inspections and child labor laws, author Irving Fisher, a Yale economist, made a eugenics to-do list, stating the theory "should be studied and gradually put into practice. This involves the prohibition of flagrant cases of marriages of the unfit, such as syphilitics, the insane, feeble-minded, epileptics, paupers, or criminals, etc." He then called for the "unsexing of rapists, criminals, idiots, and degenerates generally." In all of the twentieth century's megatons of government reports, it's unlikely one could find a more mortifying use of the abbreviation "etc." or the adverb "generally."[13]

Like Hitler, Theodore Roosevelt, a close friend of Grant's, praised his *Passing of the Great Race*. He described the contents as "the facts our people need to realize" and declared "all Americans should be sincerely grateful to you for writing it." Other pro-eugenics American presidents included Warren G. Harding and Herbert Hoover.[14]

Despite the eugenicists' phobia of Jewish immigrants and others from southern Europe, Junior and the Rockefeller philanthropies were largely sympathetic toward the plight of the newest Americans. In 1905, Senior agreed to a request from the Hebrew Technical School for Girls of New York for $15,000 (over $500,000 in 2025 dollars) for a new building where girls could learn practical skills and receive a general education in literature and other topics. Junior suggested to his father that the family contribute more, but he was overruled. Beginning in 1909, Rockefeller philanthropies supported the Society for Italian Immigrants and contributed to the North American Civic League for Immigrants which discouraged fraud against immigrants and helped recently arrived children get into schools. Other immigrant assistance organizations receiving Rockefeller funds included the Foreign Language Information Service, New York City Education Committee for non-English-speaking women, the Young Men's Hebrew Association, the Young Men's Christian Association, the Young Women's Christian Association, the International House in New York, and programs to provide better housing for single women and industrial workers.[15]

According to Eileen Rockefeller, Junior and Abby's granddaughter, Abby made a point of visiting recent immigrants employed by Standard Oil of New Jersey. She called upon the wives and children at their new homes and formally welcomed them to the United States. She asked about the traditions of their homelands and asked to see any art or heirlooms they brought with them. In her own friendly way, she encouraged them to continue celebrating their heritage as they blended into the melting pot.[16]

While proponents of eugenics pushed to protect the "superior" northern European gene pool and weed out the others, Rockefeller philanthropies were nothing if not champions of the sick, poor, and marginalized. Junior was the grandson of Harvey Buel Spelman, a member of the Ohio legislature, staunch abolitionist, and operator on the Underground Railroad. In 1854, while in high school, Senior wrote that slavery was "a violation of the laws of our country and the laws of our God."[17] As a struggling young clerk, Senior helped a Black woman in Cincinnati buy her husband out of slavery, and during the Civil War—before he became wealthy—he donated to Black orphanages, churches, and a deaf and mute society. During the late nineteenth century, Senior gave generously to Native American Bacone College in Oklahoma and had Rockefeller Hall named in his honor. Spelman College, the prestigious historically Black women's school in Atlanta, was named after Laura's (Senior's wife's) family in acknowledgment of the Rockefellers' contributions.[18]

The creation of the Rockefeller General Education Board (a much, much, much larger philanthropy than Junior's conservation projects) was spurred by the knowledge there were no high schools for Black students in the South. Between 1903 and 1910, the board fostered the creation of eight hundred southern high schools. Junior personally made significant contributions to the Fisk, Hampton, and Tuskegee institutions and chaired the United Negro College Fund's national council for ten years. Education historian James D. Anderson declared that during the first three decades of the twentieth century, the Rockefeller General Education Board had "virtual monopolistic control of educational philanthropy for the South and the Negro."[19]

In their personal lives, Junior, Abby, and their children were recognized as allies of the stricken and marginalized. In 1923, a year when membership in the Ku Klux Klan exceeded three million Americans and two hundred thousand Klansmen and supporters gathered in Kokomo, Indiana, for mass rallies, Abby decided she was overdue in addressing her three eldest sons—all of whom were away at college—on the topic of racism.[20]

In a letter, Abby stated: "For a long time I have had very much on my mind and heart a certain subject... that one of the greatest causes of evil in the world is race hatred or race prejudice." Abby pointed to the groups most often victimized as the "Jews and the Negroes." She urged her boys "to begin your lives by giving the other fellow a fair chance and a square deal. I long to have our family stand firmly for what is best and highest in life. If you older boys will do it the younger will follow."[21]

Because Junior was a very private person, there are only a few documents and anecdotes that reveal much about his personal life. One such anecdote is from the wedding of daughter Babs to Dave Milton at the Rockefeller family home in Manhattan on May 14, 1925. Although more than a thousand guests attended the extravagant (though bone-dry) reception, thousands more gathered outside the nine-story home to watch from a distance. After the wedding was over and the guests and newlyweds had departed, Junior went out on the street to chat with the uninvited onlookers who were still quietly milling about in the mild evening air. He ushered small groups into the Rockefeller home and showed them the room where the ceremony had been held and "filled their arms with flowers."[22] Within the last of his little tour groups, "he took the arm of an elderly Black woman carrying a shopping bag and personally assisted her into the house."[23] Anyone who knew Junior well might have conjectured that his generosity was at least partially motivated by his need not to let all the fresh flowers he had purchased go to waste.

In 1930 the gargantuan Rockefeller Foundation became a major funder for psychiatry and neurological sciences. One early result was "some progress in the treatment of epilepsy."[24] The General Education Board contributed more heavily to medical education than any other field and the Rockefeller Foundation has focused on public health and the elimination of disease, malnutrition, and poverty since its beginning. The massive Laura Spelman Rockefeller Memorial was tilted toward "the needs of women and children."[25]

There is not much in the record regarding any direct statements or comments by Junior on eugenics. One anecdote from a 1925 dinner conversation recalls that Junior acknowledged an interest in eugenics but "urged caution in drawing any conclusions" about assertions of deficiencies in the children of interracial couples.[26]

According to philanthropy scholar William A. Schambra, our nation's first general-purpose philanthropies "backed eugenics precisely because they considered themselves to be progressive. After all, eugenics promised to solve many of the world's problems through the application of new advances in the natural and social sciences. Foundations boasted that they could delve down to the very roots of social problems, rather than merely treating the symptoms."[27]

As to the Rockefeller Foundation's pre–World War II support for eugenics research, Lia Weintraub's award-winning paper ascertained the foundation "had not received adequate explanations of the projects they would be funding and most likely, was not alerted to the connection between their projects and the Nazi regime. The Rockefeller Foundation would not have known the injustices that its policies would fuel."[28] (The Rockefeller Foundation is currently conducting its own extensive research on the organization's past support of eugenics research.) Between 1933 and 1945, the Rockefeller Foundation also helped over three hundred scholars escape Nazi persecution.[29]

Regarding the unsettling correlation between conservationists and eugenicists of the day, some writers have suggested that Grant's, Merriam's, Eliot's, and the Osborns' desire to save the last of the greatest trees, animals, and wildlands on earth could be equated with their longing to protect the last of the beleaguered (in their minds) Nordics.[30] While this motivation is certainly plausible, a simpler explanation would be that eugenics was part of the conservationists' progressive politics and was the new and intriguing "science" of the day, especially for America's and Europe's white, well-to-do, biological scientists and scholars. Still, any lack of empathy for those considered mentally ill or deficient by such elites can only be looked upon as a cruel and inexcusable offense.

*

In the four years between Mather's recovery from depression and Grant's letter to Fosdick, the Save-the-Redwoods League slowly grew in membership and influence. Mather was one of the first major contributors to the league and one of its most effective fundraisers, calling upon his long list of upper-crust friends (especially from UC–Berkeley), while also rallying locals in redwood country. Importantly, Mather persuaded the publishers of *National Geographic* and the *Saturday Evening Post* to publish stories on threats to the coastal redwoods.[31] By 1921, the league had more than four thousand members and had raised sufficient funds to purchase small but important tracts of redwoods along the highway in Humboldt County. More importantly, they had convinced the major lumber companies to postpone cutting the best redwoods in places like Bull Creek Flat and persuaded the California legislature to approve $300,000 in matching funds to purchase private properties containing redwoods.[32]

In April of 1924, John Merriam sent a follow-up letter to Fosdick, asking for Junior's urgent help in purchasing "a famous area known as Bull

Creek Flat [200 miles north of San Francisco near Highway 101], considered by a great number of Californians as the finest of all redwood groves."[33] A few months later, Drury set off even louder alarm bells by announcing the Pacific Lumber Company was planning to build a logging railroad "through Dyerville Flat, presumably leading to Bull Creek Flat."[34]

In the fall of 1924, in response to this seemingly imminent slaughter, NPS director Cammerer contacted Junior's office with one of his thorough and persuasive memos on behalf of the national parks. Cammerer stated the U.S. government had long been interested in a redwoods national park and desired to save the best trees from becoming "grape stakes and shingles." After a number of investigations, it was decided Dyerville and Bull Creek Flat were the best candidates for national park status. Emphasizing the urgency of the situation, Cammerer warned, "It takes thousands of years to build one of these trees, and only a few hours to throw them prostrate for the mills."[35]

Having set the stage, Cammerer then made the request. It was a nice round number, $1 million ($18.5 million in 2025 dollars) to Fosdick's "principal," Junior, to purchase Dyerville Flat (where logging had already begun) and a substantial portion of Bull Creek Flat. Cammerer concluded with the assurances that "we have carefully avoided bringing his [Junior's] name into the light, wherever his interest has touched a national park or conservation measures of the sort."[36]

Although Junior took a cautious, incremental approach to contributing to projects in Acadia, Mesa Verde, and Yellowstone in 1924, he did not tarry when it came to saving the redwoods. Like his father, Junior believed in testing the philanthropic waters with a small ante before committing himself. But in this case, Junior quickly responded to Cammerer's request with a pledge of $1 million, acknowledging the urgency of the situation and the irrevocable damage about to be done.[37]

Unfortunately, the Pacific Lumber Company was not quite ready to let go of Dyerville or Bull Creek. First the lumber company offered to donate some nice redwood properties to the California state parks in exchange for a pledge from the state legislature not to use eminent domain on Bull Creek and other PLC properties. The unproductive wrangling went on through 1925 and until the Rockefeller family arrived in redwood country in the summer of 1926, following stops in Mesa Verde, Yellowstone, and New Mexico.[38]

In California, word somehow leaked to the press that the Rockefellers were in the state "traveling under cover of unusual secrecy" with Junior "incognito as John Davison."[39] Over the following days the *Los Angeles Times* crowed that the Rockefellers' "mask was pierced" and a reporter had obtained an exclusive interview. Said scoop was actually a few remarks delivered by Junior in front of the borrowed private rail car they had been traveling in. "The real reason we are traveling as we are is to allow my two younger sons . . . to see the West," Junior explained. While he posed for a few pictures and chatted with a couple of familiar faces in the crowd at the train station, Junior was adamant that his children not be photographed. "They are plain American boys and I wish to raise them as democratically as possible. Newspaper publicity might affect them at their age."[40]

Newton Drury, who would be the family's tour guide through the redwoods, feared the media leak would be pinned on him or the Save-the-Redwoods League and was reportedly "in a disturbed state of mind."[41] However, since the press failed to learn of the Rockefellers' visits to the redwoods until after the fact, it's unlikely Drury or the league were the leakers.

Drury, who was yet another University of California grad, and like Mather, had been an advertising man, became involved with the league as a promotor and fundraiser. He served as the league's executive director from 1919 to 1939 after which he became NPS director. Whenever VIPs visited redwood country, chances were good that Drury would be their guide. He also booked the visitors' accommodations in Eureka and Crescent City and arranged for touring cars and picnic lunches for the Rockefellers on their journey along the 213-mile Redwood Highway from Dyerville to Smith River.[42]

A highlight of the redwoods visit was a mass noontime picnic held at Bull Creek Flat with Drury, the Rockefellers, and carloads of locals anxious to meet the famous family. While it's fair to say there were few more reticent Americans than Junior, it is also accurate to note that he was "blown away" by the redwoods. "I have never seen redwoods like these before," he told a reporter for the *Santa Rosa Press Democrat*. "They are, to me, the most impressive of all living things."[43] A week later he declared to Merriam, "The preservation of all of them possible for the enjoyment of not only present but future generations is a task justifying any effort or sacrifice."[44]

Over the fall and winter, Drury mailed Junior maps and documents and generally kept him apprised of the redwoods situation, mainly the negotiations between the league and the timber companies. In March of 1927 Junior reiterated he was "immensely interested in the [redwoods] project and want to be helpful in any way I can." Consistent with his unflagging perfectionism, he maintained that "nothing short of the ideal thing should be considered."[45]

In August, at around the same time he was contemplating a $5 million contribution to Great Smoky Mountains National Park and more than $1 million to purchase land in Jackson Hole, Junior was likewise still all in on the redwoods. As Drury continued to cry for assistance, Junior fretted "not only will much of the land be cut over by lumbermen if there is undue delay, but also prices will tend to rise and ultimate costs be larger." Consequently, Junior resolved he was willing to contribute $5 million "and possibly a good deal more" to save the Dyerville, Bull Creek Flat, Prairie Creek, and Del Norte Coast forests. Depending on how one interprets "a good deal more" (likely the $10 million total price tag of the four crucial areas), Junior was willing to put $16 million (over $289 million in 2025 dollars) toward park conservation projects in late 1927.[46]

As it turned out, thanks to other funds raised by the league and appropriations from the state of California, Junior was only on the hook for another $1 million for the time being. Over the next few years, many of the threatened redwoods were eventually protected as state parks, including Bull Creek Flat, which Drury hailed as "unquestionably the greatest forest in the world."[47]

*

More than twenty years later, after timber companies had completed harvesting at least 90 percent of the coastal redwoods on earth (almost everything except what was officially protected in state and county parks), newspaper articles once again alerted Junior to an endangered stand of ancient trees. By this time Junior was in his late seventies and carried the scars of some of the twentieth century's most contentious conservation battles. He had also distributed much of his remaining fortune to his children and invested an enormous sum in Rockefeller Center. Yet, an article in *National Geographic Magazine* called the world's attention to the Calaveras South Grove, one of the last major redwood forests still in private hands, and home to several of the tallest trees in the world. The league had been able to dissuade the Pickering Lumber Company from harvesting the trees for a decade, but time had finally run out. Albright couldn't help but entice

Junior in a 1950 letter, "There is no question but that the South Calaveras Grove is of National Park stature."[48]

Albright, who had resigned as NPS director some seventeen years earlier, made the redwoods pitch to Junior rather sheepishly, but it's obvious he still knew how to push the Rockefeller philanthropy buttons. Earlier, in June 1949, Albright told Junior that Calaveras was "the only great natural feature conservation job to be done," and mentioned that the state could provide matching funds. He then backpedaled apologetically: "I have tried all these years not to bring new projects to your attention and I am not doing that now in the sense that I am suggesting that you take action one way or another."[49]

Junior had bristled when Albright first broached the topic of the South Calaveras Grove in 1949 (prior to settlement of the Jackson Hole controversy). "You will not misunderstand my saying to you that the high regard in which I held the National Parks when you were its head has undergone a marked change with the passing years and my enthusiasm for cooperating with the Service is considerably dampened. It is not probable that I would be interested to contribute to any further park projects, however, if you care to tell me more about the Calaveras South Grove...I will at least review the situation."[50]

One senses that Junior had a very hard time absolutely saying no to the prospect of saving big trees. The date of these rather blunt words from Junior also coincides with some definite light at the end of the Jackson Hole controversy tunnel. In fact, Junior's words could easily be interpreted as applying a little added leverage to Albright and his political allies to finally complete the expansion of Grand Teton National Park. The fifteen-year freeze (or at least hard frost) in Junior's magnanimous philanthropy for national parks was showing signs of a possible thaw.

Following the old Rockefeller philanthropy preference for partnerships, Junior ended his inertia and pledged $250,000 "on condition that the other half be secured from other interested individuals [and] preferring to be in the position of matching money which other people have agreed to give."[51] Unfortunately, the value of the trees as timber was skyrocketing and the price Pickering wanted for the lands continued to climb. So did the company's eagerness to begin cutting giant sugar pines and redwoods.

When the cutting of the Calaveras South Grove became imminent (and the expansion of Grand Teton National Park had finally become a done deal), Junior telegrammed Albright: "if entire project ... can be carried out and you so recommend will give total of whatever is necessary in securities of the approximate value of one million" to protect the spectacular grove.[52] When Junior's $1 million ($12.5 million in 2025 dollars) was paired with $1.67 million from the state of California, $130,000 in Calaveras Grove Association and Save-the-Redwoods League funds, and some U.S. Forest Service lands, it was just enough to close the deal with Pickering. The South Grove became part of the Calaveras Big Trees State Park in 1954.[53]

Albright's, Cammerer's, Junior's, and Mather's dream of a redwood national park had to wait until the late 1960s, however, after the latter three proponents had passed. Paul A. Zahl and others surveyed remote Redwood Creek in 1963 and documented significant stands of virgin redwoods and three specimens taller than 360 feet! The subsequent media coverage kick-started one of the orneriest conservation battles ever, pitting conservation group against conservation group, local against tourist, timber company against preservationist, California against the federal government, and even Governor Ronald Reagan against President Lyndon B. Johnson.

Junior and Abby's son Laurance, acting as chairman of the Citizens' Advisory Committee on Recreation and Natural Beauty, was heavily involved in the long fight. In 1968, after the vast majority of the planet's redwood trees had been milled into "grape stakes and shingles," President Johnson signed the bill for the compromised, undersized, fifty-eight-thousand-acre park. It included three existing state parks: Jedediah Smith, Del Norte Coast, and Prairie Creek, plus an additional eleven thousand acres of old-growth redwood forest. A few centuries prior to the park's establishment, indigenous peoples could have happily beheld 1.9 million acres of redwoods in the area.[54]

While the California State Park System has done a commendable job of protecting coastal redwoods, these areas, combined with the national park (now expanded to seventy thousand acres), still protect only a paltry fraction of one of the world's most magnificent ecosystems. Redwood forests are the best carbon dioxide sponges on our planet and also protect many imperiled plants and animals as well as clean water for fish and people downstream.

Redwood National Park, with its "most impressive of all living things," is mournfully only one-third the size of diminutive Shenandoah in Virginia. Although opportunities for saving significant new tracts of old-growth coastal redwood forest have passed, the prospects for adding cutover acreage for the future are excellent. Redwoods are some of the world's fastest growing trees, and by the time your great-grandchildren make their pilgrimage to northern California, second-growth redwoods will be over 150 feet tall. We should not squander our second chance to create a right-sized Redwood National Park. Or parks.

BACK IN THE GAME

The Blue Ridge Parkway, 1950–1952

It's not unusual for potential national park lands to become available adjacent to an existing national park. This is true in most of our parks today and highlights how public-private collaborations to expand our existing parks could be key to alleviating severe overcrowding in the future. Fortunately, as in the example of Linville Falls on the Blue Ridge Parkway, many owners of lands adjacent to national parks possess a strong desire to see their lands protected, greatly simplifying the process of park enlargement.

In 1950, Fritz Hossfeld, one of the last of the F. W. Hossfeld heirs—the long-time benevolent owners of Linville Falls—passed on. The land went to his sister, Giula, and rumors circulated that she was willing to sell. The family had protected the scenic gorge and refused to sell the timber for decades (though they did charge a fee—a la Niagara Falls—for the public to view the cataract). Lumber companies were already cutting the old-growth hemlocks and hardwoods in the vicinity of the Hossfeld property, and they would have been glad to acquire the Hossfeld's magnificent eleven hundred acres and harvest the ancient trees. Both the National Park Service and U.S. Forest Service were also potential buyers, as the acreage bordered the Blue Ridge Parkway and Pisgah National Forest. Neither agency, however, had the funds to make the offer.[1]

In the early summer of 1950, Junior and his children were still mourning the unexpected death of beloved wife and mother Abby. She had passed away in her bed from a heart attack at age seventy-three. The "yeast in the bread" of the Rockefeller family had departed far too soon.[2] Junior was "totally devastated by her death," granddaughter Peggy Rockefeller recalled. As the children and their spouses watched helplessly, the widower called out in unbearable pain, "Take this blow away." And though he was immediately surrounded by sympathetic family, "he was terribly lonely, hopelessly lonely."[3] Over two thousand letters and telegrams immediately poured in and friends remembered Abby Aldrich Rockefeller as "the first lady of this country."[4]

As a way to take their minds off their loss, David Rockefeller suggested a father-son road trip. Junior chose the Shenandoah Skyline Drive and Blue Ridge Parkway from Washington, DC, to Asheville, North Carolina, a five-hundred-mile cruise through sublime southern Appalachian mountains and forests. David was about to celebrate his thirty-fifth birthday and had recently embarked on his long career with Chase National Bank in New York.[5] Junior was seventy-six, could count over two dozen grandchildren, and split his time between the family estate in Pocantico Hills, the Eyrie in Maine, the old JY dude ranch in Jackson Hole, and a modest house in restored Colonial Williamsburg called Bassett Hall. The old manse on West 54th Street had been torn down and the property donated to the Museum of Modern Art.[6]

The Rockefellers plus a chauffeur started their tour in Shenandoah National Park but had to detour off Skyline Drive because of dense fog. Even in the 1950s, Junior preferred travel by train or automobile to flying. He "never was in an airplane in his life," Kenneth Chorley recalled.[7] David and Junior took the alternate route in stride and enjoyed the beauty of the Shenandoah Valley. When they finally reached the still-incomplete Blue Ridge Parkway, the weather was perfect.

During some years, on the higher elevations of the parkway, there is a short window when all three floral superstars—Catawba rhododendron, flame azalea, and mountain laurel—overlap in their periods of bloom. Mountain laurel must linger from its late-May peak, flame azalea (ranging in color from orange to cream to lava red) must also persist from May, and the big purple flowers of Catawba rhododendron must open a tad early. The Rockefellers hit the big bloom perfectly, once again managing to arrive at the right place at exactly the right time.

David, who at age seven had attended a summer class on natural history in Maine and henceforth established himself as the family's chief naturalist, took note of the area's diverse flora and fauna. He became more than a casual entomologist over the years, assembling a serious collection of over 150,000 specimens of beetles. Junior was of course captivated by the parkway's extraordinary landscape architecture. Although the early parkways of New Jersey and New York (some of which Junior helped create) had set the bar for large-scale landscape architecture projects, the Blue Ridge Parkway is recognized worldwide as a true masterpiece. The 469-mile-long scenic road was designed by a twenty-five-year-old, Stanley W. Abbott, and his associates. Abbott's vision was "to fit the Parkway into the mountains as if nature had put it there" and intersperse recreation areas and historic sites spaced "like beads on a string; the rare gems of a necklace."[8] Construction of the Blue Ridge Parkway began in 1935, jump-started by a flurry of Depression-era, New Deal make-work initiatives. At the time of the Rockefellers' 1950 visit, only about half of the Parkway had been completed.

Horace Albright, who was intensely interested in adding Linville Falls and Gorge to the Parkway, had alerted Blue Ridge Parkway superintendent Sam Weems as soon as he learned of Junior and David's excursion. Albright and Weems's best scheme for persuading Junior to purchase the Linville Falls property was predictably similar to Albright's plans decades earlier to encourage Junior to buy Jackson Hole and a grove of California redwoods. Basically, send their mark on a long scenic drive and unfurl a nice picnic lunch where the potential benefactors might gaze upon the pristine, yet endangered landscape.

Weems initially found Junior and his chauffeur sitting in their Cadillac at a scenic overlook along the parkway. Upon seeing the uniformed ranger approach, Junior asked, "Oh, are we in violation of some regulation?"[9] Weems replied to the contrary. He then listened to Junior's explanation for the missing David. The party had received an overly generous picnic lunch that morning from the hotel, and Junior, who like his father always deplored waste, proposed they share. They spotted a humble log cabin off the parkway and David trotted down through the briers to deliver the leftovers. Upon his return, David reported the residents—a widow and several children—were happy to receive the sustenance.[10]

Junior told Weems, who had come to the parkway as a project manager fifteen years earlier, his scenic road was "an extraordinary achievement. I have seen nothing finer of its kind."[11] He also complimented Weems on its maintenance. "This is the most picked-up park I've seen."[12]

The entourage spent the night in Bluffs Lodge at Doughton Park, a small, very rustic and very quiet accommodation surrounded by forest and meadows. The lodge's restaurant, famous for its fried chicken and biscuits, likely provided the next, crucial picnic lunch. In the morning, David and Junior piled into Weems's government car and had their chauffeur follow in the Caddy. Weems noticed it was about time for lunch as they neared Linville Falls and the group found a comfortable place to spread the picnic blanket beneath a grove of towering, old-growth hemlock trees. Weems invited the Rockefellers on a short hike to view the falls, but Junior declined, citing doctor's orders to avoid exertion. David, however, was game.[13]

Linville Falls could hardly be more beautiful. The Linville River is large and often grows to an absolute torrent after a typical southern Appalachian thunderstorm. The falls plummet over sheer rock walls in two steps, the second 45 feet tall, then churns through a narrow gorge replete with natural hanging fern gardens and uncommon species of rhododendron, bryophytes, and other flora. The path to the viewpoint is through 150-foot-tall white pine and hemlock, scenery so rare it was chosen by location scouts for films requiring scenes of primeval eastern forest. David noted an unusual low-growing shrub: Allegheny sand myrtle, an Ice Age relict that barely clings to a few rocky outcrops in the southern mountains.[14]

Back at the picnic, one imagines the group chewing on delicious drumsticks and digging into potato salad while Weems gives his pitch for saving the falls and its precious eleven hundred old-growth acres. He warned that lumber companies were eager to turn forest giants into two-by-fours and plywood. And just as Arno Cammerer had "filled his briefcase with all the photographs" in his meeting with Junior on the Smokies, Weems handed over a fat folder of photos, maps, and documents related to Linville Falls and Gorge.[15] Weems mentioned a ballpark figure of $100,000 ($1.35 million in 2025 dollars). Junior ended the discussion with a characteristically noncommittal "That's very interesting."[16]

*

It had been a while since Junior had engaged in any major philanthropic endeavors on behalf of national parks. In early 1947, Junior informed Chorley, "I do not find myself disposed to enter into new ways of helping the Service." He did mention, however, that if "the particular matter" (no doubt Jackson Hole) was resolved, "I may find myself more mellow toward cooperating with the Service in other ways."[17]

There is evidence that Albright's 1933 resignation from the NPS would never sit quite right with Junior, even though Albright remained highly active in park affairs long afterwards. Nearly two decades after Albright stepped down, Junior suggested the former NPS director had made a big mistake. "If you had only been able to stay at the head of the National Park Service, the history of the national parks over the last twenty years would, I feel sure, have been very different."[18] Primarily, though, Junior's disappointment was due to a whole lot of government inaction on accepting his gift of thirty-three thousand acres of land in Jackson Hole. In a 1956 letter to his children, Junior stated the Jackson Hole conflict was "burned into [his] very being."[19]

One can be certain, however, that Junior never lost his respect for Albright. "He [Albright] is one of the finest men I know," Junior told his son Nelson in 1953. "He has extraordinary public spirit and knows more about the national parks and conservation than any other man of my acquaintance."[20]

Auspiciously, around the same time that Albright and Weems were making their pitch for Linville Falls, things had finally come together in the Grand Tetons. Albright, despite his full-time job with the potash company, had never let up on working to close the Jackson Hole deal. Consequently, Junior conceded to Albright that he would likely be able to scrounge up half the $100,000 needed for Linville Falls if Albright and Weems could find matching funds. Albright and Weems replied that they were not aware of any potential matchers and reminded Junior that the clock was ticking on their option to buy the land. Once again, the NPS successfully used the prospect of falling old-growth forest to leverage Junior. He telegrammed Albright:

> if you are in agreement approve and authorize closing option of one hundred thousand dollars on blue ridge project even if i pay entire amount. j.[21]

Three months later, in January of 1952, the *New York Times* proclaimed: "The National Park Service announced today that through a grant of about $100,000 by John D. Rockefeller Jr. Linville Falls in western North Carolina would be added to the 500-mile Blue Ridge Parkway." The acquisition included the falls, gorge, and eleven hundred adjacent acres of pristine forest.[22]

Once the dust had settled in Jackson Hole, correspondence between Albright and Junior confirmed the park-rescuing dynamic duo was back. Albright discussed progress on acquiring Linville Falls, talked up the need to save the Calaveras grove of redwood trees in California, and complimented Laurance on his purchase of land for Palisades Interstate Park (making him the third generation of Rockefellers to be involved in that Hudson River conservation project). Albright never seemed to have a single unkind word regarding the New York tycoon. In a 1964 interview with writer Michael Frome, Albright stated, "I felt close to Mr. Rockefeller from 1926 until he died. He regarded me as his advisor on conservation problems. My esteem for him and my admiration for him amounted almost to making him an idol. I think he was one of the finest men I have ever known."[23]

It's important to observe that the Linville Falls expansion took barely a year and half and involved minimal gnashing of teeth. It did require a willing donor, a gracious seller, and an enthusiastic agency to tackle the challenge. Plus a decent picnic lunch. Expanding existing national parks is one of the simplest and least controversial ways to solve our overcrowding and the growing problem of loving our parks to death. Opportunities abound. Who hasn't wondered why the Rocky Mountains are so vast and Rocky Mountain National Park such a postage-stamp-size preserve? Why shouldn't the Blue Ridge Parkway be expanded outward to make it a viable protector of biodiverse southern Appalachian ecosystems (rather than a narrow ridgetop right of way) and wildlife migration corridor? Wouldn't that help limit extinctions caused by climate change? Wouldn't it be splendid if Yellowstone and Glacier could likewise be connected by a preserve wide enough to encourage the migration and intermingling of wildlife and plants?

Yes, there are a thousand reasons why this would be difficult, just as there were in Grand Teton, Yosemite, Acadia, Great Smoky Mountains, Shenandoah, the Blue Ridge Parkway, and many others.

*

In 1952, after Junior telegraphed Albright at his office in New Mexico agreeing to provide up to $1 million to save the South Calaveras grove of redwoods, Laurance S. Rockefeller, son of Junior and Abby, and his wife, Mary French Rockefeller, landed their boat *The Dauntless* on St. John, U.S. Virgin Islands.[24] Laurance and the Jackson Hole Preserve, Inc. (successor to the Snake River Land Company), soon began buying old sugarcane plantation land to create an ecotourism resort called Caneel Bay, and, following in his father's footsteps, a national park. The NPS had recommended a national park be created on St. John as early as 1939, and the 1950s project proffered by Laurance melded perfectly with the Park Service's Mission 66 ambitions to better disperse park visitors away from overcrowded areas.[25]

As Junior and Laurance corresponded about the project (mostly the son asking his father for contributions), Junior warned that the process Laurance and the Jackson Hole Preserve were pursuing could lead to another Jackson Hole-esque fiasco with the park in limbo for decades and another Rockefeller left holding the taxable bag. Then Junior concluded, "Let me say that I have been greatly pleased with what you have done in this whole matter and the fine position of leadership you have taken in so significant and public spirited an enterprise. To go along with you in it, even in a small way, gives me the utmost pleasure and satisfaction."[26]

On August 2, 1956, President Dwight D. Eisenhower signed the papers creating America's twenty-seventh national park: Virgin Islands.

The baton had been passed.

CONCLUSION

Although it is difficult to calculate an exact figure (especially with adjustments for inflation), it can safely be said that Junior contributed some $50 million to the national parks and redwoods projects during his lifetime. Roughly, this equates to over $780 million in 2025 dollars. After Junior's passing, Laurance Rockefeller and the rest of Junior and Abby's children continued to provide substantial support to national and state parks.[1] The total sum of Rockefeller philanthropy from the 1850s to 2011 has been tallied at well over $200 billion.[2]

Most scholars and ethicists maintain that government is superior to philanthropy for doing the public good. A democratically elected government's tools of taxing and spending are more equitable and egalitarian than philanthropy, in which a small group of often elite trustees control the distribution of funds. Government, however, is sometimes slow to act and its bureaucracies may hamper efficacy.[3]

The activities in this book demonstrate the efficacy of well-intentioned public-private partnerships. During the 1920s and 1930s when John D. Rockefeller Jr. conducted the bulk of his public lands' philanthropy, America's Congress was adamantly opposed to purchasing lands to create new national parks and expand existing ones. The short-sighted politicians also refused to fund most building projects in parks and were beyond draconian in staffing. In many national parks and monuments (Acadia, for example), Congress initially appropriated money for a single employee and paid that person a token salary of one dollar per month. The result was new "preserves"—especially in the West and Southwest—being overrun by pot hunters, wildlife poachers, rock-hammer-wielding souvenir collectors, and grave robbers. Much of the damage done in those years to America's treasures is, very sadly, irreversible.

Fortunately, a small number of enlightened public lands staff were wise enough to see beyond government's limited coffers and to seek out wealthy individuals who might help them protect wonderland. The vast majority of Gilded Age plutocrats of the day had little interest in supporting government projects, especially since state and federal governments were hard at work busting their trusts, regulating their industries, and taxing their income and inheritances.

Horace Albright, Arno Cammerer, George Dorr, Stephen Mather, and John D. Rockefeller Jr. were the right people at the right time to establish and augment a system of magnificent national parks—for the benefit and enjoyment of the people—that will always be the envy of the world. The top National Park Service brass were uniformly brilliant, strategic, personable workaholics who put the creation and preservation of public lands above their own career goals and private lives. In his role, Junior was more than a check writer. As a self-taught landscape architect, he helped set the standards for the park roads from which the vast majority of visitors happily experience their national parks today. He was able to contribute the skills of a giant pool of the very best lawyers, accountants, architects, landscape architects, engineers, museum planners, historic preservationists, surveyors, assessors, realtors, and scientists to help create "ideal" national parks. He provided financial and moral support to the park superintendents and NPS officers who toiled day and night at clerks' salaries. By the time of Junior's death in 1960, he and Albright had exchanged over thirteen hundred correspondences, many quite lengthy. Albright continued to communicate with the Rockefeller family for decades after Junior's death.[4]

Much of Junior's success as a conservationist can be traced to his father's achievements as the creator of Standard Oil. Senior loved efficiency and abhorred waste, attitudes that are as valuable in philanthropy as in business. Senior was detail oriented yet willing to delegate most responsibilities to top-notch managers with whom he communicated frequently and effectively. His staff and army of advisors were better than anyone else's network; they were also able to nimbly shift their skills from business to philanthropy when needed. Senior's thrift allowed the cash to pile up and become the fuel for his and Junior's munificence in education, health care, religious institutions, the arts, and conservation.

Yet Junior had his own qualities that made him productive as a conservationist. From childhood he had a love of nature and believed in the therapeutic virtues of time spent in the out-of-doors. He had an overdeveloped sense of duty; to God, to his parents, to the public, and to every righteous cause he happened to stumble upon or have presented to him. His patience was almost supernatural, and he was a perfectionist who could not even imagine tolerating half measures. How many people would have taken on the U.S Navy to avoid the slightest of imperfections on a seaside road in Acadia National Park? Most importantly (and incredibly), once he was committed to a project, he never stopped.

To be American is to long to be rich while also harboring a deep distrust of the wealthy. And for good reasons: the ultra-rich have much greater access to power than ordinary citizens and therefore tend to warrant our scrutiny. In some circumstances, the wealthy are allowed to play by a different rule book than the rest of us. They also, in some cases, get to write and edit those rules.

In terms of Junior's philanthropy for national parks, it is difficult to detect illegitimate intentions. No doubt he was motivated by a twinge of rich person's guilt, having inherited, rather than earned, his vast fortune. True, he was sensitive to improving the public's opinion of his father and the Rockefeller family name, but this is not a dishonorable cause. Yet, if burnishing the Rockefeller name had been the primary motive, he would not have done so much of his philanthropy anonymously. And he might have actually attended some of the dedication ceremonies and allowed more commemorative plaques. More importantly, Junior's conservation projects were highly controversial, attracting decades of vehement opposition, criticism, and even congressional investigations. Had Junior chosen to only write checks to hospitals, colleges, orphanages, art museums, libraries, and symphony orchestras, he could have avoided much sweat and scrutiny.

As to why Junior dedicated the majority of his life to philanthropy, the most obvious answers are likely the true ones. Junior's deep religious convictions, including his firm belief that God was the source of Rockefeller wealth and that God wanted the monies used to improve the conditions of humankind (rather than squandered on conspicuous consumption), inspired him to put the bulk of his riches to benevolent use. Junior chose protecting beautiful landscapes as one of his main charities because of his love for beauty in the natural world, because in his mind God and nature were synonymous, and because of the influence of other conservationists like Albright, Cammerer, Dorr, Eliot, Mather, and the Nusbaums. On Mount Desert Island in Maine, Junior's first step was falling in love with the place. While his initial contributions to creating the national monument were self-serving (allowing him to build more carriage roads), as time went on, he recognized that protecting parks for the "benefit and enjoyment of the people" was another way of accomplishing his family's long-time goal of enhancing "the well-being of mankind throughout the world."[5] Working with his father at Forest Hill and Kykuit, Junior had developed into an amateur landscape architect who appreciated big trees and sublime landscapes and hated to see them sullied. "I am interested in beauty," Junior once said, and as a perfectionist, he liked to see beautiful places maintained.[6] He also had sentimental recollections of happy times with his father and sisters in the out-of-doors.

Abraham Flexner, a Rockefeller General Education Board member and one of Junior's mentors, offered his (perhaps defensive) perspective on Gilded Age philanthropy: "Would it have been socially better if, instead of foundations, we had created a nobility or an aristocracy?"[7] Yet, it is worth noting that, unlike George Dorr, who spent (and borrowed) every last penny to benefit Acadia National Park and died not only broke but seriously in debt, Junior never intended to exhaust the family fortune on philanthropy. His associates continued to shrewdly invest and diversify the fruits of the Standard Oil money tree and to propagate the wealth. Long before he died, Junior passed most of his fortune on to his children and the Rockefeller foundations, as his father had, thereby prolonging the family legacy of both copious wealth and magnanimous philanthropy.

Kenneth Chorley recalled a time near Jackson, Wyoming, a few years before Junior's death in 1960. Chorley and Junior and their wives had spryly climbed the stairs into a fire tower to watch the sunset. The tower commanded a view of the south end of the Grand Teton mountain range and most of the valley called Jackson Hole. The sublime scene was enhanced by a billowing thunderstorm over the jagged peaks. Chorley recounted part of his conversation with Junior that day:

"Mr. Chorley," Junior inquired, "how much money have we spent here?"

"Well, I don't remember exactly," Chorley replied, "but certainly somewhere in the neighborhood of a million and a half or two million dollars" [before the hotels].

"Well, it was well worth it," Junior replied.

Then he turned, and the storm was at its height. It wasn't raining where we were, but you could see the lightning in the mountains, and the most beautiful cloud formations—it was just fascinating. And he [Junior] watched it for a while, and turned to me and said, "Mr. Chorley, how can they say there is no God?"[8]

*

Guided by Albright, Junior—along with his sons Laurance and John—continued their national parks crusade after the South Calaveras Grove, Linville Falls, and Virgin Islands donations. The team collaborated in raising a ruckus over the loss of Civil Service protection for key NPS officers. In 1953 Junior responded to alarming articles in *Harper's Magazine* and *Reader's Digest*, with titles like "Let's Close the National Parks," that called attention to rising visitation, deteriorating infrastructure, damage to park resources, and ridiculously expensive (though rat-infested and generally lousy) employee housing. He even went so far as to write President Eisenhower (after Albright had reviewed his draft) about the parks' crisis.[9] Eventually, the media attention and efforts of private citizens like Junior led to "Mission 66," a massive ten-year modernization of park roads, visitor centers, campgrounds, employee housing, and other facilities.

One of Junior's last great contributions to the parks was building Jackson Lake Lodge and Colter Bay Village, both in Grand Teton National Park and both designed and constructed through the nonprofit Jackson Hole Preserve, Inc. The preserve was run by Laurance Rockefeller at the time, with all income directed into the facilities' maintenance, operations, and improvement, or to general conservation projects in the parks.[10]

Even for two Rockefellers, Jackson Lake Lodge was an enormous undertaking. It was sorely needed at the time because, by 1952, park visitation became elevated to 785,343, "while accommodations were limited to 150 units. Visitors were forced to sleep in their cars on the side of the roads, trampling vegetation"[11] and "causing a very serious sanitary problem."[12]

Never one to settle for mediocre, Junior hired architect Gilbert S.

Underwood to design the lodge. Underwood had proven himself worthy of the task by creating lodges at Bryce and Zion National Parks and on the north rim of the Grand Canyon. However, when Junior received the first jaw-dropping cost estimate of $5.6 million ($67 million in 2025 dollars), he wondered if he'd hired the right person for the job.[13]

Underwood accepted the design job at a turning point for American architecture and the NPS. Modern was all the rage, and even the tradition-loving Park Service used modern architecture in most of its Mission 66 projects. Underwood decided Jackson Lake Lodge presented the perfect opportunity to blend NPS rustic with modern in the four-hundred-room facility. He called the blend "environmental," because of the way it conformed to the landscape, preserved many existing trees, and was barely visible from the main park road.[14]

Six thousand people attended Jackson Lake Lodge's open house and dedication in 1955. Laurance Rockefeller was among them; Junior was not.[15]

At the dedication, our old friend Chorley quipped, "When Mr. Rockefeller, Junior decided to build Jackson Lake Lodge I told him I thought no one in their right mind would ever invest $5 ½ million in a hotel which would only operate three months in the year. His reply was that this was not an investment, it is a gift to the American people."[16] Quite true, since concession facilities in national parks are generally owned by the federal government and leased to the concessionaire. Colter Bay Village recycled some of the old Moran cabins and tailored itself to campers and vacationers on tighter budgets.

However, after Jackson Lake Lodge, Laurance did go on to build several "eco-lodges" in beautiful areas around the world. He considered the environmentally friendly resorts that encouraged guests to immerse themselves in local nature and art to be the ideal blending of commerce and conservation.[17] His RockResorts included Caneel Bay in Virgin Islands National Park, Little Dix Bay on Virgin Gourda, the Mauna Kea Beach Hotel in Hawaii, and the Woodstock Inn in Vermont.

The most striking feature of Jackson Lake Lodge is the spacious, two-story lounge with its big windows overlooking Mount Moran, the Cathedral Group (the Grand Teton, Teewinot, and Mount Owen), and the valley of Jackson Hole. Oftentimes moose browse the willows in the marsh between the lodge and Jackson Lake. It is almost the exact same view that Junior and his family enjoyed in 1924 and 1926 after a young Horace Albright directed them to Lunch Tree Hill to sell them on a pie-in-the-sky scheme to buy Jackson Hole. Except now that view belongs to everyone.

EPILOGUE

While the view from Lunch Tree Hill was a sunny one in 1955, much has changed since. Annual visitation at little Acadia National Park has leapt from 655,000 in 1955 to around four million today.[1] Despite the success of the Island Explorer shuttlebus system, reservations are required to drive up Cadillac Mountain during the season, and finding a parking spot at popular destinations and trailheads between 8:00 a.m. and 4:00 p.m. is nearly impossible. Park officials advise summer and fall visitors to make a "backup plan" which may entail abandoning intentions of visiting the park.[2]

Likewise, in the Great Smoky Mountains there is no parking available at popular sites between 8:30 a.m. and 4:00 p.m. Some hikers now walk an extra half-mile along the busy roads from their vehicle just to reach a trailhead. Visitors to Arches, Mount Rainier, Glacier, and Rocky Mountain National Parks must purchase a timed entry pass (on top of their entry fee) to gain access to the parks' most popular areas during peak seasons. Obtaining entry passes has become a highly competitive proposition with the number of people hoping to see the parks far outstripping the availability of passes. Most reservations for Glacier's Going-to-the-Sun Road sell out for the entire summer in advance. Their website warns that traffic may cause portions of the park to be temporarily closed and suggests people consider visiting one of their park's neighbors, like Yellowstone.[3]

Yosemite officials advise people to visit before 9:00 a.m. or after 5:00 p.m. to avoid "extended traffic delays, extremely limited parking, busy trails, and no lodging or campground availability."[4] Parking for the shuttlebus into Zion National Park usually fills up by 8:00 a.m. year-round. There is parking in town, a short distance (and free shuttlebus ride) away, but it can set you back twenty dollars or more.[5] What happens when visitation rises another 10 percent?

Just as concerning as the overcrowding is the rapidly escalating cost for a national park vacation. Long gone are the days when the middle-class family piled into the overloaded station wagon for two weeks each summer to explore their national parks. Marquee park lodges like the Old Faithful Inn in Yellowstone or Jackson Lake Lodge in the Grand Tetons, which for many decades were affordable to ordinary folk, no longer fit that description. Nor is there any availability without mounting a round-the-clock internet and phone reservation campaign. The cost of accommodations in park gateway communities has also recently gone through the roof. Demand has simply outstripped supply. Although more parks will require more funding for operations, philanthropy through park Friends groups has proven to be highly effective in this regard. Also keep in mind that only one-fifteenth of one percent of the federal budget currently goes for national park operations. As former Blue Ridge Parkway superintendent and Coalition to Protect America's National Parks president Phil Francis put it, "I cannot understand why Congress refuses to invest more in our national parks when they are so important to local economies and the American people. From a purely economic standpoint, the $3 billion that we invest in national parks provides a $50.3 billion benefit to the nation's economy and supports 378,400 jobs."[6]

The National Park Service is aware of the problems and is not particularly happy about them, but their management options are few and poor. They do provide limited camping for families with the wherewithal and gear (which can be quite expensive) to do so.

If "America's Best Idea" is to endure as something the people can be proud of, we will need more national parks and larger national parks.

NOTES

Abbreviations

AAM	American Association of Museums
GPO	Government Printing Office
GSMCA	Great Smoky Mountains Conservation Association Papers, Townsend, TN
GSMNP	Archives. Great Smoky Mountains National Park Archives, Townsend, TN
JHPI	Jackson Hole Preserve, Inc., collection, Pocantico Hills, NY
JMU	James Madison University Oral History Collection
LSRM	Laura Spelman Rockefeller Memorial Archives, Pocantico Hills, NY
MEVE	Mesa Verde National Park Archives, Mesa Verde National Park, CO
NAC	National Archives Center, College Park, MD
OMR	Office of the Messrs.
RAC	Rockefeller Archives Center, Pocantico Hill, NY
YOSE	Yosemite National Park Archives, El Portal, CA

Preface

1 Scott Detrow, "Why You Might See Traffic Jams in Yosemite This Summer," *All Things Considered*, National Public Radio, July 1, 2023. See also https://www.npr.org/2023/07/01/1185658050/why-you-might-see-traffic-jams-in-yosemite-this-summer.

2 "Plan Your Trip," Various NPS Websites, NPS.gov/, 2024.

3 Holly Kays, "Smokies Seek Solutions to Overcrowding," *Smoky Mountain News* (Waynesville, NC), October 28, 2020. See also https://smokymountainnews.com/archives/item/30125-smokies-seeks-solutions-to-overcrowding.

4 Allison Pohle, "Reservations Needed at Yosemite Again," *Wall Street Journal*, December 18, 2023, A11.

5 Allison Pohle, "Rocky Mountain Park Sets Reservation System," *Wall Street Journal*, May 31, 2024, A3.
6 National Park Service Visitor Use Statistics, STATS, Social Science Program, www.Irma.nps.gov, 2024.
7 Otak, Inc., "2022 Socioeconomic Research of Great Smoky Mountains National Park," Natural Resources Report NPS/GRSM/NRR—2023/2564, University of Montana, September 1923.
8 Casey J. Wichman, "Social Media Influences National Park Visitation," *Proceedings of the National Academy of Sciences*, April 1, 2024, www.pnas.org/doi/10.1073/pnas.2310417121.
9 NPS Visitor Use Statistics, STATS, Social Science Program, www.Irma.nps.gov, 2024.
10 Robert Sterling Yard, "The People and the National Parks," *The Survey* 48, no. 13 (August 1, 1922): 547.
11 Alfred Runte, *National Parks: The American Experience* (Lanham, MD: Taylor Trade Publishing, 2010): 2.
12 Jackson Lake Lodge, GTLC.com; Yosemite Valley Lodge, Travelyosemite.com; Old Faithful Inn, usparklodging.com.
13 William L. Rice, Jennifer Thomsen, and Peter Whitney, "Exclusionary Effects of Campsite Allocation through Reservations in U.S. National Parks: Evidence from Mobile Device Location Data," *Journal of Park and Recreation Administration* 40, no. 4 (March 2022): 45–65.
14 Maxine Joselow, "Report Details How Biden Can Protect 30 Percent of U.S. Lands by 2030 without Congress," *Washington Post*, November 22, 2022. The article also notes the Center for American Progress followed up the announcement with a report suggesting sixteen national monuments and marine sanctuaries Biden could create to fulfill this pledge. Jim Robbins, "Salvation or Pipe Dream: A Movement Grows to Protect Up to Half the Planet," *Yale Environment 360* (publication of the Yale School of the Environment) (February 13, 2020), https://e360.yale.edu/features/salvation-or-pipe-dream-a-movement-grows-to-protect-up-to-half-the-planet#.
15 Personal interview with former Great Smoky Mountains National Park deputy superintendent Phillip Francis, 2022.
16 Ted Kerasote, "This Tiny Parcel of Paradise Could Be Devoured," *New York Times*, Opinion Page, December 6, 2023.
17 Courtney Lix, "The Gift: How We Gained a Mountain," *Smokies Life Magazine* 3, no. 1 (2009): 39–45.
18 "Katahdin Woods and Waters National Monument," Natural Resources Council of Maine, August 24, 2016, www.nrcm.org/programs/forests-wildlife/katahdin-national-monument. George Wuerthner, "A Maine Woods National Park?" *Wildlife News*, April 5, 2022, www.thewildlifenews.com/2022/04/05/a-maine-woods-national-park.
19 Leslie A. Richardson and Christopher Huber, "Terrestrial Carbon

Sequestration in National Parks: Values for the Conterminous United States," Natural Resources Report, pubs.usgs.gov/publication/70148512.

20 GSMNP Species Tally, Discover Life in America, dlia.org.

21 "Nature Makes You . . . ," NPS, www.nps.gov/articles/naturesbenefits.

22 Harvey Broome, "Sanctuary," film shown in Sugarlands Visitor Center, GSMNP, Harpers Ferry Design Center, circa 1970s.

23 Jason Mark, "Across the Great Divide," *Sierra* (July/August 2020): 27–32.

24 Emily Davis, NPS news release, GSMNP, August 21, 2023. Visitor Spending Effects – Economic Contributions of National Park Visitor Spending – Social Science (U.S. National Park Service) (nps.gov).

25 Phil Francis, *The Hill*, Opinion: Energy and Environment, March 16, 2024.

26 Rick Hampson, "America's Second Gilded Age," *USA Today*, May 17, 2018.

27 John D. Rockefeller Sr., *Random Reminiscences of Men and Events* (New York: Doubleday, Page & Company, 1909), 140.

Introduction

1 Peter Collier and David Horowitz, *The Rockefellers: An American Dynasty* (New York: Holt, Rinehart, and Winston, 1976), 8–10.

2 Ron Chernow, *Titan: The Life of John D. Rockefeller, Sr.* (New York: Vintage Books, 1998), 38.

3 Chernow, *Titan*, 8–9.

4 Chernow, *Titan*, 43.

5 William H. Allen, *Rockefeller; Giant, Dwarf, Symbol* (New York: New York Institute for Public Service, 1930), 74.

6 Chernow, *Titan*, 43.

7 Chernow, *Titan*, 45.

8 John D. Rockefeller Sr., *Random Reminiscences of Men and Events* (New York: Doubleday, Page & Co., 1909), 37–39.

9 Chernow, *Titan*, quote of John Wesley dictum, 55.

10 Allan Nevins, *John D. Rockefeller: A Study in Power* (New York: Charles Scribner's Sons, 1959), 9.

11 Chernow, *Titan*, 99–102.

12 Chernow, *Titan*, 80.

13 Charles R. Morris, *The Tycoons: How Andrew Carnegie, John D. Rockefeller, Jay Gould, and J. P. Morgan Invented the American Supereconomy* (New York: Henry Holt & Co., 2005), 20.

14 Collier and Horowitz, *The Rockefellers*, 22–29.

15 Morris, *Tycoons*, 84–85; also, Clarice Stasz, *The Rockefeller Women: Dynasty of Piety, Privacy, and Service* (New York: St. Martin's Press, 1995), 64.

16 Morris, *Tycoons*, 81–82.
17 Chernow, *Titan*, 289–95.
18 Chernow, *Titan*, 289.
19 Morris, *Tycoons*, 158.
20 National Register of Historic Places Inventory—Nomination Form, John D. Rockefeller Estate (Pocantico Hills), National Park Service, circa late 1970s, 3.
21 Samuel P. Hays, *Conservation and the Gospel of Efficiency: The Progressive Conservation Movement, 1890–1920* (Pittsburgh: University of Pittsburgh Press, 1999), 261.
22 Karl Zinsmeister, "Books: Five Best on Philanthropy," *Wall Street Journal*, Saturday/Sunday, April 27–28, 2024, C8.
23 National Register, John D. Rockefeller Estate, 3; Morris, *Tycoons*, 158–59.
24 National Register, John D. Rockefeller Estate, 8.
25 Raymond B. Fosdick, *John D. Rockefeller, Jr.: A Portrait* (New York: Harper & Brothers, 1956), 43.
26 Fosdick, *John D. Rockefeller, Jr.*, 84.
27 Hays, *Conservation*, 141–45.
28 Fosdick, *John D. Rockefeller, Jr.*, 403.
29 Hays, *Conservation*, 122–29.
30 Ann Rockefeller Roberts, *Mr. Rockefeller's Roads: The Untold Story of Acadia's Carriage Roads & Their Creator* (Camden, ME: Down East Books, 1990), 2.
31 Fosdick, *John D. Rockefeller, Jr.*, 418.
32 National Park Service press release, "National Park Visitation Sets New Record as Economic Engines," August 21, 2023, www.nps.gov/orgs/1207.
33 National Park Service Visitor Use Statistics, STATS, Social Science Program, www.Irma.nps.gov, 2024.

Island Idyll

1 Bernice Kert, *Abby Aldrich Rockefeller: The Woman in the Family* (New York: Random House, 1993), 128.
2 Kert, *Abby*, 128.
3 *Boston Globe*, "J. M. Sears Dies in Providence Hospital," August 13, 1908, 1.
4 *Bar Harbor Record*, "Around Town," July 8, 1908, 5; *Bar Harbor Record*, July 22, 1908, 4.
5 David Rockefeller, *Memoirs* (New York: Random House, 2002), 31.
6 Eileen Rockefeller Growald, *Being a Rockefeller, Becoming Myself* (New York: Blue Rider Press, 2013), 87.
7 Ann R. Roberts, interview by Ronald H. Epp, August 26, 2003, Southwest Harbor, ME, transcript from Ronald H. Epp papers, 4.

8 Kert, *Abby*, 138–39. All inflation-adjusted figures are from the U.S. Bureau of Labor Statistics, CPI Inflation Calculator, data.bls.gov.
9 Ron Chernow, *Titan: The Life of John D. Rockefeller, Sr.* (New York: Vintage Books, 1998), 183–86.
10 Ann Rockefeller Roberts, *Mr. Rockefeller's Roads: The Untold Story of Acadia's Carriage Roads & Their Creator* (Camden, ME: Down East Books, 1990), 29.
11 Daniel Okrent, *Great Fortune: The Epic of Rockefeller Center* (New York: Viking, 2003), 376.
12 David Rockefeller, *Memoirs*, 181.
13 Kert, *Abby*, 139.
14 Richard H. Quin, "Rockefeller Carriage Roads," Historic American Engineering Record (HAER), ME-13 (Washington, DC: National Park Service, 1997), 13.
15 Peter Collier and David Horowitz, *The Rockefellers: An American Dynasty* (New York: Holt, Rinehart, and Winston, 1976), 80.
16 Chernow, *Titan*, 124.
17 Junior to W. O. Inglis, February 19, 1918, Rockefeller Archive Center (RAC), Series H, Biography: JDR, W. O. Inglis.
18 Collier and Horowitz, *The Rockefellers*, 79.
19 Raymond B. Fosdick, *John D. Rockefeller, Jr.: A Portrait* (New York: Harper & Brothers, 1956), 58–59.
20 Andrew Wilton and Tim Barringer, *American Sublime: Landscape Painting in the United States 1820–1880*. (Princeton: Princeton University Press, 2002), 14.
21 Wilton and Barringer, *American Sublime*, 49.
22 Ralph W. Emerson, *Nature* (London: Penguin Books, 1985), 5.
23 Fosdick, *John D. Rockefeller, Jr.*, 129.
24 George W. Nichols, "Mount Desert," *Harper's New Monthly Magazine* 45 (August 1872): 336.
25 Charles W. Eliot to John D. Rockefeller Jr., February 15, 1915, RAC, III, 2.1, box 59, folder 441.
26 Fosdick, *John D. Rockefeller, Jr.*, 322.
27 Fosdick, *John D. Rockefeller, Jr.*, 110.
28 Collier and Horowitz, *The Rockefellers*, 100–4.
29 Fosdick, *John D. Rockefeller, Jr.*, 120.
30 National Register of Historic Places Inventory—Nomination Form, John D. Rockefeller Estate (Pocantico Hills), National Park Service, circa late 1970s, 8.
31 National Register, Pocantico Hills, 120.
32 Collier and Horowitz, *The Rockefellers*, 106.
33 Fosdick, *John D. Rockefeller, Jr.*, 433.
34 Charles W. Eliot to Ellen Bullard (niece), September 1, 1915, Dorr Papers, Box 1, folder 5, Bar Harbor Historical Society.

35 Roberts, interview.
36 Collier and Horowitz, *The Rockefellers*, 85.
37 Roberts, *Mr. Rockefeller's Roads*, 22.
38 Ronald H. Epp, *Creating Acadia National Park: The Biography of George Bucknam Dorr* (Bar Harbor, ME: Friends of Acadia, 2016), 12.
39 Roberts, interview.
40 Charles Eliot, "The Coast of Maine," *Garden and Forest* 3, no. 104 (February 19, 1890): 85–86.
41 Eliot, "Coast of Maine," 85–86.
42 Eliot, "Coast of Maine," 85–86.
43 Alfred Runte, *National Parks: The American Experience* (Lanham, MD: Taylor Trade Publishing International, 2010), 16.
44 Runte, *National Parks*, 2.
45 Catherine Schmitt, *Historic Acadia National Park: The Stories Behind One of America's Great Treasures* (Guilford, CT: Globe Pequot Press, 2016), 24.
46 Charles Eliot, "The Waverly Oaks," *Garden and Forest* 3, no. 104 (February 19, 1890): 86–87.
47 Schmitt, *Historic Acadia*, 168.
48 Schmitt, *Historic Acadia*, 162.
49 Epp, *Creating Acadia*, 147–49.
50 George B. Dorr, *The Story of Acadia National Park* (Bar Harbor, ME: Acadia Publishing Co. 1997), 29.
51 Dorr, *The Story of Acadia*, 29.

An Altruist Goes to Washington

1 Donald C. Swain, *Wilderness Defender: Horace M. Albright and Conservation* (Chicago: University of Chicago Press, 1970), 26.
2 Horace M. Albright as told to Robert Cahn, *The Birth of the National Park Service: The Founding Years, 1913–33* (Salt Lake City: Howe Brothers, 1985), 2.
3 Harold Fabian, interview by Ed Edwin, July 11, 1966, Rockefeller Archive Center (RAC), Oral History Collection, FA 1444, Series 4, Box 3, Folder 45.
4 Horace M. Albright, interview by Michael Frome, "Recollections and Observations by Horace M. Albright," 1964, Great Smoky Mountains National Park Library, Townsend, TN, #23575 and 23576; transcription of tapes 32 and 33, 8, 16.
5 Horace Albright and Marian Albright Schenck, *Creating the National Park Service: The Missing Years* (Norman: University of Oklahoma Press, 1999), 269.
6 Samuel P. Hays, *Conservation and the Gospel of Efficiency: The Progressive Conservation Movement, 1890–1920* (Pittsburgh: University of Pittsburgh Press, 1999), 196.

7 Act creating Yellowstone National Park, March 1, 1872, Enrolled Acts and Resolutions of Congress, 1789–1996, General Records of the United States Government, Record Group 11, National Archives Center (NAC).
8 Blair Bolles, *The Tyrant from Illinois: Uncle Joe Cannon's Experiment with Personal Power* (New York: W. W. Norton, 1951), 119–20.
9 Richard Hale Jr., *The Story of Bar Harbor* (New York: Ives Washburn, 1949), 197–98.
10 Eric J. Abrahamson, *Beyond Charity: A Century of Philanthropic Innovation* (New York: Rockefeller Foundation, 2013), 41.

Coal, Catastrophe, and a Road Toward Redemption

1 Ron Chernow, *Titan: The Life of John D. Rockefeller, Sr.* (New York: Vintage Books, 1998), 575.
2 Robin Henry, "'In Our Image, According to Our Likeness': John D. Rockefeller, Jr. and Reconstructing Manhood in Post-Ludlow Colorado." *The Journal of the Gilded Age and Progressive Era* 16 (2017): 25.
3 Thomas G. Andrews, *Killing for Coal: America's Deadliest Labor War* (Cambridge, MA: Harvard University Press, 2008), 234–35.
4 *New York Times*, October 27, 1913, 1.
5 Raymond B. Fosdick, *John D. Rockefeller, Jr.: A Portrait* (New York: Harper & Brothers, 1956), 146.
6 John D. Rockefeller Sr., *Random Reminiscences of Men and Events* (New York: Doubleday, Page & Company, 1909), 119.
7 Fosdick, *John D. Rockefeller, Jr.*, 90–91.
8 JDR Jr. to JDR Sr., November 11, 1899, in Joseph W. Ernst, ed., *"Dear Father"/"Dear Son": Correspondence of John D. Rockefeller and John D. Rockefeller, Jr.* (New York: Fordham University and Rockefeller Archive Center, 1994), 19.
9 Fosdick, *John D. Rockefeller, Jr.*, 91.
10 *The Evening World*, "Rockefeller, Jr. Babe of Finance," February 23, 1900, 11.
11 *Evansville (IN) Journal*, December 12, 1899, 4.
12 *St. Joseph (MI) Herald*, "Young Rockefeller," January 27, 1900, 5.
13 *Deseret News* (Salt Lake City), July 14, 1900, 11.
14 Allan Nevins, *John D. Rockefeller: A Study in Power* (New York: Charles Scribner's Sons, 1959, Abridgement), 179.
15 Chernow, *Titan*, 176–78.
16 Chernow, *Titan*, 575.
17 Chernow, *Titan*, 574.
18 Chernow, *Titan*, 576.
19 Fosdick, *John D. Rockefeller, Jr.*, 145.
20 Fosdick, *John D. Rockefeller, Jr.*, 147.

21 *New York Times*, April 7, 1914, 1.
22 Fosdick, *John D. Rockefeller, Jr.*, 149.
23 Andrews, *Killing for Coal*, 277.
24 *New York Times*, "45 Dead, 20 Hurt, Scores Missing," April 22, 1914, 7.
25 Andrews, *Killing for Coal*, 278–81.
26 *New York Sun*, "J. D. Rockefeller, Jr. Blamed for the Riots," April 23, 1914, 16.
27 Kirk Hallahan, "Ivy Lee and the Rockefellers' Response to the 1913–1914 Colorado Coal Strike," *Journal of Public Relations Research* 14, no. 4 (October 2002): 265–315.
28 Robert M. Dawson, *William Lyon Mackenzie King*, vol. 1 (Toronto: University of Toronto Press, 1958), 229, 232.
29 *New York Times*, "Witness Displays Spirit," May 22, 1915, 6.
30 "Statement of John D. Rockefeller, Jr. before the United States Commission on Industrial Relations, January 25, 1915," booklet, n.p., Internet Archive, https://archive.org/details/statementofjohndoorock.
31 Chernow, *Titan*, 586.
32 *New York Sun*, "Mother Jones Likes JDR Jr. She Says," January 28, 1915, 14.
33 *New York Sun*, "Rockefeller Talks to 3 Union Leaders," January 29, 1915, 14.
34 Fosdick, *John D. Rockefeller, Jr.*, 159.
35 Chernow, *Titan*, 586.
36 Ernst, *"Dear Father"/"Dear Son,"* 68.

Friends, Foes, and Favors in Our Nation's Capital

1 Charles W. Eliot to JDR Jr., February 25, 1915, Rockefeller Archive Center, OMR Rockefellers, RG III, Box 59, Folder 441.
2 John D. Rockefeller Jr., "Participation in the Formative Stage of the Park [Acadia]," Summary report, RAC, Cultural Interests, Series E, Box 78, Folder 738.
3 John D. Rockefeller, *Random Reminiscences of Men and Events* (Garden City, NJ: Doubleday, Page & Co., 1933), 183. Also, National Register of Historic Places Inventory—Nomination Form, John D. Rockefeller Estate (Pocantico Hills), National Park Service, circa late 1970s, 7.
4 Ron Chernow, *Titan: The Life of John D. Rockefeller, Sr.* (New York: Vintage Books, 1998), 320.
5 Raymond B. Fosdick, *John D. Rockefeller, Jr.: A Portrait* (New York: Harper & Brothers, 1956), 161.
6 *Denver Labor Bulletin*, September 25, 1915, 1.
7 *Denver Labor Bulletin*, September 25, 1915, 1.
8 *Raymer (CO) Enterprise*, September 23, 1915, 1.
9 *New York Tribune*, September 24, 1915, 1, 4.

10 Bernice Kert, *Abby Aldrich Rockefeller: The Woman in the Family* (New York: Random House, 1993), 150–51.
11 Fosdick, *John D. Rockefeller, Jr.*, 162–63.
12 Fosdick, *John D. Rockefeller, Jr.*, 186.
13 JDR Jr. "Labor and Capital—Partners," *Atlantic Monthly*, January 1916, 15–17.
14 Clarice Stasz, *The Rockefeller Women: Dynasty of Piety, Privacy, and Service* (New York: St. Martin's Press, 1995), 190.
15 Kert, *Abby Aldrich Rockefeller*, 152.
16 Jonathan H. Rees, *Representation and Rebellion: The Rockefeller Plan at the Colorado Fuel and Iron Company, 1914–1942* (Boulder: University Press of Colorado, 2010), xvi, 6–9.
17 Fosdick, *John D. Rockefeller, Jr.*, 167.
18 Karl Jacoby, *Crimes Against Nature: Squatters, Poachers, Thieves, and the Hidden History of American Conservation* (Berkeley: University of California Press, 2014), 2–6, 48–50.
19 Chernow, *Titan*, 590.
20 Rees, *Representation and Rebellion*, 6.
21 Fosdick, *John D. Rockefeller, Jr.*, 164.
22 *Bangor Daily News*, "Mother Jones Sends Her Congratulations," July 9, 1930, 1, 11.
23 *Bangor Daily News*, "Mother Jones," July 9, 1930, 1, 11.
24 George B. Dorr, *The Story of Acadia National Park* (Bar Harbor, ME: Acadia Publishing Co. 1997), 47.
25 Dorr, *The Story of Acadia*, 49–50.
26 Dorr, *The Story of Acadia*, 50.
27 Dorr, *The Story of Acadia*, 49.
28 *Bar Harbor Times*, August 26, 1916, 1.
29 Dorr, *The Story of Acadia*, 60–62.
30 Chernow, *Titan*, 644. Also, Philanthropy Roundtable: Hall of Fame, "John D. Rockefeller, Jr." Washington, DC, www.philanthropyroundtable.org/hall-of-fame/john-rockefeller-jr/.
31 Speech by Secretary of Interior Ray Lyman Wilbur, broadcast by WRC (NBC radio), November 2, 1931.
32 Maxine Joselow and Vanessa Montalbano, "Report Details How Biden Can Protect 30 Percent of U.S. Lands and Waters by 2030 without Congress," *Washington Post*, November 22, 2022.

A New Park and a New National Park Service

1 Andy Cerda, "Americans See Many Federal Agencies Favorably, but Republicans Grow More Critical of Justice Department," *Short Reads*, Pew Research Center, https://www.pewresearch.org/short-reads/2024/08/12/, August 12, 2024. The NPS scored the top ranking of all

federal agencies, with 76 percent of respondents in survey rating NPS as favorable.

2 Robert Shankland, *Steve Mather of the National Parks* (New York: Alfred A. Knopf, 1951), 28.

3 Richard West Sellars, *Preserving Nature in National Parks: A History* (New Haven, CT: Yale University Press, 1997), 31.

4 Horace M. Albright and Marian A. Schenck, *Creating the National Park Service: The Missing Years* (Norman: University of Oklahoma Press, 1999), 188.

5 Albright and Schenck, *Creating*, 197.

6 Donald C. Swain, *Wilderness Defender: Horace M. Albright and Conservation* (Chicago: University of Chicago Press, 1970), 65.

7 Albright and Schenck, *Creating*, 200.

8 Albright and Schenck, *Creating*, 270.

9 Richard H. Quin, "Rockefeller Carriage Roads," Historic American Engineering Record, ME-13 (Washington, DC: National Park Service, 1997), 18.

10 Anne W. Lane, ed., *The Letters of Franklin K. Lane: Personal and Political* (Boston: Houghton Mifflin, 1922), 256–58.

11 Horace M. Albright as told to Robert Cahn, *The Birth of the National Park Service: The Founding Years, 1913–33* (Salt Lake City: Howe Brothers, 1985), 85.

Drive-Through National Parks

1 Ann Rockefeller Roberts, *Mr. Rockefeller's Roads: The Untold Story of Acadia's Carriage Roads & Their Creator* (Camden, ME: Down East Books, 1990), 79.

2 JDR Jr. to George W. Pepper, August 20, 1920, Rockefeller Archive Center (RAC), Homes, FA318, Series I, Box 83, Folder 827.

3 *Bar Harbor Times*, "Mr. Rockefeller Is Doing Big Work," September 8, 1920, 4.

4 Thomas C. Vint to Horace Albright, April 26, 1955. Harper Ferry Center Library, H.C., Acadia, Box 3.

5 H. Eliot Foulds and Jeffrey Killion, "Cultural Landscape Report for the Historic Motor Road System Acadia National Park" (Boston: Olmsted Center for Landscape Preservation, Boston National Historical Park, 1993), 51.

6 Roberts, *Mr. Rockefeller's Roads*, 126–34.

7 Roberts, *Mr. Rockefeller's Roads*, 145.

8 *Bar Harbor Times*, September 8, 1920, 1, 4.

9 Foulds and Killion, "Cultural Landscape Report," 19.

10 Foulds and Killion, "Cultural Landscape Report," 19.

11 *Bar Harbor Times*, August 8, 1926, 1, 8.

12 JDR Jr. to George B. Dorr, February 8, 1915, Hon. John A. Peters Papers, Dorr Estate Correspondence, Hale & Hamlin, LLC, Ellsworth, ME (courtesy of Ronald H. Epp).
13 *Bar Harbor Times*, August 8, 1926, 1, 8.
14 Ronald H. Epp, *Creating Acadia National Park: The Biography of George Bucknam Dorr* (Bar Harbor, ME: Friends of Acadia, 2016), 195.
15 Roberts, *Mr. Rockefeller's Roads*, 90.
16 Roberts, *Mr. Rockefeller's Roads*, 108.
17 Horace M. Albright to Ray Lyman Wilbur, November 14, 1930, National Archives Center (NAC), Record Group 48, CCF 1907–36, Box 2009.
18 *Bar Harbor Times*, August 8, 1926, 1, 8.
19 Roberts, *Mr. Rockefeller's Roads*, 125.
20 Donald C. Swain, *Wilderness Defender: Horace M. Albright and Conservation* (Chicago: University of Chicago Press, 1970), 54.
21 Arno B. Cammerer, "Memorandum on a Development Plan for Lafayette National Park," NAC, Record Group 79, Acadia, Box 1, P 60, 26–27.
22 David Louter, *Windshield Wilderness: Cars, Roads, and Nature in Washington's National Parks* (Seattle: University of Washington Press, 2006), 4–5.
23 Richard H. Quin, "Rockefeller Carriage Roads," Historic American Engineering Record (HAER), ME-13 (Washington, DC: National Park Service, 1994–1995), 30.
24 Alexander Forbes to Hubert Work, May 14, 1925, NAC, Records Group 79, Cammerer Papers, Box 1, P 60.
25 J. Gresham Machen, "The Beauty of the Forest," *New York Times*, February 18, 1925, 18.
26 Richard West Sellars, *Preserving Nature in National Parks: A History* (New Haven, CT: Yale University Press, 1997), 60–61.
27 Alfred Runte, *National Parks: The American Experience* (Lanham, MD: Taylor Trade Publishing, 2010), 80.
28 Neil Maher, "Acadia National Park Motor Roads," HAER No. ME-11 (Washington, DC: National Park Service, 1994–1997), 21.
29 JDR Jr. to Cammerer, May 9, 1927, NAC, Records Group 79, Cammerer Papers, Box 1, P 60.
30 George B. Dorr, *The Story of Acadia National Park* (Bar Harbor, ME: Acadia Publishing Co., 1997), 95–101.
31 Quin, "Rockefeller Carriage Roads," 32.
32 Board of Selectmen, Town of Mt. Desert, Maine, March 20, 1924, RAC, Homes, FA318, Series I, Box 83, Folder 827.
33 John C. Clement to JDR Jr., February 18, 1924. RAC, Homes, FA318, Series I, Box 83, Folder 827.
34 Richard H. Harte to JDR Jr., October 5, 1923.
35 Epp, *Creating Acadia*, 217–18.
36 Paul S. Sutter, *Driven Wild: How the Fight Against Automobiles Launched*

the Modern Wilderness Movement (Seattle: University of Washington Press, 2002), 12.

37 Robert Shankland, *Steve Mather of the National Parks* (New York: Alfred A. Knopf, 1951), 161.

38 Roberts, *Mr. Rockefeller's Roads*, 92.

39 Brian Kevin, *Outside Magazine*, June 19, 2012.

40 Richard Perez-Pena, "Obama Designates National Monument in Maine, to Dismay of Some," *New York Times*, August 24, 2016.

41 Kevin, *Outside*.

42 Perez-Pena, *New York Times*.

43 "President Obama Designates National Monument in Maine's North Woods in Honor of the Centennial of the National Park Service," Press Release, Office of the Press Secretary, White House, August 24, 2016.

Longing for Enlightenment

1 Stephen T. Mather, "Report of the Director of the NPS to the Secretary of the Interior for the Fiscal Year ended June 30, 1920, and the Travel Season 1920" (Washington, DC: Government Printing Office, 1920).

2 Ralph H. Lewis, "Museum Curatorship in the National Park Service 1904–1982" (Washington, DC: National Park Service, Curatorial Services Division, 1993), 15.

3 Ricardo Torres-Reyes, "Mesa Verde National Park: An Administrative History, 1906–1970" (Washington, DC: National Park Service, Office of History and Historic Architecture, 1970), 91–94.

4 Lewis, "Museum Curatorship," 10.

5 Ansel F. Hall, "Statistical Analysis of the Development of Educational Activities in Yellowstone National Park, 1920–1930," Rockefeller Archive Center (RAC), LSRM, AAM, 1927–32, LS III-4, B-11, 120.

6 Francis P. Farquhar letter to Rudolf C. Bertheau of the LSRM, March 4, 1924, RAC, AAM, Yosemite, 1924–, Folder B-11.

7 Ansel F. Hall, "Report of the Park Naturalist Yosemite National Park," September 1921, Yosemite Archives, Box 80, Folder 4.

8 Carl Parcher Russell, *100 Years in Yosemite: The Story of a Great Park and Its Friends* (El Portal, CA: Yosemite Association/Heydey Books, 1992), 138.

9 Ansel F. Hall, "Report of Progress in Educational Activities in the National Parks" (Washington, DC: National Park Service, Branch of Research and Education, 1931).

10 Mather to Secretary of Interior Hubert Work, April 1925, RAC, AAM 1925–27, Folder B-11.

11 Lewis, "Museum Curatorship," 5–7.

12 Lewis, "Museum Curatorship," 8–9, 29–31.

13 Public Lands Alliance website, www.publiclandsalliance.org.

14 Beardsley Ruml to Chauncey J. Hamlin, November 22, 1922, RAC, LSRM, AAM, 1922–28, S-III-4, B-11, 117.
15 American Association of Museums, "Statement of Accomplishments 1923–25," RAC, LSRM, AAM 1922-28, S-III-4, B-11, 5-6. Also, Lewis, "Museum Curatorship," 31.
16 Arno B. Cammerer to Chauncey J. Hamlin, March 21, 1924, RAC, LSRM, S-III-4, B-11, AAM, Yosemite 1924–, 121.
17 Lewis, "Museum Curatorship," 31.
18 Ethan Carr, *Wilderness by Design: Landscape Architecture & the National Park Service* (Lincoln: University of Nebraska Press, 1998), 145.
19 Sarah Allaback, "Rustic Museums and Modern Visitor Centers: America's Most Popular Museums," in David Harmon, ed., *People, Places, and Parks*, Proceedings of the 2005 George Wright Society Conference (Hancock, MI: George Wright Society, 2005).
20 Lewis, "Museum Curatorship," 34.
21 Hall progress report to AAM, November 18, 1924, RAC, LSRM 1924–27, S-III-4, B-11, 121.
22 Hall progress report to AAM, December 20, 1924, 121.
23 Hall progress report to AAM, April 18, 1925, RAC, AAM-Yosemite, 1925–27, S-III-4, B-11, 122.
24 NPS brochure, "Cooperating Associations," Harpers Ferry Design Center Archives.
25 Hermon C. Bumpus to Cammerer and Ruml, May 17, 1926, RAC, LSRM, AAM, 1922–28, SIII-4, B-11, 117. Also, Ruml to Bumpus, March 18, 1926; Ruml to Mather, March 23, 1926; Bumpus to Ruml, April 1, 1926, RAC, LSRM, AAM, SIII-4, B-11, Trailsides 1926–29, 119.
26 Carr, *Wilderness by Design*, 143.
27 Hermon C. Bumpus Jr., *Hermon C. Bumpus: Yankee Naturalist* (Minneapolis: University of Minnesota Press, 1947), 56–57.
28 Bumpus Jr., *Hermon C. Bumpus*, 104.
29 Architectural Resources Group, "Yavapai Observation Station: Historic Structure Report" (San Francisco: Architectural Resources Group, 2001), 11.
30 Architectural Resources Group, "Yavapai Observation Station," 14.
31 Abbe Museum Archives, C. II, F.2. See also Ronald H. Epp, *Creating Acadia National Park: The Biography of George B. Dorr* (Bar Harbor, ME: Friends of Acadia, 2016), chapter 17.
32 Epp, *Creating Acadia*, 223.
33 Epp, *Creating Acadia*, 223.
34 Hubert Work to LSRM, February 9, 1928, RAC, LSRM, AAM, Yellowstone 1927–32, folder B-11.
35 JDR Jr. to Hermon C. Bumpus, circa 1928, NAC, Records Group 79, Horace M. Albright, Box 3, Entry P 59.
36 Bumpus to Rockefeller Foundation, April 23, 1931, RAC, AAM, Yellowstone 1927–32, Folder B-11.

37 Kiki L. Rydell and Mary S. Culpin, *Managing the Matchless Wonders: A History of Development in Yellowstone* (Yellowstone National Park, WY: Yellowstone Center for Resources, 2006), 110.
38 Rydell and Culpin, *Managing the Matchless Wonders*, 110.
39 Bumpus to Albright, August 19, 1929, NAC, Records Group 79, Horace M. Albright, Box 3, Entry P 59.
40 Flora S. McHarg to Herbert Maier (acting regional NPS director), November 22, 1938, RAC, FA314, Records Group 2, Series E, Box 77, Folder 734.
41 Laura S. Harrison, National Register of Historic Places Inventory—Nomination Form, Norris, Madison, and Fishing Bridge Museums, National Park Service, 1986, 8.
42 Rydell and Culpin, *Managing the Matchless Wonders*, 103.
43 "History of the NPS Arrowhead," NPS, www.nps.gov/wrst/learn/historyculture/history-of-the-nps-arrowhead.htm.
44 Linda Flint McClelland, *Presenting Nature: The Historic Landscape Design of the National Park Service 1916 to 1942* (Washington, DC: National Park Service, 1993), 234–40.
45 Ray L. Wilbur to Rockefeller Foundation, May 6, 1931, RAC, AAM Yellowstone 1927–32, Folder B-11.
46 Bumpus to Thomas B. Appleget, August 22, 1932, RAC, AAM Yellowstone 1927–32, Folder B-11.
47 Appleget to Bumpus, September 14, 1932, RAC, AAM Yellowstone 1927–32, Folder B-11.
48 Raymond B. Fosdick, *John D. Rockefeller, Jr.: A Portrait* (New York: Harper & Brothers, 1956), 420.
49 Fosdick, *John D. Rockefeller, Jr.*, 418.
50 Andras Szanto, *Rockefeller Philanthropy: A Selected Guide* (Sleepy Hollow, NY: Rockefeller Archive Center, 2011), 20.

Colorado High Desert High

1 JDR Jr. to J. F. Welborn, President Colorado Fuel & Iron Co., April 24, 1924, Rockefeller Archive Center (RAC), JDR Jr. Personal Papers, Travel, U.S., Glacier National Park 1924–26, Record Group 2, Box 43, Folder 392.
2 Jesse Nusbaum to Freeman Tilden and Horace M. Albright, April 1954, Mesa Verde National Park archives (MEVE); MEVE 1328-017, Article II.
3 Nusbaum to Tilden and Albright, 2.
4 Nusbaum to Tilden and Albright, 2.
5 *New York Daily News*, "A Lot of Guessing," June 21, 1924, 59.
6 Ricardo Torres-Reyes, "Mesa Verde National Park: An Administrative History, 1906–1970" (Washington, DC: National Park Service, Office of History and Historic Architecture, 1970), 91.

7 Torres-Reyes, "Mesa Verde National Park," 91.
8 Horace M. Albright, as told to Robert Cahn, *The Birth of the National Park Service: The Founding Years, 1913–33* (Salt Lake City: Howe Brothers, 1985), 159.
9 Duane A. Smith, *Mesa Verde National Park: Shadows of the Centuries* (Boulder: University of Colorado Press, 2002), 108.
10 Jesse L. Nusbaum, interview by Herb Evison, 1962, MEVE Archives, Mesa Verde National Park, Rosemary Talley/Jesse L. Nusbaum Papers, Series 001, Folder 030, ACC MEVE-01328, CAT MEVE-96900, 10.
11 Smith, *Mesa Verde*, 108.
12 Nusbaum to Tilden and Albright, 8.
13 Torres-Reyes, "Mesa Verde National Park," 6.
14 T. Mitchell Pruden, "The Prehistoric Ruins of the San Juan Watershed in Utah, Arizona, Colorado, and New Mexico," *American Anthropologist* 5, no. 2 (April–June 1903): 237.
15 Jesse Nusbaum to JDR Jr., February 23, 1926, RAC, Cultural Interests, Mesa Verde National Park, 1924–34, Series E, Box 82, Folder 767.
16 Nusbaum to Tilden and Albright, 8.
17 Nusbaum to Tilden and Albright, 8
18 Nusbaum to Tilden and Albright, 4.
19 Nusbaum to Tilden and Albright, 3.
20 Nusbaum to Tilden and Albright, 4.
21 Nusbaum to Arno B. Cammerer, January 29, 1924, NAC, Records Group 79, CF, Mesa Verde, Box 097, Entry P 9.
22 Jesse Nusbaum, "Notes on Contributions to Fund Museum Development at Mesa Verde," August 4, 1944, MEVE Archives, Mesa Verde National Park, Rosemary Talley/Jesse L. Nusbaum Papers, Series 001, Folder 030, ACC MEVE-01328, CAT MEVE-96900.
23 Nusbaum to Tilden and Albright, 8–9.
24 Nusbaum to Cammerer, June 18, 1924, NAC, Horace Albright, Record Group 79, Box 3, Entry P 59.
25 JDR Jr. to Abby A. Rockefeller, July 5, 1924, RAC, JDR Jr. Personal, Series II, Travel Records, U.S., Box 4, Folder 392.
26 Patti Bell, "Mrs. Nusbaum's Plays," unpublished, 2010.
27 Nusbaum to Tilden and Albright, 8.
28 Aileen Nusbaum to JDR Jr., July 1924, RAC, Series E, FA314, Organizations and Parks, Box 82, Mesa Verde 1924–1934.
29 Nusbaum to Tilden and Albright, 9.
30 Peter Collier and David Horowitz, *The Rockefellers: An American Dynasty* (New York: Holt, Rinehart, and Winston, 1976), 138.
31 Ron Chernow, *Titan: The Life of John D. Rockefeller, Sr.* (New York: Vintage Books, 1998), 637.
32 Collier and Horowitz, *The Rockefellers*, 156.
33 Nusbaum to Tilden and Albright, 9.

34 Personal interview with Kathy Fiero, former Mesa Verde employee, May 12, 2021.
35 Nusbaum to Tilden and Albright, 8.
36 Nusbaum to Tilden and Albright, 5–6.

Long Trip Through Wonderland

1 Horace M. Albright, as told to Robert Cahn, *The Birth of the National Park Service: The Founding Years*, 1913–33 (Salt Lake City: Howe Brothers, 1985), 159.
2 Albright, *The Birth of the National Park Service*, 158.
3 Albright, *The Birth of the National Park Service*, 89.
4 Albright, *The Birth of the National Park Service*, 89.
5 Horace M. Albright, interview by Michael Frome, "Recollections and Observations by Horace M. Albright," 1964, Great Smoky Mountains National Park library, Townsend, TN, #23575 and 23576; transcription of tapes 32 and 33, p. 16.
6 Albright, *The Birth of the National Park Service*, 160.
7 Diary of JDR III, Rockefeller Archive Center (RAC), FA 108, 1920, Trips, Diary, Box 86, Folder 725.
8 Albright interview, "Recollections and Observations," 16.
9 Albright interview, "Recollections and Observations," 16.
10 Albright, *The Birth of the National Park Service*, 160.
11 Albright, *The Birth of the National Park Service*, 160.
12 Albright, *The Birth of the National Park Service*, 160.
13 Robert V. Goss, "Creating Art from Yellowstone Resources," *Geyser Bob's Yellowstone Park Historical Service*, 2020, www.geyserbob.com/ole-anderson.
14 Aubrey L. Haines, *The Yellowstone Story* (Yellowstone National Park, WY: Yellowstone Library and Museum Association, 1977), 1:207.
15 Haines, *The Yellowstone Story*, 1:200.
16 Wallace Stegner, oral remarks on national parks, 1983, www.nps.gov/parkhistory/hisnps/npsthinking/famousquotes.htm.
17 Mark David Spence, *Dispossessing the Wilderness: Indian Removal and the Making of the National Parks* (New York: Oxford University Press, 1999), 62–70.
18 Albright, *The Birth of the National Park Service*, 160. See also Haines, *The Yellowstone Story*, 2:354.
19 J. E. Haynes, *Haynes New Guide*: *Yellowstone National Park*, 35th ed. (St. Paul, MN: J. E. Haynes, 1923), 68.
20 Senator Tasker L. Oddie to JDR Jr., January 20, 1925, RAC, FA314, Records Group 2, Series E, Box 77, Folder 734.
21 Haynes, *Haynes New Guide*, 138.

22 Diary of JDR Jr., July 12, 1924, RAC Travel Records, 1924–25, Series 2, Box 43, Folder 395.
23 Diary of JDR Jr., July 13, 1924.
24 Albright, *The Birth of the National Park Service*, 161.
25 Albright, *The Birth of the National Park Service*, 161.
26 Albright, *The Birth of the National Park Service*, 161.
27 Ron Chernow, *Titan: The Life of John D. Rockefeller, Sr.* (New York: Vintage Books, 1998), 236.
28 Chernow, *Titan*, 626–27.

The Landscape Architect

1 Alfred Runte, *National Parks: The American Experience* (Lanham, MD: Taylor Trade Publishing, 2010), 80–85.
2 JDR Jr. to Horace M. Albright, August 15, 1924, Rockefeller Archive Center (RAC), FA314, Series E, Box 82, Folder 769.
3 JDR Jr. to Albright, August 15, 1924.
4 JDR Jr. to Albright, August 15, 1924.
5 JDR Jr. to Charles J. Kraebel, August 15, 1924, RAC, OMR Rockefellers, Records Group 2, Box 43, Folder 392, 1.
6 JDR Jr. to Kraebel, August 15, 1924.
7 Stephen T. Mather to JDR Jr., September 24, 1924, RAC, FA 314, Records Group 2, Series E, Box 77, Folder 734.
8 Kraebel to JDR Jr. September 22, 1924, RAC, Record Group 2, Cultural Interests, Glacier National Park, Box 43, Folder 392.
9 Robert Shankland, *Steve Mather of the National Parks* (New York: Alfred A. Knopf, 1951), 209–10.
10 JDR Jr. to Henry C. Wallace, September 6, 1924, RAC, FA314, Record Group 2, Series E, Box 77, Folder 734.
11 Wallace to JDR Jr., September 18, 1924.
12 JDR Jr. to Wallace, September 26, 1924.
13 Albright to JDR Jr., September 2, 1924, RAC, FA314, Series E, Box 82, Folder 769.
14 Stephen T. Mather, Report of the Director of the NPS (Washington, DC: National Park Service, Government Printing Office, 1925), 24.
15 Albright to JDR Jr., September 2, 1924, RAC, FA314, Series E, Box 82, Folder 769.
16 Blair Bolles, *The Tyrant from Illinois: Uncle Joe Cannon's Experiment with Personal Power* (New York: W. W. Norton, 1951), 119. See also Horace M. Albright, as told to Robert Cahn, *The Birth of the National Park Service: The Founding Years, 1913–33* (Salt Lake City: Howe Brothers, 1985), 162.
17 Albright to JDR Jr., September 2, 1924, RAC, FA314, Series E, Box 82, Folder 769.

18 Ronald H. Epp, *Creating Acadia National Park: The Biography of George Bucknam Dorr* (Bar Harbor, ME: Friends of Acadia, 2016), 218.
19 Albright to JDR Jr., September 15, 1924, RAC, Record Group 2, FA314, Series E, Box 82, folder 769.
20 Albright to JDR Jr., September 15, 1924.
21 JDR Jr. to Albright, October 16, 1924.
22 Albright to JDR Jr., October 17, 1924.
23 Albright to JDR Jr., November 4, 1924.
24 Horace M. Albright and Marian Albright Schenck, *Creating the National Park Service: The Missing Years* (Norman: University of Oklahoma Press, 1999), 274–77.
25 Aubrey L. Haines, *The Yellowstone Story* (Yellowstone National Park, WY: Yellowstone Library and Museum Association, 1977), 1:272–73.
26 Albright to JDR Jr., "Report of Roadside Clean-Up, Autumn 1925," RAC, Cultural Interests—Yellowstone, Series E, Box 82, Folder 770.
27 Charles A. Heydt to JDR Jr., December 5, 1925, RAC, Cultural Interests—Yellowstone, Series E, Box 82, Folder 770.
28 Albright to JDR Jr., January 18, 1926, RAC, Record Group 2, FA314, Series E, Box 82, Folder 769.
29 Linda Flint McClelland, *Presenting Nature: The Historic Landscape Design of the National Park Service 1916 to 1942* (Washington, DC: National Park Service, 1993), 108.
30 Yellowstone National Park, "Report of Roadside Cleanup," October 18, 1927, RAC, Cultural Interests—Yellowstone, Series E, Box 82, Folder 770.
31 Arno B. Cammerer to JDR Jr., September 20, 1926, RAC, Record Group 2, FA314, Series E, Box 82, Folder 769.
32 Yellowstone National Park, "Report of Roadside Cleanup."
33 Albright to JDR Jr., August 13, 1928, RAC, Cultural Interests—Yellowstone, Series E, Box 82, Folder 770.
34 Albright to JDR Jr., December 27, 1928.
35 Albright to JDR Jr., November 13, 1936, RAC, FA314, Records Group 2, Series E, Box 77, Folder 734.
36 Joseph W. Ernst, ed., *Worthwhile Places: Correspondence of John D. Rockefeller, Jr. and Horace M. Albright* (New York: Fordham University Press/ Rockefeller Archive Center, 1991), 1.

Country Mice, City Mice

1 Jesse L. Nusbaum to Horace M. Albright, March 24, 1925, Mesa Verde National Park Archives (MEVE), Nbr: 001-02, RC-4-70, Drawer 2, Folder 82-b.
2 Bernice Kert, *Abby Aldrich Rockefeller: The Woman in the Family* (New York: Random House, 1993), 142.

3 Nusbaum to Albright, March 24, 1925.
4 Nusbaum to Albright, March 24, 1925.
5 Nusbaum to JDR Jr., October 31, 1924, MEVE Archives, Nbr: 001-02, RC-4-70, Drawer 1, Folder 59.
6 Jesse Nusbaum to JDR Jr., Nov. 29, 1924, Rockefeller Archive Center (RAC), MVNP –Jesse Nusbaum 1924–27, MEVE Acc. 1459, 2.
7 JDR Jr. to Jesse Nusbaum, August 22, 1924, MEVE Archives, Nusbaum Papers, 96900.
8 Jesse Nusbaum to JDR Jr., September 8, 1924.
9 Jesse Nusbaum to Albright, March 24, 1925, 1.
10 Nusbaum to Albright, March 24, 1925, 1.
11 Nusbaum to Albright, March 24, 1925, 2.
12 Nusbaum to Albright, March 24, 1925, 3.
13 Nusbaum to Albright, March 24, 1925, 3.
14 Nusbaum to Albright, March 24, 1925, 3.
15 Nusbaum to Albright, March 24, 1925, 4.
16 Nusbaum to Albright, March 24, 1925, 5.
17 Jesse Nusbaum to Freeman Tilden and Horace M. Albright, April 1954, MEVE Archives; MEVE 1328-017, Article II, 11–12.
18 "Memorandum of Personal Gifts to Mr. Jesse L. Nusbaum," 1935, RAC, OMR Rockefellers, FA 317, Friends & Services, Series H Friends and Relations, Jessie L. Gifts 1925–37, Box 95.
19 Robert Shankland, *Steve Mather of the National Parks* (New York: Alfred A. Knopf, 1951), 166–67.

The View from Lunch Tree Hill

1 David Rockefeller, *Memoirs* (New York: Random House, Inc., 2002), 41.
2 JDR Jr. to Horace M. Albright, September 6, 1926, Rockefeller Archive Center (RAC), Records Group II, Personal, Travel, Series II, Box 43, Folder 404.
3 Horace M. Albright, as told to Robert Cahn, *The Birth of the National Park Service: The Founding Years, 1913–33* (Salt Lake City: Howe Brothers, 1985), 163.
4 Albright to JDR Jr., July 12, 1949, RAC, Cultural Interests, FA 314, Series E, Box 80, Folder 751.
5 Albright, *The Birth of the National Park Service*, 163–64.
6 John Daugherty, *A Place Called Jackson Hole: The Historic Resource Study of Grand Teton National Park* (Moose, WY: Grand Teton National Park, National Park Service, 1999), 302.
7 Maxwell Struthers Burt to Albright, February 10, 1927, National Archives Center, Records Group 79, Cammerer Papers, Box 10, Entry P 60.
8 Albright, *The Birth of the National Park Service*, 154.

9 Robert W. Righter, *Crucible for Conservation: The Struggle for Grand Teton National Park* (Moose, WY: Grand Teton Association, 1982), 34.
10 Daugherty, *A Place Called Jackson Hole*, 304, 222–25.
11 Maxwell Struthers Burt, *The Diary of a Dude-Wrangler* (New York: Charles Scribner & Sons, 1924), 327.
12 Maxwell Struthers Burt, "The Battle of Jackson's Hole," *The Nation* 122, No. 3165 (March 3, 1926): 225–26.
13 Albright, *The Birth of the National Park Service*, 164–65.

That Isn't What I Had in Mind at All

1 Raymond B. Fosdick, *John D. Rockefeller, Jr.: A Portrait* (New York: Harper & Brothers, 1956), 372–73.
2 Fosdick, *John D. Rockefeller, Jr.*, 374–79.
3 Ron Chernow, *Titan: The Life of John D. Rockefeller, Sr.* (New York: Vintage Books, 1998), 90. Also, "Wet Plank Urged by Rockefeller, Jr.," *Buffalo Evening News*, June 7, 1932, 4.
4 Fosdick, *John D. Rockefeller, Jr.*, 215–17.
5 Fosdick, *John D. Rockefeller, Jr.*, 353–54.
6 Fosdick, *John D. Rockefeller, Jr.*, 354–55.
7 Fosdick, *John D. Rockefeller, Jr.*, 356.
8 Fosdick, *John D. Rockefeller, Jr.*, 420.
9 Fosdick, *John D. Rockefeller, Jr.*, 403.
10 Forest Hills Homeowners, Inc., "The Rockefellers and Forest Hill," Cleveland Heights, OH, www.fhho.org.
11 Joseph W. Ernst, *"Dear Father"/"Dear Son": Correspondence of John D. Rockefeller and John D. Rockefeller, Jr.* (New York: Fordham University Press/Rockefeller Archive Center, 1994), 84.
12 Eric J. Abrahamson, *Beyond Charity: A Century of Philanthropic Innovation* (New York: The Rockefeller Foundation, 2013), 47.
13 Chernow, *Titan*, 624.
14 Donna Bordeaux, "Rich as a Rockefeller," *Calculated Moves*, www.CalculatedMoves.com, March 29, 2017.
15 Shamus Khan, "The Rich Haven't Always Hated Taxes," *Time Magazine*, September 18, 2012, https://ideas.time.com/2012/09/18/the-rich-havent-always-hated-taxes/.
16 Chernow, *Titan*, 673.
17 Khan, "The Rich."
18 Ernst, *"Dear Father"/"Dear Son,"* 85–86.
19 Bernice Kert, *Abby Aldrich Rockefeller: The Woman in the Family* (New York: Random House, 1993), 137.
20 Hugh J. McCauley, "Visions of Kykuit," *Hudson River Valley Regional Review* 10, no. 2 (Summer 1993): 38–39.
21 Chernow, *Titan*, 556.

22 Ida M. Tarbell, "John D. Rockefeller: A Character Study, Part II," *McClure's Magazine* 25, no. 3 (July 1905): 387.
23 Horace M. Albright to JDR Jr., September 29, 1926, Rockefeller Archive Center (RAC), Correspondence—Horace M. Albright, Organizations and Parks, Yellowstone National Park, 1924–27.
24 Albright to JDR Jr., September 29, 1926.
25 JDR Jr. to Albright, November 9, 1926.
26 Fosdick, *John D. Rockefeller, Jr.*, 415.
27 Ann Rockefeller Roberts, *Mr. Rockefeller's Roads: The Untold Story of Acadia's Carriage Roads & Their Creator* (Camden, ME: Down East Books, 1990), 21.
28 Albright to JDR Jr., September 16, 1926, RAC, Record Group 2, RG FA314, Series E, Box 82, Folder 769.
29 Albright, *The Birth of the National Park Service*, 166.
30 Albright, *The Birth of the National Park Service*, 166.
31 Albright, *The Birth of the National Park Service*, 167.
32 Donald C. Swain, *Wilderness Defender: Horace M. Albright and Conservation* (Chicago: University of Chicago Press, 1970), 160.
33 Albright, *The Birth of the National Park Service*, 165–66.
34 Albright, *The Birth of the National Park Service*, 167.
35 JDR Jr. to Albright, December 22, 1926.
36 Albright to JDR Jr., January 4, 1927.
37 Albright to JDR Jr., January 4, 1927.
38 Joseph W. Ernst, ed., *Worthwhile Places: Correspondence of John D. Rockefeller, Jr. and Horace M. Albright* (New York: Fordham University Press/Rockefeller Archive Center, 1991), 75.
39 Arno B. Cammerer to JDR Jr., January 27, 1927, RAC, Record Group 2, RG FA314, Series E, Box 82, Folder 769.
40 Albright, *The Birth of the National Park Service*, 167.

Clouds of Dust, Honks of Horns . . . and Smells of Gasoline

1 Edward K. Dunham, "Early Years of the Seal Harbor Village Improvement Society," August 20, 1980, Seal Harbor [Maine] Library Archives, 3. See also various letters of acknowledgment of contributions, Rockefeller Archives Center (RAC), OMR Homes, Bar Harbor, Series 1, FA318, Box 79, Folders 801–3.
2 Ronald H. Epp, *Creating Acadia National Park: A Biography of George Bucknam Dorr* (Bar Harbor, ME: Friends of Acadia, 2016), 175.
3 National Park Service, Acadia, "Where Science Makes History," www.nps.gov/acad/. Also, various letters of acknowledgement of contributions, RAC.
4 Epp, *Creating Acadia National Park*, 229.
5 Epp, *Creating Acadia National Park*, 269.

6 Ann Rockefeller Roberts, *Mr. Rockefeller's Roads: The Untold Story of Acadia's Carriage Roads & Their Creator* (Camden, ME: Down East Books, 1990), 97.
7 Epp, *Creating Acadia National Park*, 235.
8 Roberts, *Mr. Rockefeller's Roads*, 98.
9 Epp, *Creating Acadia National Park*, 236.
10 H. Eliot Foulds and Jeffrey Killion, "Cultural Landscape Report for the Historic Motor Road System Acadia National Park" (Boston: Olmsted Center for Landscape Preservation, Boston National Historical Park), 21.
11 George B. Dorr, *The Story of Acadia National Park* (Bar Harbor, ME: Acadia Publishing Company, 1997), 115.
12 "Summer Resident Worth Having," *Bangor (ME) Daily News*, September 17, 1930, 4.
13 John H. Edwards (acting secretary of the interior) to JDR Jr., September 11, 1930, National Archives Center (NAC), Records Group 48, CCF 1907–36, Box 2009.
14 *Bangor (ME) Daily News*, September 17, 1930, 4.
15 Epp, *Creating Acadia National Park*, 241–43.
16 *Bar Harbor (ME) Times*, December 3, 1930, 1.
17 Paul S. Sutter, *Driven Wild: How the Fight Against Automobiles Launched the Modern Wilderness Movement* (Seattle: University of Washington Press, 2002), 12.
18 Epp, *Creating Acadia National Park*, 214.
19 Robert Marshall, "The Problem of the Wilderness," *Scientific Monthly* 30 (February 1930): 142–45.
20 Sutter, *Driven Wild*, 3–9, 47.
21 Foulds and Killion, "Cultural Landscape Report," 26.
22 *Bar Harbor Times*, January 28, 1931, 1.
23 *Bar Harbor Times*, January 28, 1931, 2.
24 Cammerer to Wilbur, September 4, 1930, NAC, Records Group 48, CCF 1907-36, Box 2009.
25 Albright to Wilbur, November 7, 1930.
26 *Bar Harbor Times*, February 4, 11, 18, 1931, 1–8.
27 "Unemployment News," *Daily Mirror*, circa 1930, RAC, OMR Rockefeller, Seal Harbor, FA318, Box 114, Folder 1144.
28 *Bar Harbor Times*, February 4, 1931, 6.
29 Sutter, *Driven Wild*, x–xi.
30 Dr. Francis G. Peabody to Editor of *Bar Harbor Times*, January 20, 1931, RAC, OMR Rockefellers, Seal Harbor, FA 318, Box 114, Folder 1144.
31 A. F. Sherman to JDR Jr., January 31, 1931, RAC, OMR Rockefellers, Seal Harbor, FA318, Box 114, Folder 1144.
32 Neil Maher, "The Acadia National Park Roads and Bridges Recording Project," Historic American Engineering Record, HAER No. ME-11 (Washington, DC: National Park Service, 1994–97), 17.

33 Stephen T. Mather, "Report of the Director of the NPS to the Secretary of the Interior" (Washington, DC: Government Printing Office, 1923), 6–11.
34 Roberts, *Mr. Rockefeller's Roads*, 107.
35 Dorr, *The Story of Acadia*, 104–11.
36 Horace M. Albright, as told to Robert Cahn, *The Birth of the National Park Service: The Founding Years, 1913–33* (Salt Lake City: Howe Brothers, 1985), 228.
37 JDR Jr. to Cammerer, October 1929, NAC, Record Group 79, Acadia, Box 1, P 60.
38 Dorr, *The Story of Acadia*, 113–14.
39 James J. Lee III, "U.S. Naval Radio Station—Apartment Building #1 Historic Structure Report" (Lowell, MA: National Park Service, 2009), 23.
40 Dorr, *The Story of Acadia*, 114.
41 Captain S. C. Hooper to Chief of Naval Operations, October 24, 1932, NAC, Records Group 79, Cammerer Papers, Box 1, P 60.
42 Dorr, *The Story of Acadia*, 117.
43 JDR Jr. to Ray Lyman Wilbur, May 28, 1931, NAC, Records Group 48, CCF 1907-36, Box 2009.
44 Dorr, *The Story of Acadia*, 117–18.
45 Hooper to Chief of Naval Operations, October 24, 1932.
46 Lee, "U.S. Naval Radio Station," 25–26, 28–29.
47 Epp, *Creating Acadia National Park*, 238–39.
48 Dorr, *The Story of Acadia*, 108.
49 Horace M. Albright, *Report of the Director of the National Park Service* (Washington, DC: National Park Service, 1929), 4.
50 Epp, *Creating Acadia National Park*, 368.
51 *Bangor Daily News*, "Hope for Start of Scenic Road in Acadia Park Is Brightened," November 8, 1932, 1.
52 *Bangor Daily News*, January 2, 1933, 12.
53 Grace M. Oakes, *Portland Press Herald*, October 22, 1933, 35.
54 Oliver G. Taylor to Arthur E. Demaray, February 1, 1934, NAC, Records Group 48, CCF 1907–36, Box 2009.
55 Ron Chernow, *Titan: The Life of John D. Rockefeller, Sr.* (New York: Vintage Books, 1998), 667.
56 JDR Jr. to Arno B. Cammerer, January 16, 1934, NAC, Records Group 48, CCF 1907–36, Box 2009.
57 Cammerer to JDR Jr., February 9, 1934.
58 Cammerer to JDR Jr., March 7, 1932.
59 Schoodic Institute, https://schoodicinstitute.org/about-us/history/.
60 Richard H. Quin, "Cadillac Mountain Road," Acadia National Park Roads & Bridges, Historic American Engineering Records, No. ME-58 (Washington, DC: National Park Service, 1994), 7–8, 11–12.

61 Ronald H. Epp, author of *Creating Acadia National Park: The Biography of George Bucknam Dorr*, phone interview with Steve Kemp, 2022.
62 Maher, "Acadia Roads and Bridges," 15–16.
63 Maher, "Acadia Roads and Bridges," 16.
64 Albright to JDR Jr., June 13, 1932, in Joseph W. Ernst, *Worthwhile Places: Correspondence of John D. Rockefeller, Jr. and Horace M. Albright* (New York: Fordham University Press/Rockefeller Archive Center, 1991), 125–26.
65 Quin, "Cadillac Mountain Road," 22.
66 *Bar Harbor Times*, July 27, 1932, 1.
67 Ray L. Wilbur to JDR Jr., June 27, 1932. RAC, Homes, FA318, Series I, Box 83, Folder 827.
68 Maher, "Acadia Roads and Bridges," 5.
69 *Kennebec Journal*, July 25, 1932, 1.
70 David Louter, *Windshield Wilderness: Cars, Roads, and Nature in Washington's National Parks* (Seattle: University of Washington Press, 2006), 36.
71 Ernst, *Worthwhile Places*, 5, 249–50.
72 Andras Szanto, *Rockefeller Philanthropy: A Selected Guide* (Sleepy Hollow, NY: Rockefeller Archive Center, 2011), 8.

Both Sorts of Green

1 STATS, NPS, Irma.nps.gov/stats.
2 *Knoxville Journal*, April 5, 1953, 50.
3 "Save Our Mountains," *Knoxville News*, August 28, 1923, 4.
4 W. P. Davis to W. S. Richardson, September 22, 1924, Rockefeller Archive Center (RAC), OMR Rockefellers, Cultural Interests, Series E, Box 92, Folder 851.
5 Dr. Arthur Kendall to Junior's officers, October 15, 1924, GSMNP timeline, RAC, Great Smoky Mountains 1924–30, Series E, Box 92, Folder 851.
6 Richardson to Kendall, October 18, 1924.
7 Richardson to Major W. A. Welch, September 17, 1924.
8 Margaret Lynn Brown, *The Wild East: A Biography of the Great Smoky Mountains* (Gainesville: University of Florida Press, 2000), 88.
9 Stephen T. Mather, "Report of the Director of the NPS to the Secretary of the Interior" (Washington, DC: National Park Service, 1923), 14.
10 Cammerer to Kenneth Chorley, January 27, 1927, National Archives Center (NAC), Records Group 79, Horace Albright, Box 3, Entry P 59.
11 Mather, "Report of the Director of the NPS," 14.
12 Mather to JDR Jr., December 13, 1924.
13 Mather to JDR Jr., December 12, 1924.
14 John D. Rockefeller, Sr., *Random Reminiscences of Men and Events* (New York: Doubleday, Page & Company, 1909), 183.

15 Mather to JDR Jr., December 13, 1924.
16 Carlos C. Campbell, *Birth of a National Park in the Great Smoky Mountains* (Knoxville: University of Tennessee Press, 1960), 22.
17 Campbell, *Birth*, 23.
18 Daniel S. Pierce, *The Great Smokies: Natural Habitat to National Park* (Knoxville: University of Tennessee Press, 2000), 70–71.
19 Paul J. Adams, *Mount Le Conte* (Knoxville: University of Tennessee Press, 2016), 23–27.
20 Adams, *Mount Le Conte*, 27–28.
21 Carson Brewer, *Hiking Trails of the Smokies* (Gatlinburg, TN: Great Smoky Mountains Natural History Association, 2001), 419.
22 Adams, *Mount Le Conte*, 29.
23 Adams, *Mount Le Conte*, 30.
24 Pierce, *Great Smokies*, 71.
25 Pierce, *Great Smokies*, 71.
26 Paul M. Fink, *Backpacking Was the Only Way* (Johnson City, TN: Research Advisory Council, Eastern Tennessee State University, 1975), 65.
27 JDR Jr. to Major W. A. Welch, September 26, 1927, NAC, Cammerer Personal File—Frank Maloney, Records Group 79, Box 6, Entry P 60.
28 Pierce, *Great Smokies*, 75.
29 H.W. Temple, Glen S. Smith, W. A. Welch, Harlan P. Kelsey, and William C. Gregg to Hubert Work, January 29, 1925, "Providing for the Acquisition of Land in the Southern Appalachian Mountains for Park Purposes," U.S. House of Representatives, 68th Congress, 2nd Session, Report No. 1320.
30 Pierce, *Great Smokies*, 75–76.
31 Mather to JDR Jr., December 13, 1924, RAC, OMR Rockefellers, Cultural Interests, Series E, Box 92, Folder 851.
32 Pierce, *Great Smokies*, 75.
33 Pierce, *Great Smokies*, 76.
34 Theodore Catton, *Mountains for the Masses: A History of Management Issues in Great Smoky Mountains National Park* (Gatlinburg, TN: Great Smoky Mountains Association, 2014), 22.
35 Speech by Ray Lyman Wilbur, November 2, 1931, Broadcast on WRC, NBC, from the Secretary's office.
36 Brown, *Wild East*, 97–98.
37 Herb Trentham oral history, GSMNP Archives.
38 Brown, *Wild East*, 96.
39 Frank Maloney to Arno B. Cammerer, September 13, 1926, NAC, Records Group 79, Cammerer, Box 12, Entry CCP 60.
40 NPS Land acquisition guidelines from Cammerer, NAC, Records Group 79, Arno B. Cammerer, Box 6, Entry P 60, 4.
41 Dennis Drabelle, *The Power of Scenery: Frederick Law Olmsted and the*

Origins of National Parks (Lincoln: University of Nebraska Press, 2021), 22–24.
42 Drabelle, *The Power of Scenery*, 188.
43 NPS, *A Journey of Injustice*, 2023, www.nps.gov/trte/learn/historyculture/index.htm.
44 Brown, *Wild East*, 58, 96–99.
45 Pierce, *Great Smokies*, 97–98.
46 Pierce, *Great Smokies*, 140.
47 Pierce, *Great Smokies*, 103.
48 Pierce, *Great Smokies*, 97–98.
49 Pierce, *Great Smokies*, 112–14.
50 Pierce, *Great Smokies*, 78–79.
51 *Knoxville News* and *Knoxville Journal*, March 16, 1925.
52 Denise Aday and Michael Aday, *Letters from the Smokies* (Gatlinburg, TN: Great Smoky Mountains Association, 2023), 67.
53 Michael Frome, *Strangers in High Places: The Story of the Great Smoky Mountains* (Knoxville: University of Tennessee Press, 1966), 186.
54 Pierce, *Great Smokies*, 80.
55 Catton, *Mountains for the Masses*, 25, 28.
56 Congressman Joseph G. Cannon, U.S. Speaker of the House, quoted by Blair Bolles, *Tyrant from Illinois: Uncle Joe Cannon's Experiment with Personal Power* (New York: W. W. Norton, 1951), 119. See also Catton, *Mountains for the Masses*, 27.
57 *Asheville (NC) Times*, November 22, 1925, 17. Though wildly optimistic at the time, Sprague's estimates have proved to be grossly understated. In 2024, GSMNP hosted over 12 million visitors who helped generate over $3 billion in revenue.
58 Pierce, *Great Smokies*, 89–91.
59 Campbell, *Birth*, 38.
60 Pierce, *Great Smokies*, 99–105.
61 Campbell, *Birth*, 41–42.
62 David Chapman, *Knoxville Journal*, February 21, 1926, 1.
63 *Knoxville Journal*, March 16, 1926, editorial, 1.
64 Aday and Aday, *Letters from the Smokies*, 73.
65 Pierce, *Great Smokies*, 105.
66 Cammerer to JDR Jr, July 17, 1940 (extract from other material), RAC, 108, C, 2.

Rich People Are Temperamental

1 Theodore Catton, *Mountains for the Masses: A History of Management Issues in Great Smoky Mountains National Park* (Gatlinburg, TN: Great Smoky Mountains Association, 2014), 21.

2 Daniel S. Pierce, *The Great Smokies: Natural Habitat to National Park* (Knoxville: University of Tennessee Press, 2000), 118–21.
3 Cammerer to Kenneth Chorley, February 10, 1927, National Archive Center (NAC), Records Group 79, Horace M. Albright, Box 3, Entry P 59.
4 Chorley to Arthur Woods, March 9, 1928, Rockefeller Archives Center (RAC), LSRM, 1926–28, 143; S III-4, B-13.
5 Pierce, *Great Smokies*, 118–21.
6 Cammerer to Mark Squires, April 28, 1927, NAC, Records Group 79, Horace M. Albright, Box 3, Entry P 59.
7 Pierce, *Great Smokies*, 158–64.
8 Frank Maloney to Cammerer, October 14, 1927, NAC, Records Group 79, Cammerer Papers, Box 12, Entry P 60.
9 Carlos C. Campbell, *Birth of a National Park in the Great Smoky Mountains* (Knoxville: University of Tennessee Press, 1960), 65.
10 Raymond B. Fosdick, *John D. Rockefeller, Jr.: A Portrait* (New York: Harper & Brothers, 1956), 322.
11 David Chapman to Cammerer, September 13, 1927, GSMNP Archives, GSMCA, XI-8.
12 Mark Squires to Cammerer, July 18, 1927, NAC, Records Group 79, Cammerer Papers, Box 6, Entry P 60.
13 Squires to Chapman, January 8, 1928, GSMNP Archives, GSMCA, IX-9-9.
14 Cammerer to Squires, July 20, 1927, NAC, Records Group 79, Cammerer Papers, Box 6, Entry P 60.
15 Cammerer to JDR Jr, September 13, 1927, RAC, OMR Rockefellers, Cultural Interests, Series E, Box 92, Folder 851.
16 Cammerer to Chapman, January 28, 1928, GSMNP Archives, GSMCA, XI-10.
17 Cammerer to Chapman, January 14, 1928.
18 Squires to Cammerer, August 3, 1927, NAC, Records Group 79, Cammerer Papers, Box 6, Entry P 60.
19 Squires to Cammerer, August 3, 1927.
20 Cammerer to Frank Maloney, September 14, 1926.
21 Cammerer to JDR Jr., May 28, 1927, RAC, OMR Rockefellers, Cultural Interests, Series E, Box 92, Folder 851.
22 Cammerer to Chapman, September 14, 1927.
23 Campbell, *Birth of a National Park*, 60.

Sight Unseen

1 Horace M. Albright and Marian Albright Schenck, *Creating the National Park Service: The Missing Years* (Norman: University of Oklahoma Press, 1999), 314.

2 Horace M. Albright, interview by Michael Frome, "Recollections and Observations," 1964, Great Smoky Mountains National Park (GSMNP) Library, 23575–23576, 21.
3 Edwin C. Bearss, *Arno B. Cammerer*, Biographical Vignette, NPS, www.nps.gov/parkhistory/online_books/sontag/cammerer.htm. See also Michael Frome, *Strangers in High Places* (Knoxville: University of Tennessee Press, 1966), 227.
4 Cammerer to Chapman, January 11, 1928, GSMNP archives, GSMCA, XI-10.
5 Cammerer to Chorley, December 15, 1928, National Archives Center (NAC), Records Group 79, Horace Albright, Box 3, Entry P 59.
6 Cammerer to JDR Jr., November 12, 1929, NAC, Records Group 48, CCF 1907–36, Box 2009.
7 Ron Chernow, *Titan: The Life of John D. Rockefeller, Sr.* (New York: Vintage Books, 1998), 237.
8 F. Roger Miller to JDR Jr., December 1, 1926, Rockefeller Archive Center (RAC), OMR Rockefellers, Cultural Interests, Series E, Box 92, Folder 851.
9 Thomas B. Appleget to Miller, December 13, 1926.
10 Major W. A. Welch to Miller, December 9, 1926, GSMNP archives, GSMCA, XIV-25.
11 JDR Jr. to Cammerer, July 11, 1939, RAC, Office of Messrs. Rockefeller, Series H (FA317), Friends and Relations, Box 52, Cammerer illness letters.
12 Chorley to Cammerer, March 14, 1928, NAC, Records Group 79, Horace M. Albright, Box 3, Entry P 59.
13 Cammerer to Chorley, February 10, 1927.
14 JDR Jr. to Cammerer, May 17, 1927, NAC, Records Group 79, Cammerer Papers, Box 01, Entry P 60.
15 Cammerer to Mark Squires, August 5, 1927, GSMNP Archives, GSMCA, XI-8.
16 Daniel Okrent, *Great Fortune: The Epic of Rockefeller Center* (New York: Viking, 2003), front cover flap.
17 Peter Collier and David Horowitz, *The Rockefellers: An American Dynasty* (New York: Holt, Rinehart, and Winston, 1976), 154.
18 Riverside Church website, 2023, www.trcnyc.org, 1.
19 Philip Kopper, *Colonial Williamsburg* (New York: Harry N. Abrams, 1986), 161.
20 Raymond B. Fosdick, *John D. Rockefeller, Jr.: A Portrait* (New York: Harper & Brothers, 1956), 290.
21 Kert, *Abby Aldrich Rockefeller*, 231. See also "Colonial Williamsburg," Rockefeller Brothers Fund, Timeline, www.rbf.org/about/our-history/timeline/colonial-williamsburg.
22 Collier and Horowitz, *The Rockefellers*, 165–66.
23 Collier and Horowitz, *The Rockefellers*, 165–66.
24 Fosdick, *John D. Rockefeller, Jr.*, 239.

25 Okrent, *Great Fortune*, 129, 191.
26 Kenneth Chorley, Oral History Research Office, Columbia University, 1973, RAC, Jackson Hole Preserve Project, FA 1444, Box 3, Folder 44, 9.
27 Cammerer to Mark Squires, August 5, 1927, Oral History Research Office, Columbia University, 1973, RAC, Jackson Hole Preserve Project, FA 1444, Box 3, Folder 44.
28 Chernow, *Titan*, 180–81.

Angels Can't Do More

1 Cammerer to JDR Jr., August 12, 1927, Rockefeller Archive Center (RAC), Cultural Interests: Great Smoky Mountains 1924–30, Box 92, Folder 169.
2 Cammerer to JDR Jr., August 12, 1927.
3 Paul S. Sutter, *Driven Wild: How the Fight Against Automobiles Launched the Modern Wilderness Movement* (Seattle: University of Washington, 2002), 12.
4 Sutter, *Driven Wild*, 12.
5 Ann Rockefeller Roberts, *Mr. Rockefeller's Roads: The Untold Story of Acadia's Carriage Roads & Their Creator* (Camden, ME: Down East Books, 1990), 12.
6 Sutter, *Driven Wild*, 279.
7 Cammerer to David Chapman, January 12, 1928, GSM Archives, GSMCA, XI-10.
8 JDR Jr. to W. A. Welch, September 26, 1927, National Archive Center (NAC), Records Group 79, Box 6, Entry P 60.
9 JDR Jr. to Cammerer, September 3, 1927, RAC, Cultural Interests: Great Smoky Mountains 1924–30, Box 92, Folder 169.
10 Notes of Great Smoky Mountains Conservation Association (GSMCA) meeting, April 8, 1953, Knoxville, TN, GSMNP Archives, GSMCA, XI-8.
11 Cammerer to Laura Spelman Rockefeller Memorial (LSRM), February 11, 1928, RAC, LSRM, S-III-4, B-13, 143, 2.
12 JDR Jr. to Cammerer, April 30, 1940, NAC, Records Group 79, Cammerer Papers, Box 10, Entry P 60.
13 Daniel S. Pierce, *The Great Smokies: From Natural Habitat to National Park* (Knoxville: University of Tennessee Press, 2000), 126.
14 Kenneth Chorley to Colonel Woods, March 9, 1928, RAC, LSRM, S-III-4, B-13, 143, 3.
15 Chorley to Woods, March 9, 1928.
16 Chorley to Woods, March 9, 1928.
17 *Knoxville News Sentinel*, March 6, 1928, 1.
18 Francis Christy, "GSMNP: An Episode in Its Creation," RAC, Cultural Interests, Series E, FA314, Box 93, Folder 854.

19 Horace M. Albright to JDR Jr., May 28, 1930, RAC, OMR Rockefellers, Cultural Interests, Series E, Box 92, Folder 851.
20 JDR Jr. to children, "Motor Trip in the South," 1934 travel diary, RAC, OMR Rockefellers, JDR Jr., FA335, Series Z, Box 45, Folder 424, 6–8.
21 *Asheville (NC) Citizen*, September 3, 1940, 1.
22 JDR Jr. to Cammerer, April 30, 1940, NAC, Records Group 79, Cammerer Papers, Box 10, Entry P 60.
23 JDR Jr. to Harold L. Ickes, July 11, 1939, RAC, OMR Rockefellers, JDR Jr., Series Z, FA335, Box 76, Folder 269.
24 Michael Frome, *Strangers in High Places: The Story of the Great Smoky Mountains* (Knoxville: University of Tennesee Press, 1966), 205.
25 Albright, interview by Michael Frome, "Recollections and Observations," 1964, GSMNP Library, 20.
26 George B. Dorr to Cammerer, November 8, 1935, NAC, Records Group 79, Cammerer Papers, Box 12, Entry P 60.
27 John S. Mullings to JDR Jr., July 6, 1953, RAC, OMR Rockefellers, Series E, FA314, Box 78, Folder 738.
28 Norris Farm to JDR Jr., October 17, 1956.
29 J. J. Goodacre to JDR Jr., November 3, 1958.
30 Notes of GSMCA meeting, April 8, 1953, GSMNP archives, GSMCA, XI-8.
31 Joseph W. Ernst, *"Dear Father"/"Dear Son": Correspondence of John D. Rockefeller and John D. Rockefeller, Jr.* (New York: Fordham University Press/Rockefeller Archive Center, 1994), 226.
32 Raymond B. Fosdick, *John D. Rockefeller, Jr.: A Portrait* (New York: Harpers & Brothers, 1956), 432.
33 Lawrence K. Frank to Beardsley Ruml, February 3, 1926, RAC, Great Smoky Mountains Memorial Fund, LSRM, SIII-4, B-13, Folder 143.
34 Emily Davis, GSMNP news release, NPS, August 21, 2023, www.nps.gov/grsm/learn/news.
35 Frome, *Strangers in High Places*, 213.

The Lingering Sorrow of the Dispossessed

1 John Bradley, interview by Dorothy Noble Smith, June 12, 1978, SdArch SNP-15, Shenandoah National Park Oral History Collection, 1963–1999, Special Collections, Carrier Library, James Madison University, 42.
2 Bradley interview, 41.
3 Darwin Lambert, *The Undying Past of Shenandoah National Park* (Boulder, Colorado: Roberts Rinehart, Inc., 1989), 289–302.
4 Lambert, *The Undying Past*, 241–42.
5 Lambert, *The Undying Past*, 289–90.
6 Darwin Lambert, "Administrative History Shenandoah National Park, 1924–1971" (Washington, DC: National Park Service, 1979), 95.

7 Lambert, "Administrative History," 135–39.
8 *Richmond Times-Dispatch*, October 5, 1935, 6.
9 Lambert, "Administrative History," 8–9. Also, Audrey J. Horning, "When Past Is Present: Archaeology of the Displaced in Shenandoah National Park," paper presented at 2001 Society of Historical Archaeology Conference, Long Beach, CA, 2001.
10 Lambert, "Administrative History," 20–22.
11 Lambert, "Administrative History," 61–68.
12 Lambert, "Administrative History," 136.
13 Harry F. Byrd to Arno B. Cammerer, March 8, 1928, National Archive Center (NAC), Records Group 79, Cammerer Papers, Box 11, Entry P 60.
14 Cammerer to Dr. Frank Bohn, September 7, 1928.
15 Cammerer to Mrs. J. Ogden Armour, January 23, 1929.
16 Albright to JDR Jr., October 13, 1932, Rockefeller Archive Center (RAC), Cultural Interests, Series E, FA 314, Box 93, Folder 856. Also, JDR Jr. to Edsel Ford, March 2, 1928, RAC, Cultural Interests, Great Smoky Mountains National Park (GSMNP) 1924–30, FA 314, Series E, box 92, Folder 851.
17 Lambert, "Administrative History," 92.
18 Cammerer to Ray L. Wilbur, October 15, 1929, NAC, Records Group 79, Cammerer Papers, Box 11, Entry P 60.
19 Laurel A. Racine, "Rapidan Camp: 'The Brown House,'" Historic Furnishings Report (Washington, DC: National Park Services, Northeast Museum Services, 2001), 17.
20 Racine, "Rapidan Camp," 6, 17–18.
21 Lambert, *The Undying Past*, 219, 232.
22 Lambert, *The Undying Past*, 254.
23 Congressional Record, 75th Congress, 1st session, 1937, 7966. See also Lambert, "Administrative History," 231.
24 Lambert, "Administrative History," 233.
25 Lambert, "Administrative History," 237.
26 Lambert, "Administrative History," 246–47.
27 *Staunton (VA) News-Leader*, September 18, 1935, 1.
28 Dorothy H. Housh, interview by Dennis Carter and Dorothy Noble Smith, March 30, 1978, SdArch SNP-15, Shenandoah National Park Oral History Collection, 1963–1999, Special Collections, Carrier Library, JMU.
29 Lambert, "Administrative History," 125.
30 Lambert, *The Undying Past*, 233.
31 Lambert, *The Undying Past*, 246.
32 James G. Burner Sr., interview by Dorothy Noble Smith, September 16, 1977, SdArch SNP-19, Shenandoah National Park Oral History Collection, 1963–1999, Special Collections, Carrier Library, JMU, 45.

33 Ralph Cave, interview by Nancy Smith, September 21, 1977, SdArch SNP-25, Shenandoah National Park Oral History Collection, 1963–1999, Special Collections, Carrier Library, JMU, 33.
34 Miriam Sizer to John Bohn, September 13, 1928, NAC, Record Group 79, Cammerer, Box 12, Entry P 60.
35 Burner Sr. interview, 1.
36 Philip Kopper, *Colonial Williamsburg* (New York: Harry N. Abrams, 1986), 164.
37 "TVA," Tennessee Historical Society, www.tennesseehistory.org.
38 *Richmond Times-Dispatch*, December 22, 1935, 42.
39 Claire Comer, NPS News Release, August 31, 2023, www.nps.gov/shen/learn/news.
40 Horace M. Albright to JDR Jr., October 13, 1932, NAC, Records Group 79, Cammerer Papers, Box 11, Entry P 60.

The Sugar Pines' Last Stand

1 JDR Jr. to Cammerer, February 28, 1928, Rockefeller Archive Center (RAC), Cultural Interests, Yosemite National Park, FA 314, Series E, Box 93, Folder 858.
2 "To Sell 20 Miles of Yosemite Trees," *New York Times*, February 21, 1928, 3.
3 "The Yosemite Trees," *New York Times*, February 22, 1928, 20.
4 "Pleads for Forests in Yosemite Park," *New York Times*, February 25, 1928.
5 Willard G. Van Name, *Vanishing Forest Reserves: Problems of the National Forests and National Parks* (Boston: R. G. Badger, 1929), 114–16.
6 Letter from Horace Albright to Duncan McDuffie, May 15, 1928, YOSE, Yosemite Lumber Company, File 610-09, 1925–28; Box 33, F 102.
7 Carl Bachem to Kenneth Chorley, July 10, 1929, YOSE, Rockefeller Purchase, 1918–1935, File 610, 4.
8 Van Name, *Vanishing Forest Reserves*, 118–22.
9 Albright to JDR Jr., June 28, 1928, RAC, Cultural Interests, FA314, Series E, Box 93, Folder 857.
10 Carl Bachem to Superintendent Lewis, September 25, 1925, YOSE, Yosemite Lumber Co., Box 33, F 103_C.
11 Personal letter to California state senator Sanborn Young, June 10, 1926, YOSE, Yosemite Lumber Company, File 610-09, 1925–28; Box 33, F 103_B.
12 Quoted in *Great Falls Tribune*, December 26, 1926, 21.
13 Albright to JDR Jr., April 16, 1928, National Archive Center (NAC), Records Group 79, Cammerer Papers, Box 10, Entry P 60.
14 Horace M. Albright and Marian Albright Schenck, *Creating the National*

Park Service: The Missing Years (Norman: University of Oklahoma Press, 1999), 293.

15 Albright to A. Crawford Greene, February 3, 1930, YOSE, File 610, BRN44372-005673.

16 Stephen T. Mather to Albright, June 11, 1928, YOSE, Box 33, Folder 102.

17 Mather to George A. Ball, June 15, 1928.

18 Albright to Mather, May 3, 1928.

19 Mather to Ball, June 15, 1928.

20 Mather to Albright, May 3, 1928.

21 William O. Ingles, "Suggestions for a Biography of John D. Rockefeller," 1918, RAC, Ingles Papers, 11.

22 John D. Rockefeller Sr., *Random Reminiscences of Men and Events* (New York: Doubleday, Page, and Co., 1908), 21–22.

23 Dyana Z Furmansky, "Role Model for a Conservationist," RAC Research Reports, RAC, 2013, 7, https://rockarch.issuelab.org/resources/28024/28024.pdf.

24 Arno B. Cammerer to JDR Jr., March 12, 1928, RAC, Cultural Interests, Yosemite National Park 1928–29, Series E, Box 93, Folder 858.

Friends in High Places

1 Hubert Work to William M. Jardine, April 14, 1928, Rockefeller Archive Center (RAC), Cultural Interests, Yosemite National Park 1928–29, FA 314, Series E, Box 93, Folder 857.

2 Work to Jardine, April 14, 1928.

3 Russell Franzen, "Louis C. Cramton," *Michigan Political History Society News* 2, no. 2 (May 1995): 3.

4 Louis C. Cramton, "The Grand Canyon of the Colorado," speech to the U.S. House of Representatives, March 3, 1924, 89293-323, 3.

5 "A New Yosemite Threat," *New York Times*, May 24, 1928.

6 Arthur N. Pack, "Saving the Yosemite," *New York Times*, May 24, 1928.

7 "A New Yosemite Threat."

8 Cammerer to JDR Jr., June 28, 1928, RAC, Cultural Interests, Yosemite N. P. 1928–29, FA 314, Series E, Box 93, Folder 858.

9 Cammerer to JDR Jr., June 28, 1928.

10 Cammerer to JDR Jr., June 28, 1928.

11 Arthur Woods to JDR Jr., November 26, 1929, RAC, Cultural Interests, Organizations and Parks—Yosemite N.P., Series E, Box 93, Folder 858.

12 Albright to Mather, September 11, 1928, YOSE, Box 33, Folder 101.

13 Albright to Mather, July 13, 1928, YOSE, Box 33 F, Folder 102.

14 Arthur Woods to JDR Jr., August 20, 1928, RAC, Cultural Interests, Organizations and Parks—Yosemite N.P., Series E, Box 93, Folder 858.

15 JDR Jr. to Chorley, August 29, 1928, RAC, Cultural Interests, Organizations and Parks—Yosemite N.P., Series E, Box 93, Folder 857.
16 Chorley summary memo of conversation with Cramton, November 14, 1929, RAC, Cultural Interests, Organizations and Parks—Yosemite N.P., Series E, Box 93, Folder 857.
17 Hand-written code key, YOSE, BRN44372_005792, Rockefeller Purchase, #610, 1918–35.
18 Vanderbilt Webb to John Ball, November 1928, RAC, Cultural Interests, Organizations and Parks—Yosemite N.P., Series E, Box 93, Folder 857.
19 Chorley to T. M. Debevoise, November 21, 1928.
20 Albright to Chorley, July 3, 1929.
21 Albright to Vanderbilt Webb, July 1, 1929.
22 Albright to JDR Jr., February 16, 1927, National Archive Center (NAC), Records Group 79, Cammerer Papers, Box 10, Entry P 60.
23 Thomas M. Debevoise letter to Chorley, Oct 31, 1928. RAC, Cultural Interests, Series E, Box 93, Folder 858.
24 Congressional Record—House, December 14, 1928, 652.
25 Albright to Chorley, November 17, 1928, NAC, Records Group 79, Albright Papers, Box 3, Entry P 59.
26 "Yosemite Trees in Danger," *New York Times*, February 5, 1929.
27 Albright to A. Crawford Greene, January 4, 1930, YOSE, BRN44372_005673, Rockefeller Purchase, File No. 610, 1918–35, 3.
28 Albright to Chorley, March 1, 1930, and Albright to Greene, January 4, 1930, 3.
29 Arthur E. Demaray to JDR Jr., Memo: "Yosemite Valley Timber," June 19, 1929. RAC, Cultural Interests, FA 314, Series E, Box 93, 857.
30 Albright to Vanderbilt Webb, June 30, 1929.
31 Harold W. Clark, editorial, *The Live Oak* 2, no. 1 (September 1928): 1–2.
32 Arthur Woods to JDR Jr., April 16, 1929, RAC, Cultural Interests, FA 314, Series E, Box 93, 857.
33 JDR Jr. to Arthur Woods, December 18, 1929.
34 JDR Jr. to Mather, July 18, 1929, RAC, FA 314, Records Group 2, Series E, Box 77, Folder 734.
35 Albright to W. C. Deming, March 7, 1930, NAC, Records Group 79, Horace M. Albright, Box 3, Entry P 59. Both Albright and Cammerer received $25,000 in Mather's will. Joseph W. Ernst, ed., *Worthwhile Places: Correspondence of John D. Rockefeller, Jr. and Horace M. Albright* (Fordham University Press/Rockefeller Archive Center, 1991), 92.
36 Arthur Woods to JDR Jr., May 17, 1929, RAC, Cultural Interests, FA 314, Series E, Box 93, Folder 858.
37 Albright to Chorley, July 3, 1929.
38 Memo from Chorley for files, September 24, 1929.
39 _______strom [illegible] to Frank McKee [California Chamber of Commerce], April 16, 1930. Also, Albright to Superintendent

C. G. Thomson, April 22, 1930. See also, Albright to A. Crawford Greene, April 5, 1930, YOSE, Rockefeller Purchase, 1918–1935, BRN44372_005673, File No. 610.

40 Albright to Greene, March 27, 1930.

41 Burnham Enersen & Carole Hicke, Practicing Law with the McCutchen Law Firm Since 1930, Regional Oral History Office, the Bancroft Library, University of California, Berkeley, 1995, 295.

42 Letter in Sunday *New York Times* from Arthur N. Pack, President of American Nature Association, December 16, 1928, Section E, 5.

43 NPS, National Register of Historic Places Registration Form, Thomas J. Walsh Lodge, OMB no. 1024-0018, D-599, October 1990. Also, Chris Peterson, "Burton Wheeler Cabin," *Hungry Horse (MT) News*, January 20, 2023.

44 "Yosemite National Park* Interest," summary for files, May 27, 1930, RAC, Cultural Interests, FA 314, Series E, Box 93, Folder 858.

45 Albright to JDR Jr., June 9, 1930.

46 C. G. Thomson to JDR Jr., April 14, 1930, YOSE, BRN44372_005673, Rockefeller Purchase, 1918–35, File No. 610.

47 *New York Times*, May 30, 1930, 18.

48 Albright to Archie Stevenot, manager of Hotel Tioga, April 15, 1930, YOSE, BRN44372_005673, Rockefeller Purchase, 1918–35, File No. 610.

49 Albright, "FY 1930 Annual Report from the Director of the NPS" (Washington, DC: National Park Service, 1930).

50 JDR Jr. to Albright, April 29, 1930. Ernst, *Worthwhile Places*, 100–1.

51 Arthur N. Pack to Ball, July 3, 1928, YOSE, 1001, Box 33 F, Folder 102.

A Hell Made from Heaven

1 Albright to JDR Jr., February 16, 1927, National Archive Center (NAC), Records Group 79, Cammerer Papers, Box 10, Entry P 60.

2 Donald C. Swain, *Wilderness Defender: Horace M. Albright and Conservation* (Chicago: University of Chicago Press, 1970), 161.

3 Philip Kopper, *Colonial Williamsburg* (New York: Harry N. Abrams, 1986), 164.

4 Robert W. Righter, *Crucible for Conservation: The Struggle for Grand Teton National Park* (Moose, WY: Grand Teton Association, 1982), 35–36.

5 Douglas Brinkley, *Rightful Heritage: Franklin D. Roosevelt and the Land of America* (New York: Harper Collins, 2016), 178.

6 Kenneth Chorley interview, Oral History Research Office, Columbia University, 1973, Rockefeller Archive Center (RAC), Jackson Hole Preserve Project, FA 1444, Box 3, Folder 44, 19.

7 "Park Inquiry Eyes Rockefeller Gift," *New York Times*, February 19, 1933, 4C.

8 Albright to Kenneth Chorley, July 18, 1927, NAC, Records Group 79, Albright Papers, Box 3, Entry P 59.
9 John Daugherty, *A Place Called Jackson Hole: The Historic Resource Study of Grand Teton National Park* (Moose, WY: Grand Teton National Park, National Park Service, 1999), 140–41.
10 Harold P. Fabian, interview by Ed Edwin, July 11, 1966, RAC, Oral History Collection, FA 1444, Series 4, Box 3, Folder 45, 12.
11 Ramona Ruhl and Richard Young, "National Register of Historic Places nomination sheet," Geraldine Lucus Homestead, Teton County, Wyoming, Grand Teton National Park, NPS, 1998.
12 Ruhl and Young, "National Register of Historic Places nomination sheet."
13 Harold P. Fabian, interview by Ed Edwin no. 2, July 18, 1966, RAC, Oral History Collection, FA 1444, Series 4, Box 3, Folder 45, 94.
14 Fabian, interview no. 2, 94.
15 Fabian, interview no. 2, 95.
16 *Jackson Hole News*, "Richards Honored by Rotary," February 22, 1981.
17 *Jackson Hole Guide*, February 15, 1989, C-3.
18 Righter, *Crucible*, 62.
19 Righter, *Crucible*, 62.
20 *Jackson Hole News*, February 22, 1981.
21 *Jackson Hole News*, February 25, 1991.
22 *Jackson Hole Guide*, June 8, 1978, 14.
23 Homer C. Richards oral history, Teton County Library, Jackson, Wyoming, 5.
24 Righter, *Crucible*, 62.
25 *Jackson Hole News*, February 22, 1981.
26 *Jackson Hole Guide*, circa 1970s, photo caption, courtesy Teton County Library.
27 Richards oral history, 5.
28 *Jackson Hole Guide*, January 24, 1974, 18.
29 Fabian, interview no. 2, 4.
30 Fabian, interview no. 2, 13.
31 Righter, *Crucible*, 55–56. Also, Daugherty, *A Place Called Jackson Hole*, 309–10.
32 Daugherty, *A Place Called Jackson Hole*, 309–10.
33 Daugherty, *A Place Called Jackson Hole*, 140.
34 Barry Goldberg, November 4, 2019, "John D. Rockefeller, Jr. Creates a National Park," RAC, Resource, https://resource.rockarch.org/story/john-d-rockefeller-jr-creates-a-national-park.
35 Righter, *Crucible*, 35–36.
36 John Ise, *Our National Park Policy* (Baltimore, MD: Johns Hopkins Press, 1961), 329.
37 *Jackson's Hole Courier*, April 6, 1930, 2.
38 *Cody (WY) Enterprise*, January 28, 1931.

39 *Cody Enterprise*, November 27, 1930.
40 *Deseret News* (Salt Lake City, UT), October 9, 1930, 2.
41 *Jackson's Hole Courier*, November 6, 1930.
42 *Grand Teton*, May 31, 1932, RAC, Kenneth Chorley Papers, FA067, Series 1, Box 21, Folder 181.
43 *Casper Tribune-Herald*, April 13, 1930.
44 *Jackson's Hole Courier*, June 5, 1930.
45 Maxwell Struthers Burt to Albright, February 10, 1927, NAC, Records Group 79, Cammerer Papers, Box 10, Entry P 60.
46 Daugherty, *A Place Called Jackson Hole*, 220–53.
47 Daugherty, *A Place Called Jackson Hole*, 234–35.
48 "John D. Jr. Park Quiz Slated," *San Francisco Call Bulletin*, June 23, 1933.
49 "Land 'Scandal' to Be Aired in Wyoming Quiz," *Washington Herald*, July 16, 1933, 8-B.
50 Peter Norbeck to Albright, July 21, 1933, NAC, Records Group 48, CCF: 1907-36, Box 2013.
51 *Great Falls (MT) Tribune*, August 11, 1933, 8.
52 *Great Falls Tribune*, August 10, 1933, 12.
53 A. E. Wall, *Billings (MT) Gazette*, August 9, 1933, 1–2.
54 *Billings Gazette*, August 10, 1933.
55 A. E. Wall, *Billings Gazette*, August 9, 1933, 1.
56 Associated Press, August 11, 1933.
57 *Rocky Mountain News*, August 12, 1933.
58 *Grand Teton*, June 27, 1933.
59 Swain, *Wilderness Defender*, 215.
60 Horace M. Albright and Marian Albright Schenck, *Creating the National Park Service: The Missing Years* (Norman: University of Oklahoma Press, 1999), 314.
61 Albright, "Statement to NPS Personnel Upon His Resignation as Director," 1933. www.nps.gov/articles/000/horace-albright-s-national-park-service-legacy.htm.
62 Joseph W. Ernst, ed., *Worthwhile Places: Correspondence of John D. Rockefeller, Jr. and Horace M. Albright* (New York: Fordham University Press/Rockefeller Archive Center, 1991), 153–54.
63 Ernst, *Worthwhile Places*, 154–55.

Throwing in the Towel

1 E. J. Fleming, *The Fixers: Eddie Mannix, Howard Strickling, and the MGM Publicity Machine* (Jefferson, NC: McFarland & Company, 2005), 176.
2 Newton B. Drury, speech to Commonwealth Club of San Francisco, July 25, 1947, Rockefeller Archive Center (RAC), Record Group 2, FA 314, Series E, Box 78, Folder 738.

3 Conrad L. Wirth, interview by Ed Edwin, 1966, RAC, Oral History Collection, Jackson Hole Preserve, Inc., FA 1444, Series 4, Box 3, Folder 50, 27.
4 *Jackson's Hole Courier*, May 6, 1943, 1.
5 *Salt Lake City Tribune*, May 19, 1943, 6.
6 Westbrook Pegler, *New York World Telegram*, June 16, 1943.
7 Conrad Wirth interview.
8 Conrad Wirth interview, 25–26.
9 Joseph W. Ernst, ed., *Worthwhile Places: Correspondence of John D. Rockefeller, Jr. and Horace M. Albright* (New York: Fordham University Press/ Rockefeller Archive Center, 1991), 207.
10 Ernst, *Worthwhile Places*, 207.
11 Kenneth Chorley oral history, Columbia University, Jackson Hole Preserve, Inc. Project, RAC, FA 1444, Box 3, folder 44, 72–73.
12 Conrad Wirth interview, 58.
13 Robert W. Righter, *Crucible for Conservation: The Creation of Grand Teton National Park* (Moose, WY: Grand Teton Association, 1982), 107–8. Also, John Daugherty, *A Place Called Jackson Hole: The Historic Resource Study of Grand Teton National Park* (Moose, WY: Grand Teton National Park, National Park Service, 1999), 316.
14 Chorley oral history, Columbia University, Jackson Hole Preserve, Inc. Project, RAC, FA 1444, Box 3, folder 44, 73.
15 Ann Rockefeller Roberts, *Mr. Rockefeller's Roads: The Untold Story of Acadia's Carriage Roads & Their Creator* (Camden, ME: Down East Books, 1990), 107.
16 Conrad Wirth interview, 58.
17 Robert W. Righter, "National Monuments to National Parks," NPS History eLibrary, 1989, http://npshistory.com/publications/righter/index.htm.
18 Douglas Brinkley, *Rightful Heritage: Franklin D. Roosevelt and the Land of America* (New York: Harper Collins, 2016), 544–48.
19 Chorley to Harold L. Ickes, November 8, 1937, National Archive Center (NAC), Records Group 79, Albright Papers, Box 3, Entry P 59.
20 Brinkley, *Rightful Heritage*, 244–45.
21 Horace M. Albright, as told to Robert Cahn, *The Birth of the National Park Service: The Founding Years, 1913–33* (Salt Lake City: Howe Brothers, 1985), 284.
22 Albright, *Birth of the National Park Service*, 296.
23 Anne M. Whisnant, *Super-Scenic Motorway: A Blue Ridge Parkway History* (Chapel Hill: University of North Carolina Press, 2006), 36.
24 Brinkley, *Rightful Heritage*, 562.
25 Richard W. Sellars, *Preserving Nature in National Parks: A History* (New Haven, CT: Yale University Press, 1997), 13.
26 Neil M. Maher, *Nature's New Deal: The Civilian Conservation Corps and*

the Roots of the American Environmental Movement (New York: Oxford University Press, 2008), 43.

27 Brinkley, *Rightful Heritage*, 546.
28 Ernst, *Worthwhile Places*, 209.
29 Brinkley, *Rightful Heritage*, 547.
30 Brinkley, *Rightful Heritage*, 145.
31 Conrad Wirth interview, 58.
32 "Gun Play," *Time Magazine*, no. 41, May 17, 1943, 21.
33 Righter, *Crucible*, 117–19.
34 Brinkley, *Rightful Heritage*, 564.
35 Righter, "National Monuments to National Parks."
36 Righter, *Crucible*, 7.
37 Daughterly, *A Place Called Jackson Hole*, 317.
38 Righter, *Crucible*, 123.
39 Righter, *Crucible*, 105.
40 Chorley to Harold Fabian, February 5, 1947, RAC, OMR Rockefellers, FA 314, Series E, Box 92, Folder 845.
41 Chorley to Harold Fabian, February 5, 1947.
42 Chorley to Harold Fabian, February 5, 1947.
43 Wyoming Governor Lester C. Hunt to Secretary of the Interior J. A. Krug, June 17, 1947, RAC, OMR Rockefellers, Cultural Interests, Series E, Box 92, Folder 845.
44 Olaus J. Murie, "The Jackson Hole National Monument," *National Parks Magazine*, no. 75 (October–December 1943): 9.
45 Katharine Boyd, "Heard About Jackson Hole?," *Atlantic Monthly*, April 1945, 106.
46 Newton B. Drury to JDR Jr., September 17, 1945, RAC, JHPI, FA 476, Record Group 8, Series 1, Box 30.
47 Walter L. Doty, "Destination: Our Land," *Sunset* (May 1945): 2–6.
48 Drury to JDR Jr., September 17, 1945, JHP RAC, JHPI, FA 476, Record Group 8, Series 1, Box 30.

Never Say Never

1 *Grand Teton* (Jackson, WY), May 31, 1932.
2 *Casper (WY) Tribune-Herald*, April 13, 1930.
3 Robert W. Righter, *Crucible for Conservation: The Struggle for Grand Teton National Park* (Moose, WY: Grand Teton Association, 1982), 59. See also John Daugherty, *A Place Called Jackson Hole: The Historic Resource Study of Grand Teton National Park* (Moose, WY: Grand Teton National Park, National Park Service, 1999), 311.
4 Righter, *Crucible*, 98.
5 Maxwell Struthers Burt, *Jackson's Hole Courier*, June 5, 1930, 1, 8.

6 Raymond B. Fosdick, *John D. Rockefeller, Jr.: A Portrait* (New York: Harper & Brothers, 1956), 432.
7 Robert Shankland, *Steve Mather of the National Parks* (New York: Alfred A. Knopf, 1951), 114–27.
8 Alliston Boyer to Laurance S. Rockefeller and others, September 15, 1948, Rockefeller Archive Center (RAC), OMR Rockefellers, Cultural Interests, Series E, Box 92, Folder 845.
9 George B. Dorr, *The Story of Acadia National Park* (Bar Harbor, ME: Acadia Publishing Company, 1997), 9. Also, Joseph W. Ernst, ed., *Worthwhile Places: Correspondence of John D. Rockefeller, Jr. and Horace M. Albright* (New York: Fordham University Press/Rockefeller Archive Center, 1991), 88.
10 Elmo S. Watson, *Jackson's Hole Courier*, March 8, 1928, 6.
11 Alfred Runte, *National Parks: The American Experience* (Lanham, MD: Taylor Trade Publishing, 2010), 38.
12 Peter Morrison, "NPS Cultural Landscapes Inventory: Cadillac Mountain Summit," rev. ed. (Washington, DC: National Park Service, Acadia National Park, 2008), 7.
13 Horace M. Albright, as told to Robert Cahn, *The Birth of the National Park Service: The Founding Years, 1913–33* (Salt Lake City: Howe Brothers, 1985), 163–64.
14 Elizabeth Flint, "Beneath the Jagged Tetons: Gilbert Stanley Underwood and the Architecture of Jackson Lake Lodge," Master's thesis, University of Virginia, 2009, 66.
15 Harold P. Fabian to Ray L. Wilbur, February 1, 1930, National Archive Center (NAC), Records Group 48, CCF 1907–36, Box 2013.
16 Elizabeth Flint, "Beneath the Jagged Tetons," 32.
17 Albright to JDR Jr., July 14, 1945, RAC, OMR Rockefellers, Cultural Interests, Series E, Box 92, Folder 845.
18 Director of Investigations [signature illegible] to Secretary of the Interior [Harold L. Ickes], January 24, 1936, NAC, Records Group 48, CCF, Box 2013.
19 Cammerer to Harold L. Ickes, October 29, 1935, NAC, Records Group 48, CCF 1907–36, Box 2013.
20 Elizabeth Engle, "Cultural Resources in a 'Natural' Park: Early Preservation Efforts at Menor's Ferry in Grand Teton National Park" (Moose, WY: National Park Service, 2013), 3.
21 Harold Fabian, interview by Ed Edwin, July 11, 1966, RAC, Oral History Collection, FA 1444, Series 4, Box 3, Folder 45, 60.
22 Fabian interview, 61.
23 *Jackson's Hole Courier*, August 25, 1949, 4; September 8, 1949.
24 *Billings Gazette*, August 20, 1949, 5.
25 *Jackson's Hole Courier*, September 8, 1949, 4.
26 Fabian interview, 61.
27 Kenneth Chorley to Albright, Laurance Rockefeller, and others,

January 13, 1948, RAC, OMR Rockefellers, Cultural Interests, Series E, Box 92, Folder 845.

28 Albright to Laurance S. Rockefeller, May 7, 1947.

29 JDR Jr. to Albright, August 1, 1947.

30 Righter, *Crucible*, 137–38.

31 Chorley to Laurance S. Rockefeller, March 22, 1949.

32 Floy Tonkin, "Riding the Range . . . ," *Jackson's Hole Courier*, July 7, 1949, 8.

33 Newton B. Drury to JDR Jr., July 18, 1946.

34 JDR Jr. to Chorley, August 24, 1946.

35 Righter, *Crucible*, 134.

36 "A Report by the National Park Service on the Proposal to Extend the Boundaries of Grand Teton National Park, Wyoming" (Washington, DC: National Park Service, 1937), cited by Righter, *Crucible*, 93.

37 Kenneth Chorley, Oral History Research Project, Columbia University, 1973, RAC, Jackson Hole Preserve Project, FA 1444, Box 3, Folder 44, 102–3.

38 Righter, *Crucible*, 139.

39 "An Act to Establish a New Grand Teton National Park," September 14, 1950, U.S. House of Representatives (64 Stat. 849).

40 Righter, *Crucible*, 140.

41 Oscar L. Chapman to JDR Jr., December 16, 1949, RAC, OMR Rockefellers, Cultural Interests, Series E, Box 92, Folder 847.

42 Oscar L. Chapman to JDR Jr., December 16, 1949.

43 JDR Jr. to Albright, February 2, 1950. Ernst, *Worthwhile Places*, 266.

44 Albright to JDR Jr., December 23, 1949, RAC, OMR Rockefellers, Cultural Interests, Series E, Box 92, Folder 847.

45 John S. McLaughlin to JDR Jr., December 22, 1949.

46 JDR Jr. to Governor Leslie A. Miller, September 28, 1950, RAC, OMR Rockefellers, FA 314, Series E, Box 92, Folder 895.

47 JDR Jr. to Oscar L. Chapman, December 21, 1950, RAC, OMR Rockefellers, Cultural Interests, Series E, Box 92, Folder 847.

48 Kenneth Chorley, Oral History Research Project, Columbia University, 1973, RAC, Jackson Hole Preserve Project, FA 1444, Box 3, Folder 44, 91, 102–3.

49 Chorley, Oral History Research Project, 75.

50 Terry Tempest Williams, *The Hour of Land* (New York: Picador, 2016), 27–28.

51 Fabian interview, 73.

Of Trees and Men

1 Horace M. Albright, as told to Robert Cahn, *The Birth of the National Park Service: The Founding Years, 1913–33* (Salt Lake City: Howe Brothers, 1985), 63.

2 Philip Weiss, "Masters of the Universe Go to Camp: Inside the Bohemian Club," *Spy Magazine* (November 1989): 59–76.
3 "Save the Redwoods League," *Bancroftiana: Newsletter of the Friends of the Bancroft Library*, University of California, Berkeley, no. 150 (Summer, 2019): 1.
4 "Save the Redwoods League," 3.
5 Horace M. Albright and Frank J. Taylor, "How We Saved the Big Trees," *Saturday Evening Post*, February 7, 1953.
6 John C. Merriam, "Forest Windows," *Scribner's Magazine* 83, no. 6 (June 1928): 733.
7 "Save the Redwoods League," 3–4. Also, James C. Shirley, *The Redwoods of the Coast and Sierra* (Berkeley: University of California Press, 1940), Size of the Redwoods chapter.
8 Albright and Taylor, "How We Saved the Big Trees."
9 Madison Grant to Raymond B. Fosdick, February 29, 1924, Rockefeller Archive Center (RAC), Organizations and Parks, 1924–31, Materials from R. B. Fosdick office, 1924–31, Series E, Box 82, Folder 762.
10 Adam Serwer, "White Nationalism's Deep American Roots," *Atlantic* (April 2019): 1.
11 Michael Cook, "Buck v. Bell, One of the Supreme Court's Worst Mistakes," *Bioethics News*, Kennedy Institute of Bioethics, Georgetown University, February 27, 2016.
12 Adam S. Cohen, "Harvard's Eugenics Era," *Harvard Magazine* (March–April 2016): 48. Also, Charles Wohlforth, "Conservation and Eugenics," *Orion Magazine* (June 2010).
13 Irving Fisher, "Report on National Vitality: Its Wastes and Conservation," report prepared for the National Conservation Commission (Washington, DC: Government Printing Office, 1909), 129.
14 Theodore Roosevelt blurb in *Scribner's Magazine* 62, no. 6 (December 1917): 32d.
15 Robert F. Zeidel, "Pursuit of Human Brotherhood: Rockefeller Philanthropy and American Immigration, 1900–33," *New York History* 90, no. 1/2 (Winter–Spring 2009): 85–106. http://www.jstor.org/stable/23185099.
16 Eileen Rockefeller, *Being a Rockefeller, Becoming Myself: A Memoir* (New York: Blue Rider Press, 2013), 12–13.
17 Ron Chernow, *Titan: The Life of John D. Rockefeller, Sr.* (New York: Vintage Books, 1998), 39.
18 Chernow, *Titan*, 238–39, 421. Also, Zeidel, "Pursuit of 'Human Brotherhood,'" 85–106.
19 James D. Anderson, "Northern Foundations and the Shaping of Black Rural Education, 1902–35," *History of Education Quarterly* 18, no. 4 (Winter 1978): 379. See also Fosdick, *John D. Rockefeller, Jr.*, 374–77.
20 "Malfalfa Park, Kokomo," Digital Civil Rights Museum, 2024, https://

www.digitalresearch.bsu.edu/digitalcivilrightsmuseum/items/show/110. See also, Bernice Kert, *Abby Aldrich Rockefeller: The Woman in the Family* (New York: Random House, 1993), 207.

21 Kert, *Abby*, 208.

22 *New York Times*, May 15, 1925, 1.

23 Kert, *Abby*, 224.

24 Andrew Scull, "Report on Research on the Rockefeller Foundation and American Psychiatry," RAC Reports, 2020, https://rockarch.issuelab.org/resources/36738/36738.pdf.

25 John D. Rockefeller, Jr. to John D. Rockefeller, April 9, 1926, in Joseph W. Ernst, ed., *"Dear Father"/"Dear Son": Correspondence of John D. Rockefeller and John D. Rockefeller, Jr.* (New York: Fordham University Press/Rockefeller Archive Center, 1994), 146.

26 Zeidel, "Pursuit of 'Human Brotherhood,'" 103.

27 William A. Schambra, "Philanthropy's Original Sin," *New Atlantis* 39 (Summer 2013).

28 Lia N. Weintraub, "The Link between the Rockefeller Foundation and Racial Hygiene in Nazi Germany" (Medford, MA: Tufts Archival Research Center, Tufts University, 2012).

29 Andras Szanto, *Rockefeller Philanthropy: A Selected Guide* (Sleepy Hollow, NY: Rockefeller Archive Center, 2011), 20.

30 Susan R. Schrepfer, *The Fight to Save the Redwoods: A History of Environmental Reform, 1917–1978* (Madison: University of Wisconsin Press, 1983), 43–45.

31 Robert Shankland, *Steve Mather of the National Parks* (New York: Alfred A. Knopf, 1951), 196–97.

32 Shankland, *Steve Mather*, 198–99.

33 John C. Merriam to Raymond B. Fosdick, February 17, 1925, RAC, Organizations and Parks, Materials from Fosdick Office, 1924–31, Series E, Box 82, Folder 762.

34 Newton B. Drury to Grant, October 17, 1924.

35 Arno B. Cammerer to Fosdick, November 21, 1924.

36 Arno B. Cammerer to Fosdick, November 21, 1924.

37 Barbara Shubinski, Liesel Vink, and James Allen Smith, "Photo Essay: The Rockefellers, National Parks, and Public Lands," June 16, 2021, resource.rockarch.org.

38 John C. Merriam to Fosdick, April 8, 1924, and February 17, 1925, RAC, Organizations and Parks, Materials from Fosdick Office, 1924–31, Series E, Box 82, Folder 762.

39 *Los Angeles Times*, June 21, 1926, 17.

40 *Los Angeles Times*, June 26, 1926, 21.

41 Merriam to Fosdick, June 28, 1926, RAC, Organizations and Parks, Materials from Fosdick Office, 1924–31, Series E, Box 82, Folder 762.

42 Schrepfer, *The Fight to Save the Redwoods*, 23. Also, Newton B. Drury

to Robert W. Gumbel, June 4, 1926, RAC, Series II, JDR Jr. Personal Papers, Travel Records–United States, 1926, Western Trip.

43 *Santa Rosa Press Democrat*, July 8, 1926, 9.

44 JDR Jr. to Merriam Webster, July 16, 1926, RAC Organizations and Parks, Materials from Fosdick Office, 1924–31, Series E, Box 82, Folder 762.

45 JDR Jr. to Fosdick, March 8, 1927.

46 JDR Jr. to Fosdick, August 9, 1927.

47 Newton B. Drury, oral history, Parks and redwoods, 1919–1971: oral history transcript / and related material, 1959–1972. / Newton Drury; tape recorded interview conducted by Amelia R. Fry and Susan Schrepfer in 1959–1970, BANC MSS 73/157, Regents of the University of California, Bancroft Library, University of California, Berkeley, 140.

48 Albright to JDR Jr., November 22, 1950. Joseph W. Ernst, ed., *Worthwhile Places: Correspondence of John D. Rockefeller, Jr. and Horace M. Albright* (New York: Fordham University Press/Rockefeller Archive Center, 1991), 270.

49 Albright to JDR Jr., June 22, 1949, RAC, FA 314, Records Group 2, Series E, Box 77, Folder 734.

50 JDR Jr. to Albright, June 28, 1949, RAC, Cultural Interests, FA 314, Series E, Box 80, Folder 751.

51 JDR Jr. to Albright, April 2, 1951.

52 JDR Jr. to Albright, May 19, 1952.

53 "Rockefeller Gift Spares Redwoods," *New York Times*, April 10, 1954, 17. The story also appeared in *Landscape Architecture* magazine in April 1954.

54 Edwin C. Bearss, "Redwood National Park History Basic Data," section XIV b, "The Establishment of Redwood National Park" (Washington, DC: National Park Service, Office of Archaeology and Historic Preservation, 1969).

Back in the Game

1 Ashton Chapman, *Johnson City Press-Chronicle*, February 3, 1952, 6.

2 Bernice Kert, *Abby Aldrich Rockefeller: The Woman in the Family* (New York: Random House, 1993), 89.

3 Kert, *Abby*, 472–73.

4 Kert, *Abby*, 472–73.

5 David Rockefeller, *Memoirs* (New York: Random House, 2002), 182.

6 "John D. Rockefeller, Jr. Residence, 10 West 54th St., New York, NY 10019," NYCago.org.

7 Kenneth Chorley, Oral History Research Office, Cornell University, 1973, Jackson Hole Preserve Project, Rockefeller Archive Center (RAC), FA 1444, Box 3, Folder 44.

8 Leslie Gignaux, "Stanley Abbott and the Design of the Blue Ridge

Parkway," *Proceedings of the BRP Golden Anniversary Conference* (Boone, NC: Appalachian Consortium Press, 1986), 36–37.

9 Michael Frome, *Strangers in High Places* (Knoxville: University of Tennessee Press, 1980), 214.

10 Frome, *Strangers*, 214.

11 JDR Jr. to Sam P. Weems, June 14, 1950, RAC, OMR Rockefeller, Organizations and Parks, FA 314, Series E, Box 80, Folder 753.

12 Frome, *Strangers*, 214.

13 Frome, *Strangers*, 215.

14 Frome, *Strangers*, 215.

15 Cammerer to Mark Squires, August 5, 1927, GSMNP Archives, GSMCA papers.

16 Frome, *Strangers*, 215.

17 JDR Jr. to Chorley, February 5, 1947, RAC, OMR Rockefellers, Cultural Interests, Series E, Box 92, Folder 845.

18 JDR Jr. to Albright, January 10, 1950, RAC, OMR Rockefellers, Cultural Interests, Series E, Box 92, Folder 847.

19 JDR Jr. to children, February 28, 1956, RAC, Record Group 2, Series 2E, Box 90, Folder 831.

20 JDR Jr. to Nelson A. Rockefeller, August 24, 1953, RAC, FA 314, Records Group 2, Series E, Box 77, Folder 735.

21 JDR Jr. to Albright, October 19, 1951, RAC, Record Group 2, FA 314, Series E, Box 80, Folder 753.

22 "U.S. Parkway Gains Scenic Falls Area," *New York Times*, January 13, 1952, 61.

23 Horace M. Albright, interview by Michael Frome, "Recollections and Observations by Horace M. Albright," 1964, Great Smoky Mountains National Park library, Townsend, TN, nos. 23575 and 23576; transcription of tapes 32 and 33, 16.

24 JDR Jr. to Albright, Telegram, May 19, 1952, RAC, FA 314, Series E, Box 80, Folder 751.

25 Robin W. Winks, *Laurance S. Rockefeller: Catalyst for Conservation* (Washington, DC: Island Press, 1997), 62–65.

26 JDR Jr. to Laurance Rockefeller, December 17, 1954, RAC, Record Group 2, FA 314, Series F, Box 86, Folder 801.

Conclusion

1 Joseph W. Ernst, ed., *Worthwhile Places: Correspondence of John D. Rockefeller, Jr. and Horace M. Albright* (New York: Fordham University Press/ Rockefeller Archive Center, 1991), 16–17. Also, "CPI Inflation Calculator," U.S. Bureau of Labor Statistics, data.bls.gov.

2 Andras Szanto, *Rockefeller Philanthropy: A Selected Guide* (New York: Rockefeller Archive Center, 2011), 4.

3 Beth Breeze, *In Defence of Philanthropy* (Newcastle, UK: Agenda Publishing, 2021), 151–59.
4 Ernst, *Worthwhile Places*, 343.
5 John D. Rockefeller Jr. to John D. Rockefeller, April 9, 1926, in Joseph W. Ernst, ed., *"Dear Father"/"Dear Son,": Correspondence of John D. Rockefeller and John D. Rockefeller, Jr.* (New York: Fordham University Press/Rockefeller Archive Center, 1994), 147.
6 Peter Collier and David Horowitz, *The Rockefellers: An American Dynasty* (New York: Holt, Rinehart, and Winston, 1976), 146.
7 Abraham Flexner, *An Autobiography* (New York: Simon and Schuster, 1960), 126.
8 Kenneth Chorley oral history, Columbia University, Jackson Hole Preserve, Inc. Project, Rockefeller Archive Center (RAC), FA 1444, Box 3, folders 44, 79.
9 Ernst, *Worthwhile Places*, 300–8. See also Bernard A. DeVoto, "Let's Close the National Parks," *Harper's Magazine* (October 1953).
10 Elizabeth Flint, "Beneath the Jagged Tetons: Gilbert Stanley Underwood and the Architecture of Jackson Lake Lodge," Master's thesis, University of Virginia, 2009, 12. Also, Robin W. Winks, *Laurance S. Rockefeller: Catalyst for Conservation* (Washington, DC: Island Press, 1997), 60.
11 Flint, "Beneath the Jagged Tetons," 27.
12 Chorley oral history, Columbia University, 82.
13 Flint, "Beneath the Jagged Tetons," 29.
14 Flint, "Beneath the Jagged Tetons," 54.
15 Flint, "Beneath the Jagged Tetons," 46.
16 Flint, "Beneath the Jagged Tetons," 46.
17 Michael T. Kaufman, Laurance S. Rockefeller obituary, *The New York Times,* July 12, 2004, A 18.

Epilogue

1 All park visitation statistics from NPS, Integrated Resource Management Applications, IRMA.nps.gov.
2 NPS, nps.gov/acadia.
3 NPS, nps.gov/glac.
4 NPS, nps.gov/yose.
5 NPS, nps.gov/zion; Town of Springdale, Utah Parking, www.springdaletown.com/435/Parking-Rates.
6 Phil Francis, "Funding Cuts Are Pushing Our National Parks to the Breaking Point," *The Hill, Opinion: Energy and the Environment e-newsletter*, March 16, 2024.

BIBLIOGRAPHY

Archival Collections

Cammerer, Arno B. Papers, NAC, Records Group 79, College Park, MD.
Dorr, George Bucknam. Papers. Bar Harbor Historical Society, Bar Harbor, ME.
Epp, Ronald H. Papers. Jessup Memorial Library, Bar Harbor, ME.
Harpers Ferry Design Center Archives, Harpers Ferry, WV.
National Archives Center, College Park, MD.
Peters, John A. Papers. Dorr Estate Correspondence, Hale & Hamlin, LLC, Ellsworth, ME.
Rockefeller Archive Center (RAC), Sleepy Hollow, NY.
Seal Harbor Library Archives, Seal Harbor, ME.
Shenandoah National Park Oral History Collection, 1964–1999, Special Collections, Carrier Library, James Madison University, Harrisonburg, VA.
Talley, Rosemary, and Jesse L. Nusbaum. Papers. Mesa Verde National Park Archives, Mesa Verde National Park, CO.
Yosemite Archives, El Portal, CA.

Secondary Sources

Abrahamson, Eric J. *Beyond Charity: A Century of Philanthropic Innovation.* New York: Rockefeller Foundation, 2013.
Adams, Paul J. *Mount Le Conte.* Knoxville: University of Tennessee Press, 2016.
Aday, Denise, and Michael Aday. *Letters from the Smokies.* Gatlinburg, TN: Great Smoky Mountains Association, 2023.
Albright, Horace M. "July 26, 1961." Interview by S. Herbert Evison. National Park Service Oral History Collection (HFCA 1817), 1961.
Albright, Horace M. "Recollections and Observations." Interview by Michael Frome. Great Smoky Mountains National Park Library, 23575–23576, 1961.

Albright, Horace M. *Report of the Director of the National Park Service*. Washington, DC: National Park Service, 1929.

Albright, Horace M., and Marian Albright Schenck. *Creating the National Park Service: The Missing Years*. Norman: University of Oklahoma Press, 1999.

Albright, Horace M., and Frank J. Taylor. "How We Saved the Big Trees." *Saturday Evening Post*, February 7, 1953.

Albright, Horace M., as told to Robert Cahn. *The Birth of the National Park Service: The Founding Years, 1913–33*. Salt Lake City: Howe Brothers, 1985.

Allaback, Sarah. "Rustic Trailside Museums and Modern Visitor Centers: America's Most Popular Museums." In *People, Places, and Parks*, edited by David Harmon. Proceedings of the 2005 George Wright Society Conference. Hancock, MI: George Wright Society, 2005.

Allen, William H. *Rockefeller: Giant, Dwarf, Symbol*. New York: New York Institute for Public Service, 1930.

Anderson, James D. "Northern Foundations and the Shaping of Black Rural Education, 1902–35." *History of Education Quarterly* 18, no. 4 (Winter 1978): 371–96.

Andrews, Thomas G. *Killing for Coal: America's Deadliest Labor War*. Cambridge, MA: Harvard University Press, 2008.

Architectural Resources Group. "Yavapai Observation Station: Historic Structure Report." San Francisco: Architectural Resources Group, 2001.

Bearss, Edwin C. *Arno B. Cammerer*. Biographical Vignette, National Park Service, www.nps.gov/parkhistory/online_books/sontag/cammerer.htm.

Bearss, Edwin C. "Redwood National Park History Basic Data." Section XIV b, "The Establishment of Redwood National Park." Washington, DC: National Park Service, Office of Archaeology and Historic Preservation, 1969.

Bolles, Blair. *The Tyrant from Illinois: Uncle Joe Cannon's Experiment with Personal Power*. New York: W. W. Norton, 1951.

Bordeaux, Donna. "Rich as Rockefeller." *Calculated Moves* (blog), March 29, 2017. https://www.calculatedmoves.com/post/rich-as-rockefeller.

Boyd, Katharine. "Heard About Jackson Hole?" *Atlantic Monthly* 175, no. 4 (April 1945): 102–6.

Breeze, Beth. *In Defence of Philanthropy*. Newcastle, England: Agenda Publishing, 2021.

Brewer, Carson. *Hiking Trails of the Smokies*. Gatlinburg, TN: Great Smoky Mountains Natural History Association, 2001.

Brinkley, Douglas. *Rightful Heritage: Franklin D. Roosevelt and the Land of America*. New York: Harper Collins, 2016.

Brown, Margaret Lynn. *The Wild East: A Biography of the Great Smoky Mountains*. Gainesville: University of Florida Press, 2000.

Bumpus, Hermon Carey, Jr. *Hermon Carey Bumpus: Yankee Naturalist*. Minneapolis: University of Minnesota Press, 1947.

Burt, Maxwell Struthers. "The Battle of Jackson's Hole." *The Nation* 122, no. 3165 (March 3, 1926): 225–26.

Burt, Maxwell Struthers. *The Diary of a Dude-Wrangler*. New York: Charles Scribner & Sons, 1924.

Campbell, Carlos C. *Birth of a National Park in the Great Smoky Mountains*. Knoxville: University of Tennessee Press, 1960.

Carr, Ethan. *Wilderness by Design: Landscape Architecture & the National Park Service*. Lincoln: University of Nebraska Press, 1998.

Catton, Theodore. *Mountains for the Masses: A History of Management Issues in Great Smoky Mountains National Park*. Gatlinburg, TN: Great Smoky Mountains Association, 2014.

Chernow, Ron. *Titan: The Life of John D. Rockefeller, Sr*. New York: Vintage Books, 1998.

Clark, Harold W. Editorial. *The Live Oak* 2, no. 1 (September 1928): 1–2.

Cohen, Adam S. "Harvard's Eugenics Era." *Harvard Magazine* 116 (March–April 2016): 48–52. www.harvardmagazine.com/2016/03/harvard-eugenics-era.

Collier, Peter, and David Horowitz. *The Rockefellers: An American Dynasty*. New York: Holt, Rinehart and Winston, 1976.

Cramton, Louis C. "The Grand Canyon of the Colorado." Speech in the U.S. House of Representatives (March 3, 1924).

Daugherty, John. *A Place Called Jackson Hole: The Historic Resource Study of Grand Teton National Park*. Moose, WY: Grand Teton National Park, National Park Service, 1999.

Dawson, Robert M. *William Lyon Mackenzie King*. Vol. 1. Toronto: University of Toronto Press, 1958.

Dorr, George B. *The Story of Acadia National Park*. Bar Harbor, ME: Acadia Publishing, 1997.

Doty, Walter L. "Destination: Our Land." *Sunset* 94, no. 5 (May 1945): 2–6.

Drabelle, Dennis. *The Power of Scenery: Frederick Law Olmsted and the Origins of National Parks*. Lincoln: University of Nebraska Press, 2021.

Eliot, Charles. "The Coast of Maine." *Garden and Forest* 3, no. 104 (February 19, 1890): 85–86.

Eliot, Charles. "The Waverly Oaks." *Garden and Forest* 3, no. 104 (February 19, 1890): 86–87.

Emerson, Ralph W. *Nature*. London: Penguin Books, 1985.

Enerson, Burnham. "Practicing Law with the McCutcheon Law Firms Since 1930." Interview. Regional Oral History Office, Bancroft Library, University of California, Berkeley, 1995.

Engle, Elizabeth. "Cultural Resources in a 'Natural' Park: Early Preservation Efforts at Menor's Ferry in Grand Teton National Park." Moose, WY: Rockefeller Archive Center, 2013.

Epp, Ronald H. *Creating Acadia National Park: The Biography of George Bucknam Dorr*. Bar Harbor, ME: Friends of Acadia, 2016.

Ernst, Joseph W., editor. *"Dear Father"/"Dear Son": Correspondence of John D. Rockefeller and John D. Rockefeller, Jr.* New York: Fordham University Press/Rockefeller Archive Center, 1994.

Ernst, Joseph W., editor. *Worthwhile Places: Correspondence of John D. Rockefeller, Jr. and Horace M. Albright.* New York: Fordham University Press/Rockefeller Archive Center, 1991.

Fink, Paul M. *Backpacking Was the Only Way.* Johnson City, TN: Research Advisory Council, Eastern Tennessee State University, 1975.

Fisher, Irving. "Report on National Vitality: Its Wastes and Conservation." Report prepared for the National Conservation Commission. Washington, DC: Government Printing Office, 1909.

Fleming, E. J. *The Fixers: Eddie Mannix, Howard Strickling, and the MGM Publicity Machine.* Jefferson, NC: McFarland & Company, 2005.

Flexner, Abraham. *Abraham Flexner: An Autobiography.* New York: Simon and Schuster, 1960.

Flint, Elizabeth (Engle). "Beneath the Jagged Tetons: Gilbert Stanley Underwood and the Architecture of Jackson Lake Lodge." Master's thesis, University of Virginia, 2009.

Fosdick, Raymond B. *John D. Rockefeller, Jr.: A Portrait.* New York: Harper & Brothers, 1956.

Foulds, H. Eliot, and Jeffrey Killion. "Cultural Landscape Report for the Historic Motor Road System, Acadia National Park." Boston: Olmsted Center for Landscape Preservation, Boston National Historical Park.

Franzen, Russell. "Louis C. Cramton." *Michigan Political History Society News* 2, no. 2 (May 1995): 1, 3.

Frome, Michael. *Strangers in High Places: The Story of the Great Smoky Mountains.* Knoxville: University of Tennessee Press, 1966.

Furmansky, Dyanna Z. "Role Model for a Conservationist: John D. Rockefeller's Relationship to Nature." Rockefeller Archive Center Reports, 2013.

Gignaux, Leslie. "Stanley Abbott and the Design of the Blue Ridge Parkway." In *Proceedings of the BRP Golden Anniversary Conference.* Boone, NC: Appalachian Consortium Press, 1986.

Growald, Eileen Rockefeller. *Being a Rockefeller, Becoming Myself.* New York: Blue Rider Press, 2013.

Haines, Aubrey L. *The Yellowstone Story.* 2 vols. Yellowstone National Park, WY: Yellowstone Library and Museum Association, 1977.

Hale, Richard, Jr. *The Story of Bar Harbor.* New York: Ives Washburn, 1949.

Hall, Ansel F. "Report of Progress in Educational Activities in the National Parks." Washington, DC: National Park Service, Branch of Research and Education, 1931.

Harrison, Laura S. "National Register of Historic Places Inventory Nomination: Norris, Madison, and Fishing Bridge Museums (Yellowstone National Park)." Washington, DC: National Park Service, 1986.

Haynes, J. E. *Haynes New Guide: Yellowstone National Park*. Saint Paul, MN: J. E. Haynes Publisher, 1923.

Hays, Samuel P. *Conservation and the Gospel of Efficiency: The Progressive Conservation Movement, 1890–1920*. Pittsburgh: University of Pittsburg Press, 1999.

Henry, Robin. "'In Our Image, According to Our Likeness': John D. Rockefeller, Jr. and Reconstructing Manhood in Post-Ludlow Colorado." *Journal of the Gilded Age and Progressive Era* 16 (2017): 24–33.

Horning, Audrey J. "When Past Is Present: Archaeology of the Displaced in Shenandoah National Park." Paper presented at 2001 Society of Historical Archaeology Conference, Long Beach, CA, 2001.

Ickes, Harold L. *The Secret Diary of Harold L. Ickes: The First Thousand Days 1933-36*. New York: Simon and Schuster, 1953.

Ise, John. *Our National Park Policy: A Critical History*. Baltimore, MD: Johns Hopkins University Press, 1961.

Jacoby, Carl. *Crimes Against Nature: Squatters, Poachers, Thieves, and the Hidden History of American Conservation*. Berkeley: University of California Press, 2014.

Kert, Bernice. *Abby Aldrich Rockefeller: The Woman in the Family*. New York: Random House, 1993.

Khan, Shamus. "The Rich Haven't Always Hated Taxes." *Time Magazine*, September 18, 2012.

Kopper, Phillip. *Colonial Williamsburg*. New York: Harry N. Abrams, 1986.

Lambert, Darwin. "Administrative History Shenandoah National Park: 1924–1976." Washington, DC: National Park Service, 1979.

Lambert, Darwin. *The Undying Past of Shenandoah National Park*. Boulder, CO: Roberts Rinehart, 1989.

Lane, Anne W., editor. *The Letters of Franklin K. Lane: Personal and Political*. Boston: Houghton Mifflin, 1922.

Lee, James J., III. "U.S. Naval Radio Station—Apartment Building #1 Historic Structure Report." Lowell, MA: National Park Service, 2009.

Lewis, Ralph H. "Museum Curatorship in the National Park Service: 1904–1982." Washington, DC: National Park Service, Curatorial Services Division, 1993.

Lix, Courtney. "The Gift: How We Gained a Mountain." *Smokies Life Magazine* 3, no. 1 (2009): 39–45.

Louter, David. *Windshield Wilderness: Cars, Roads, and Nature in Washington's National Parks*. Seattle: University of Washington Press, 2006.

Maher, Neil. "Acadia National Park Motor Roads." Historic American Engineering Record, no. ME-11. Washington, DC: National Park Service, 1997.

Maher, Neil M. *Nature's New Deal: The Civilian Conservation Corps and the Roots of the American Environmental Movement*. New York: Oxford University Press, 2008.

Mark, Jason. "Across the Great Divide." *Sierra* (July/August 2020): 27–32.

Marshall, Robert. "The Problem of the Wilderness." *Scientific Monthly* 30 (February 1930): 142–45.

Mather, Stephen T. "Report of the Director of the National Park Service to the Secretary of the Interior." Washington, DC: Government Printing Office, 1916–1918; 1920; 1922–1924; 1926–1927.

McCauley, Hugh J. "Visions of Kykuit: John D. Rockefeller's House at Pocantico Hills, Tarrytown, New York." *Hudson River Valley Review* 10, no. 2 (Summer 1993): 1–49.

McClelland, Linda Flint. *Presenting Nature: The Historic Landscape Design of the National Park Service 1916 to 1942.* Washington, DC: National Park Service, 1993.

Merriam, John C. "Forest Windows." *Scribner's Magazine* 83, no. 6 (June 1928): 733–37.

Morris, Charles R. *The Tycoons: How Andrew Carnegie, John D. Rockefeller, Jay Gould, and J. P. Morgan Invented the American Supereconomy.* New York: Holt Paperbacks, 2005.

Morrison, Peter. "NPS Cultural Landscapes Inventory: Cadillac Mountain Summit." Revised edition. Washington, DC: National Park Service, Acadia National Park, 2008.

Murie, Olaus J. "The Jackson Hole National Monument." *National Parks Magazine*, no. 75 (October–December 1943): 5–9.

Nevins, Allan. *John D. Rockefeller: The Heroic Age of American Enterprise.* New York: Charles Scribner's Sons, 1940.

Nevins, Allan. *John D. Rockefeller: A Study in Power.* New York: Charles Scribner's Sons, 1959.

Nichols, George W. "Mount Desert." *Harper's New Monthly Magazine* 45, no. 267 (August 1872): 321–41.

Okrent, Daniel. *Great Fortune: The Epic of Rockefeller Center.* New York: Viking, 2003.

Pierce, Daniel S. *The Great Smokies: From Natural Habitat to National Park.* Knoxville: University of Tennessee Press, 2000.

Pruden, T. Mitchell. "The Prehistoric Ruins of the San Juan Watershed in Utah, Arizona, Colorado, and New Mexico." *American Anthropologist* 5, no. 2 (April–June 1903): 224–88.

Quin, Richard H. "Cadillac Mountain Road." Historic American Engineering Record, No. ME-58. Washington, DC: National Park Service, 1994.

Quin, Richard H. "Rockefeller Carriage Roads." Historic American Engineering Record, No. ME-44. Washington, DC: National Park Service, 1997.

Racine, Laurel A. "Rapidan Camp: 'The Brown House'" Historic Furnishings Report. Washington, DC: National Park Service, Northeast Museum Services, 2001.

Rees, Jonathon H. *Representation and Rebellion: The Rockefeller Plan at the Colorado Fuel and Iron Company, 1914–1942.* Boulder: University of Colorado Press, 2010.

Rice, William L., Jennifer Thomsen, and Peter Whitney. "Exclusionary Effects of Campsite Allocation through Reservations in U.S. National Parks: Evidence from Mobile Device Location Data." *Journal of Park and Recreation Administration* 40, no. 4 (March 2022): 45–65.

Righter, Robert. *Crucible of Conservation: The Creation of Grand Teton National Park*. Moose, WY: Grand Teton Association, 1982.

Roberts, Ann Rockefeller. *Mr. Rockefeller's Roads: The Untold Story of Acadia's Carriage Roads & Their Creator*. Camden, ME: Down East Books, 1990.

Rockefeller, David. *Memoirs*. New York: Random House, 2002.

Rockefeller, Eileen. *Being a Rockefeller, Becoming Myself*. New York: Blue Rider Press, 2013.

Rockefeller, John D., Jr. "This I Believe." Speech to USO parents of soldiers, July 8, 1941. Rockefeller Archive Center online, http://www.rockarch.org/bio/jdrjr.php.

Rockefeller, John D., Sr. *Random Reminiscences of Men and Events*. New York: Doubleday, Page & Co., 1909.

Runte, Alfred. *National Parks: The American Experience*. Lanham, MD: Taylor Trade Publishing, 2010.

Russell, Carl Parcher. *100 Years in Yosemite: The Story of a Great Park and Its Friends*. El Portal, CA: Yosemite Association/Heydey Books, 1992.

Rydell, Kiki L., and Mary S. Culpin. *Managing the "Matchless Wonders": A History of Administrative Development in Yellowstone National Park, 1872–1965*. Yellowstone National Park, WY: National Park Service, 2006.

Save the Redwoods League. "Save the Redwoods League's Centennial Year." *Bancroftiana* 150 (Summer 2019).

Schambra, William A. "Philanthropy's Original Sin." *New Atlantis* 39 (Summer 2013): 3–21.

Schmitt, Catherine. *Historic Acadia National Park: The Stories Behind One of America's Great Treasures*. Guilford, CT: Globe Pequot Press, 2016.

Schrepfer, Susan R. *The Fight to Save the Redwoods: A History of Environmental Reform, 1917–78*. Madison: University of Wisconsin Press, 1983.

Sellars, Richard W. *Preserving Nature in the National Parks: A History*. New Haven, CT: Yale University Press, 1997.

Serwer, Adam. "White Nationalism's Deep American Roots." *Atlantic* 323, no. 3 (April 2019): 84–94.

Shankland, Robert. *Steve Mather of the National Parks*. New York: Alfred A. Knopf, 1951.

Shirley, James C. *The Redwoods of the Coast and Sierra*. Berkeley: University of California Press, 1940.

Smith, Duane A. *Mesa Verde National Park: Shadows of the Centuries*. Boulder: University of Colorado Press, 2002.

Spence, Mark D. *Dispossessing the Wilderness: Indian Removal and the Making of the National Parks*. New York: Oxford University Press, 1999.

Stasz, Clarice. *The Rockefeller Women: Dynasty of Piety, Privacy, and Service.* New York: St. Martin's Press, 1995.

Sutter, Paul S. *Driven Wild: How the Fight Against Automobiles Launched the Modern Wilderness Movement.* Seattle: University of Washington Press, 2002.

Swain, Donald C. *Wilderness Defender: Horace M. Albright and Conservation.* Chicago: University of Chicago Press, 1970.

Szanto, Andras. *Rockefeller Philanthropy: A Selected Guide.* Sleepy Hollow, NY: Rockefeller Archive Center, 2011.

Tarbell, Ida M. "John D. Rockefeller: A Character Study, Part II." *McClure's Magazine* 25, no. 4 (August 1905): 386–97.

Torres-Reyes, Ricardo. "Mesa Verde National Park: An Administrative History, 1906–1970." Washington, DC: National Park Service, Office of History and Historic Architecture, 1970.

Van Name, Willard G. *Vanishing Forest Reserves: Problems of the National Forests and National Parks.* Boston: R. G. Badger, 1929.

Weintraub, Lia N. "The Link between the Rockefeller Foundation and Racial Hygiene in Nazi Germany." Medford, MA: Tufts Archival Research Center, Tufts University, 2012.

Weiss, Philip. "Masters of the Universe Go to Camp: Inside the Bohemian Club." *Spy Magazine* (November 1989): 59–76.

Whisnant, Anne M. *Super-Scenic Motorway: A Blue Ridge Parkway History.* Chapel Hill: University of North Carolina Press, 2006.

Williams, Terry Tempest. *The Hour of Land.* New York: Picador, 2016.

Wilton, Andrew, and Tim Barringer. *American Sublime: Landscape Painting in the United States, 1820–1880.* Princeton: Princeton University Press, 2002.

Winks, Robin W. *Laurance S. Rockefeller: Catalyst for Conservation.* Washington, DC: Island Press, 1997.

Wohlforth, Charles. "Conservation and Eugenics." *Orion Magazine* (June 2010): 20–29.

Yard, Robert Sterling. "The People and the National Parks." *The Survey* 48, no. 13 (August 1, 1922): 546–53, 583.

Zeidel, Robert F. "Pursuit of Human Brotherhood: Rockefeller Philanthropy and American Immigration, 1900–33." *New York History* 90, no. 1/2 (Winter–Spring 2009): 85–106.

INDEX

Italicized page numbers indicate illustrations or material in the front matter.